Molecula
and Translationa

PROGRESS IN

Molecular Biology and Translational Science

Mechanisms of DNA Repair

edited by

Paul W. Doetsch,

Ph.D. Professor Departments of Biochemistry,
Radiation Oncology and Hematology & Medical Oncology,
Distinguished Chair of Cancer Research,
Associate Director for Basic Research, Winship Cancer Institute

Volume 110

ELSEVIER

AMSTERDAM • BOSTON • HEIDELBERG • LONDON
NEW YORK • OXFORD • PARIS • SAN DIEGO
SAN FRANCISCO • SINGAPORE • SYDNEY • TOKYO
Academic Press is an imprint of Elsevier

Academic Press is an imprint of Elsevier
The Boulevard, Langford Lane, Kidlington, Oxford, OX51GB, UK
Radarweg 29, PO Box 211, 1000 AE Amsterdam, The Netherlands
225 Wyman Street, Waltham, MA 02451, USA
525 B Street, Suite 1900, San Diego, CA 92101-4495, USA

This book is printed on acid-free paper. ∞

Library of Congress Cataloging-in-Publication Data
A catalog record for this book is available from the Library of Congress

British Library Cataloguing in Publication Data
A catalogue record for this book is available from the British Library

ISBN: 978-0-12-387665-2
ISSN: 1877-1173

For information on all Academic Press publications
visit our website at store.elsevier.com

Printed and bound by CPI Group (UK) Ltd, Croydon, CR0 4YY

Transferred to digital print 2012

Contents

Contributors

Numbers in parentheses indicate the pages on which the authors' contributions begin.

Ravindra Amunugama, Biophysics Graduate Program, and Department of Molecular Virology, Immunology, and Medical Genetics, Human Cancer Genetics, The Ohio State University Medical Center and Comprehensive Cancer Center, The Ohio State University, Columbus, Ohio, USA (155)

Pranshu Bansal, Department of Medicine, University of New Mexico, Albuquerque, New Mexico, USA (207)

Nicholas C. Bauer, Department of Biochemistry, and Graduate Program in Genetics and Molecular Biology, Emory University School of Medicine, Atlanta, Georgia, USA (93)

Anna L. Chambers, MRC Genome Damage and Stability Centre, University of Sussex, Falmer, Brighton, United Kingdom (229)

Anita H. Corbett, Department of Biochemistry, and Winship Cancer Institute, Emory University School of Medicine, Atlanta, Georgia, USA (93)

Paul W. Doetsch, Department of Biochemistry; Winship Cancer Institute; Department of Radiation Oncology, and Hematology and Medical Oncology, Emory University School of Medicine, Atlanta, Georgia, USA (93)

Sylvie Doublié, Department of Microbiology and Molecular Genetics, The Markey Center for Molecular Genetics, University of Vermont, Burlington, Vermont, USA (71)

Jessica A. Downs, MRC Genome Damage and Stability Centre, University of Sussex, Falmer, Brighton, United Kingdom (229)

Richard Fishel, Department of Molecular Virology, Immunology, and Medical Genetics, Human Cancer Genetics, The Ohio State University Medical Center and Comprehensive Cancer Center, and Physics Department, The Ohio State University, Columbus, Ohio USA (155)

Ann Ganesan, Department of Biology, Stanford University, Stanford, California, USA (25)

Alba Guarné, Department of Biochemistry and Biomedical Sciences, McMaster University, Hamilton, Ontario, Canada (41)

Philip C. Hanawalt, Department of Biology, Stanford University, Stanford, California, USA (25)

Muralidhar L. Hegde, Department of Biochemistry and Molecular Biology, University of Texas Medical Branch, Galveston, Texas, USA (123)

Robert Hromas, Department of Medicine, University of Florida, Gainesville, Florida, USA (207)

Tadahide Izumi, Graduate Center for Toxicology, University of Kentucky, Lexington, Kentucky, USA (123)

Neil M. Kad, Department of Biological Sciences, University of Essex, Colchester, Essex, United Kingdom (1)

Sankar Mitra, Department of Biochemistry and Molecular Biology, University of Texas Medical Branch, Galveston, Texas, USA (123)

Sandra Morandell, Departments of Biology and Biological Engineering, David H. Koch Institute for Integrative Cancer Research, Massachusetts Institute of Technology, Cambridge, Massachusetts, USA (289)

Aishwarya Prakash, Department of Microbiology and Molecular Genetics, The Markey Center for Molecular Genetics, University of Vermont, Burlington, Vermont, USA (71)

Brendan D. Price, Division of Genome Stability and DNA Repair, Department of Radiation Oncology, Dana-Farber Cancer Institute, Harvard Medical School, Boston, Massachusetts, USA (263)

Graciela Spivak, Department of Biology, Stanford University, Stanford, California, USA (25)

Daniel B. Swartzlander, Department of Biochemistry, and Graduate Program in Genetics and Molecular Biology, Emory University School of Medicine, Atlanta, Georgia, USA (93)

Bennett Van Houten, Hillman Cancer Center, Department of Pharmacology and Chemical Biology, University of Pittsburgh School of Medicine & University of Pittsburgh Cancer Institute, University of Pittsburgh, Pittsburgh, Pennsylvania, USA (1)

Susan S. Wallace, Department of Microbiology and Molecular Genetics, The Markey Center for Molecular Genetics, University of Vermont, Burlington, Vermont, USA (71)

Elizabeth A. Williamson, Department of Medicine, University of Florida, Gainesville, Florida, USA (207)

Justin W. Wray, Department of Medicine, University of New Mexico, Albuquerque, New Mexico, USA (207)

Chang Xu, Division of Genome Stability and DNA Repair, Department of Radiation Oncology, Dana-Farber Cancer Institute, Harvard Medical School, Boston, Massachusetts, USA (263)

Ye Xu, Division of Genome Stability and DNA Repair, Department of Radiation Oncology, Dana-Farber Cancer Institute, Harvard Medical School, Boston, Massachusetts, USA (263)

Michael B. Yaffe, Departments of Biology and Biological Engineering, David H. Koch Institute for Integrative Cancer Research, Massachusetts Institute of Technology, Cambridge, Massachusetts, USA (289)

Preface

Recent advances in our knowledge of DNA repair mechanisms and DNA damage responses have led to understanding in great detail how these processes operate at several different organizational levels ranging from single molecules to chromatin supra-macromolecular complexes. This volume comprises 11 chapters that span a range of topics addressing a number of major issues in the areas of DNA repair and the orchestration of the DNA damage response. The chapters and topics are presented as a progression from prokaryotic to eukaryotic cellular models and experimental strategies designed to interrogate the operational features of the major DNA damage management systems.

Kad and Van Houten, in the first chapter, illustrate the successful application of single molecule approaches to unravel how several hundred copies of nucleotide excision repair (NER) proteins can scan an entire bacterial (*Escherichia coli*) genome and target damage with amazing specificity. This is followed by an up-to-date review by Ganesan *et al.* of how *E. coli* transcription-coupled repair, a subpathway of NER, achieves its function to prioritize the repair of various classes of DNA damage in actively expressed genes. We move from here to mismatch repair (MMR) in *E. coli* where Guarné reviews recent functional and structural studies aimed at defining the multiple functions of the MutL endonuclease and how its activity is regulated by other repair factors. In the fourth chapter by Prakash *et al.*, the subject switches to the base excision repair (BER) pathway in bacteria where the key initiating DNA glycosylases of the Fpg/Nei family are highlighted in terms of their structural features, substrate specificities, and how they search for their DNA damage targets.

The chapters that follow then review recent advances in knowledge of DNA damage management in eukaryotic systems. Swartzlander *et al.*, in the fifth chapter, address the issue of BER regulation in *Saccharomyces cerevisiae* via dynamic localization from cytoplasm to nucleus and mitochondria of an initiating DNA glycosylase (Ntg1) in response to oxidative stress. The theme of BER regulation continues in the sixth chapter by Hegde *et al.*, where the DNA glycosylases operating in the early steps of BER and single strand break repair are reviewed in terms of their structurally disordered regions and posttranslational modifications. Another critically important repair pathway involves the utilization of homologous recombination (HR) for repair of DNA double strand breaks. Amunugama and Fishel, in the seventh chapter, provide a timely review of the components and mechanics of HR in eukaryotes.

Following the seventh chapter, the topics move toward discussions of how chromatin structure and modifications impact DNA repair and DNA damage response processes. In the eighth chapter, Williamson *et al*. provide a succinct review of histone modifications associated with the major DNA repair processes mentioned above including NER, BER, MMR, HR, and nonhomologous end joining (NHEJ). This sets the stage for the ninth chapter by Chambers and Downs where they address the issue of how chromatin remodeling enzymes (RSC and INO80) in *S. cerevisiae* pave the way for sites of DNA damage to be accessed by the repair machinery in the context of HR and NHEJ for repair of DNA double strand breaks. In the tenth chapter, by Xu *et al*., the focus of attention moves to the DNA damage response circuits that involve the master regulator of dsb repair, the ATM protein kinase, as well as histone methylation codes in human cells. These events, when corrupted, have important implications for the acquisition of genomic instability and the development of cancers. Finally, Morandell and Yaffe, in the eleventh chapter, discuss several recent examples of how synthetic lethal interactions (with an emphasis on p53) between DNA damage signaling and cell cycle checkpoint control can be exploited to kill tumor cells, illustrating how knowledge of damage response and repair pathways is being exploited for new therapies in cancer patients.

Thus, the complex landscapes of DNA repair processes and DNA damage responses are punctuated with timely examples of how the latest investigations are informing a detailed understanding of mechanistic processes.

Paul W. Doetsch
Atlanta, 2012

Dynamics of Lesion Processing by Bacterial Nucleotide Excision Repair Proteins

Neil M. Kad* and Bennett Van Houten†

*Department of Biological Sciences, University of Essex, Colchester, Essex, United Kingdom

†Hillman Cancer Center, Department of Pharmacology and Chemical Biology, University of Pittsburgh School of Medicine & University of Pittsburgh Cancer Institute, University of Pittsburgh, Pittsburgh, Pennsylvania, USA

Single-molecule approaches permit an unrivalled view of how complex systems operate and have recently been used to understand DNA–protein interactions. These tools have enabled advances in a particularly challenging problem, the search for damaged sites on DNA. DNA repair proteins are present at the level of just a few hundred copies in bacterial cells to just a few thousand in human cells, and they scan the entire genome in search of their specific substrates. How do these proteins achieve this herculean task when their targets may differ from undamaged DNA by only a single hydrogen bond? Here we examine, using single-molecule approaches, how the prokaryotic nucleotide excision repair system balances the necessity for speed against specificity. We discuss issues at a theoretical, biological, and technical level and finally pose questions for future research.

Progress in Molecular Biology
and Translational Science, Vol. 110
DOI: 10.1016/B978-0-12-387665-2.00001-8

I. Structural Insights of Bacterial Nucleotide Excision Repair

A. Overview of the Process

Bacterial nucleotide excision repair (NER) is mediated by six proteins[1] (Fig. 1). Initial damage recognition is performed by UvrA, which acts as a dimer. The cocrystal structure of a UvrA dimer bound to a DNA fragment containing a fluorescein-modified thymine indicates that the damaged DNA fits snugly into a channel formed by the UvrA dimer.[2] It is believed that positively charged amino acids along this channel help to stabilize the binding of DNA to UvrA.[2,3] Damage recognition is apparently mediated by the C-terminal zinc fingers.[4] Most recently, it has been shown that these zinc fingers help to communicate with the ATPase sites within the protein.[5] ATP binding promotes UvrA dimerization, whereas ATP hydrolysis promotes monomerization. However, ATP hydrolysis is required for damage discrimination, as ATP-γ-S abrogates damage recognition by increasing the nonspecific binding of UvrA.[6,7] These ATP-binding sites are also important for allowing damage verification by UvrB,[5] as hydrolysis by UvrA is believed to promote UvrA release during loading of UvrB at the site of damage.[8,9] UvrB has been determined to melt 4–6 bases by inserting a beta-hairpin through the two DNA strands containing the DNA lesion.[10] Tyr96 at the base of the beta-hairpin is essential for damage verification.[11,12] UvrB resembles monomeric helicases, and it is believed that ATP hydrolysis of UvrB's helicase-like fold is essential for damage verification and proper lesion positioning for binding and incision by UvrC.[8]

UvrC contains two nuclease domains in the N-terminal and C-terminal regions, required for dual incisions 3′ and 5′ on the damaged nucleotide, respectively.[13,14] Once UvrC performs both endonucleolytic cuts, two additional proteins, UvrD and DNA polymerase I (Pol I), are required to displace the postincision complex, fill in the repair patch, and allow UvrB and UvrC to turn over.[15] The final step in prokaryotic NER is the sealing of the remaining nick by DNA ligase I.

B. Dynamics of the $UvrA_2B_2$–DNA Complex

Determination of the stoichiometry of the UvrAB complex has been controversial, but two definitive experiments indicate that the working complex consists of two subunits each of UvrA and UvrB.[16,17] There are three important problems regarding the dynamics of the interaction of $UvrA_2B_2$ with damaged DNA: (1) the way the damage is handed off from the UvrA dimer to UvrB; (2) the nature of the search mechanism; and (3) the rate-limiting step in damage processing by these two proteins. Regarding the first question of damage handoff, one alternative is that the damage site slides from the central location of the two zinc fingers at the UvrA dimer interface to the beta-hairpin of UvrB.[2,18,19] Another more likely

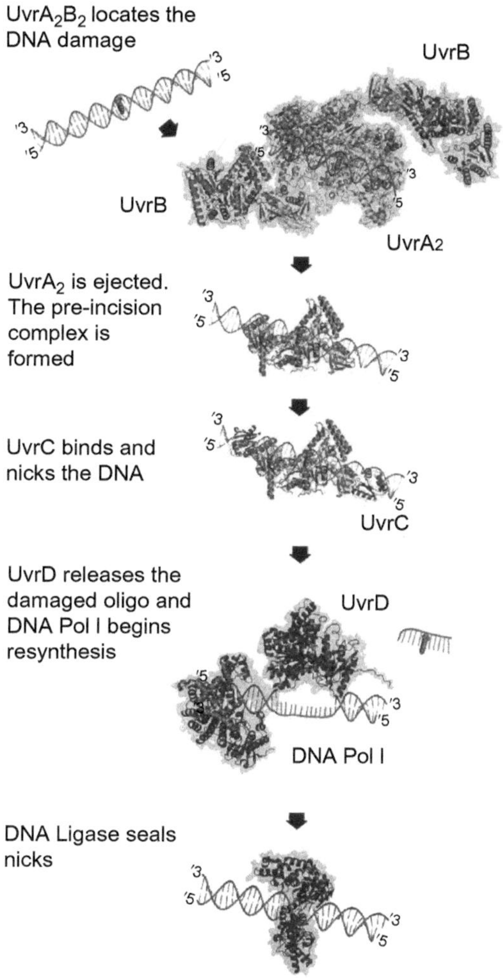

FIG. 1. Structural model of bacterial nucleotide excision repair mediated by six proteins. The process of NER is a complex multiprotein cascade of events. Each step requires the recruitment of another protein to the lesion, with UvrB remaining at the lesion site as it interacts with each component of the reaction. Remarkably, despite this central role, UvrB is incapable of binding to the lesion site directly, requiring loading by $UvrA_2$. (See Color Insert.)

alternative is that the $UvrA_2UvrB_2$–DNA complex and interaction interface between the proteins is flexible enough to allow UvrB to clamp down onto the DNA face opposite $UvrA_2$ in order to be positioned for a direct handoff of the lesion.[2,18,19] In this model, $UvrA_2$ may bend the DNA to facilitate localized unwinding of the helix to allow easy insertion of UvrB's beta-hairpin.[1] Still unresolved is how the UvrA/UvrB interface is weakened to allow dissociation of the UvrA dimer during loading of UvrB. One plausible explanation is that the conformation of this interface is altered by damage-induced hydrolysis of ATP by both UvrA and UvrB.[20] This review describes how single-molecule techniques offer a new approach toward understanding these problems. In particular, we focus on novel imaging platforms that have recently been developed to allow the process of NER to be examined one molecule at a time without the influence of flow or surfaces and to highlight the use of current protein-tagging technologies, particularly quantum dots.

C. Kinetic Proofreading as Part of a Dynamic DNA Damage Recognition Process: Role of ATP

As explained in the previous section, DNA damage recognition is a dynamic, ATP-requiring process. Damage recognition, damage verification, and finally DNA damage processing leading to incision, all require a high degree of specificity, which is probably achieved by what has been termed "kinetic proofreading".[21–23] It is postulated that to cope with the frequent small differences in binding energy between target and nontarget sites during NER, high-energy protein–substrate intermediate complexes are created using ATP; however, if the target is not confirmed, the complex can dissociate. This process utilizes energy to generate a time gap between initial protein binding and subsequent catalysis during which the complex can dissociate and the catalysis (DNA incision) is aborted. By including multiple proofreading steps prior to catalysis, damage specificity is improved. For NER, we envision that once $UvrA_2$ senses damage, $UvrA_2$ utilizes ATP hydrolysis to position UvrB for damage verification, representing the first kinetic proofreading step. Following this, in the second kinetic proofreading step, UvrB hydrolyses ATP and releases $UvrA_2$, making reversal of this process improbable; this also helps to impose a specific conformation on the damaged DNA, promoting efficient incision[11] (see Fig. 1). In this way, UvrC is probably "capturing" this preincision UvrB–DNA complex during an ATPase cycle on UvrB.[11] This occurs prior to the enzymatic commitment step, which in this case is the dual incision of the DNA that alters the integrity of the genome.[21–23]

II. So Few DNA Repair Proteins, So Much DNA: Defining the Big Problem

A. Challenge of Repair Inside a Bacterial Cell

The average size of an *Escherichia coli* cell is about 1 μm in diameter. Packed inside this volume of $\sim 1 \times 10^{-15}$ L is 4.6×10^6 bp of genomic DNA, making the base pair concentration in the millimolar range. An undamaged

E. coli cell contains about 20, 50, and 20 copies of UvrA, UvrB, and UvrC respectively, and their concentrations are approximately equivalent to 20, 50, and 20 nM, respectively. During DNA damage-induced SOS induction, which takes ~20–40 min, the levels of UvrA and UvrB proteins may rise about 10-fold, whereas the *uvrC* gene is not under SOS control. Assuming UvrC employs the same search mechanism as UvrA and UvrB, it is likely that UvrC's search process for the UvrB–DNA preincision complex would be rate-limiting for NER. In addition to global NER, transcription-coupled repair occurs when damage blocks the progression of transcribing RNA polymerase (RNAP).[24] The MFD or transcription-coupling protein, which contains a fold similar to UvrB, helps recruit the repair machinery to the site of the damage-induced stalled RNAP.[25] *E. coli* K12 WT strain showed about 10% survival after 22 J/m^2 of 254 nm UV light (UVC),[26] and quantitative polymerase chain reaction (PCR) analysis of the damaged *E. coli* indicated that UVC induced cyclobutane pyrimdine dimers (CPD) and 6–4 photoproducts at a ratio of 2.1:1, respectively.[27] A fluence of 100 J/m^2 of UVC produced about 0.4 photoproducts/kb, of which about 16.5% and 36.7% of these lesions were repaired in 15 min on the nontranscribed and transcribed strands of the *lacI* gene, respectively.[27,28] Thus, assuming an initial noninduced state, some 10 $UvrA_2B_2$ complexes were capable of processing about 1000 photoproducts in 15 min or about 6.6 lesions/complex/min. If, however, there was ample induction of UvrA and UvrB proteins, this rate of repair could be up to 10 times slower. Using a complementary approach of monitoring removal of ~4 CPD/pBR322 in *E. coli*, it was found that the UvrABC system could remove about 0.045 dimers/plasmid (4.4 kb)/min.[29] As pBR322 exists at about 10 copies per cell and *E. coli* has 4.6×10^6 bp genome, also irradiated in this experiment, that is equivalent to a total of about 45 CPD/min being processed by roughly 10 $UvrA_2B_2$ complexes or about 4.5 CPD/complex/min. Together these conditions give a repair rate of 1 photoproduct/1–2 kbp in roughly 10–100 s/$UvrA_2B_2$ complex. How can such a small number of proteins search through a vast sea of undamaged DNA to quickly repair that quantity of DNA damage? This fundamental problem in NER is discussed in more detail in the following section.

B. Potential Modes of Damage Site Location

The question of how proteins find their cognate binding sites in the presence of excess DNA has been of great interest for over 50 years and several search mechanisms have been proposed (reviewed in Ref. 30). Despite the large number of proteins packed into the small volume of a bacterial cell, proteins diffuse rapidly. The green fluorescence protein (GFP), MW of 30 kDa, was found to have a three-dimensional (3D) diffusion constant of about 7.7 μm^2/s.[31] This rapid Brownian motion is caused by collisions with water molecules such that a single GFP molecule can diffuse the width of a bacterium in ~22 ms.[32] However, the likelihood of a $UvrA_2B_2$ complex landing directly on a lesion in DNA has a low probability, and a large number of binding and dissociation steps would be required to adequately find DNA damage during a period of cellular division using a 3D searching strategy.

This problem is exacerbated by the presence of a vast excess of undamaged DNA for which the $UvrA_2B_2$ complex has affinity. In reality, however, the presence of excess nontarget sequences has been shown to accelerate target acquisition[33], suggesting that such a 3D searching strategy is an inadequate explanation.

Alternative mechanisms of target search include DNA sliding or hopping, which reduces the search space from three dimensions to one. Using techniques described below, researchers have found several proteins to slide various distances on DNA (see Table I). It should be noted, however, that the one-dimensional (1D) diffusion constants laid out in Table I are all slower than those expected considering the hydrodynamics of motion. The helical structure of DNA offers a unique challenge to a one-dimensionally diffusing molecule. Consider by analogy a modern rollercoaster ride, in which there are corkscrew sections analogous to the helical groove of DNA. The rollercoaster car spins as it traverses these sections whereas over the linear sections the car slides rapidly unimpeded by any spiraling. In direct analogy, the drag experienced by a protein that slides along DNA without being confined to the helical structure of the DNA is much less than that by a protein corkscrewing around the DNA along the grooves. Let us first consider the case of a sphere the size of a protein (r) sliding in one dimension. Ignoring the DNA helix, we calculate the maximum possible diffusion constant ($=\kappa_B T/6\pi\eta r$) and compare this to the measured value from Table I. This difference in diffusion constant provides information on the magnitude of the energy barrier to diffusion. In all cases, the value is extraordinarily large. Instead, if we calculate the maximum diffusion constant based on the protein spiraling around the DNA, we find that the drag due to rotation (effective friction factor$=1+4/3[2\pi]^2+[r/3.4\times10^{-9}]^2$) has a major effect on the diffusion constant,[53] thus lowering the theoretical maximum value. In turn, this lowers the difference between the expected and measured diffusion constants and thus the associated energy barrier to reasonable values ranging from near zero to just a few $\kappa_B T$.

Early studies looking at the processivity of UvrABC-mediated repair of DNA damage in plasmids suggested that the $UvrA_2B_2$ complex is capable of limited 1D DNA scanning. These studies showed that following a lag period after UV-induced damage, during which plasmids were bound, $UvrA_2B_2$ utilized some type of scanning mechanism to search for and repair lesions.[29] This study also implied that incision and gap filling were highly coupled, preserving the topological state of the plasmid with no accumulation of nicked plasmids.

TABLE I

Single-Molecule Analysis of Protein Movement on DNA

Protein	Process	MSD ($\mu m^2/s$)	Mode	References
EcoRI	Host restriction	3.5×10^{-3}	Sliding	34
EcoRV	Host restriction	1.1×10^{-2}	Sliding and jumping	35
EcoRV	Host restriction	3.15×10^{-3}	Sliding	36
Fpg	BER	2.3×10^{-2}	Sliding	37
Lac I repressor	Transcription	2.3×10^{-4}–1.3×10^{-1}	Sliding	38
Lac I repressor	Transcription	4.6×10^{-2}	Sliding	39
Lac I repressor	Transcription	0.4	Sliding	39
Lac I repressor	Transcription	3[a]	3D search	39
Mlh1	MMR	1.37×10^{-1}	Hopping (?)	40
Mlh1-Pms1	MMR	0.2–9.9×10^{-1}	Hopping[b]	40
Msh2-Msh6	MMR	1.2×10^{-2}	Sliding	41
MutM	BER	4.0×10^{-2}	Sliding	42
MutS	MMR	3.6×10^{-2}	Sliding/paused[c]	43
Nei	BER	3.4×10^{-2}	Sliding	37
Nth	BER	5.8×10^{-2}	Sliding	37
Ogg1	BER	5.8×10^{-1}	Sliding	42
PCNA	Replication	1.16	Sliding (direct and rotational)	44
p53	DDR	1.62×10^{-1}	Complex	45
p53 (core)	DDR	2.39×10^{-2}	Hopping	45
p53 (C-ter)	DDR	7.76×10^{-1}	Sliding	45
Rad51	Recombination	4.2×10^{-2}		46
RecBCD	Recombination	1250 bp/s[d]	Directed motion	47,48
RNAP (T7)	Transcription	6.1×10^{-3}–4.3×10^{-1}	Sliding	49
RNAP (*Escherichia coli*)	Transcription	1.0×10^{-2}	Sliding	50
T7 DNA pol	Replication	1.86 (34 mM)[e]	Hopping	51
T7 DNA pol/ thioredoxin	Replication	4×10^{-1}	Sliding	51
Uracil glycosylase	BER	1.15×10^{-3}	Sliding	52
UvrA	NER	None	Jumping	16
UvrAB	NER	4.4×10^{-4}	Sliding	16
UvrAB	NER	1.3×10^{-3} ($\mu m/s$)[f]	Paused, directed	16

Conversions:

$1 \times 10^{-15}\ m^2/s = 1 \times 10^{-11}\ cm^2 = 1 \times 10^{-3}\ \mu m^2/s = 8.6 \times 10^3\ bp^2/s$

BER, base excision repair; DDR, DNA damage response; Fpg, formamidopyrimidine DNA glycosylase; MMR, mismatch repair; Nei, endonuclease VIII; Nth, endonuclease III; NER, nucleotide excision repair.

[a] 3D diffusion constant.

[b] D increased with increasing salt, evidence for hopping.

[c] short sliding events, followed by long ATP-induced pauses.

[d] rate for translocating through lambda DNA.

[e] at this ionic strength.

[f] directed motion, therefore units = μm/s.

It is valid to assume that the distance a protein could slide before encountering another protein bound to DNA is very short given the large numbers of DNA-binding proteins. Thus, an alternative to DNA sliding has been proposed, DNA hopping, in which the protein performs microdissociation and reassociation without undergoing macro-dissociation away from the DNA helix (see Fig. 2). This interaction is thought to be mediated by an ionic interaction between the protein and DNA; most DNA-binding proteins possess a positively charged DNA-binding cleft such as that found at the dimer interface of UvrA.[2] In order to discriminate between sliding and hopping, researchers use buffers containing different ionic strengths.[54] The rate of sliding is believed to be independent of ionic strength in a purely ionic-mediated system because there is no change in the number of ions bound as the proteins move. However, hopping is greatly affected by the ionic strength as the ionic-mediated association of the protein to the DNA will be shielded with increasing salt concentrations, leading to shorter attached lifetimes and potentially more rapid diffusion.[55] One extreme form of hopping is jumping, in which the $UvrA_2B_2$ complex could rebind to a second DNA helix at a great sequence distance from the initial DNA-binding site. This could occur inside a living cell where the local DNA concentration is very high. However, regardless of the two types (sliding or hopping) of movement on DNA, with no input of external energy, the probability of stepping in one direction versus the other is equal and therefore there is no net

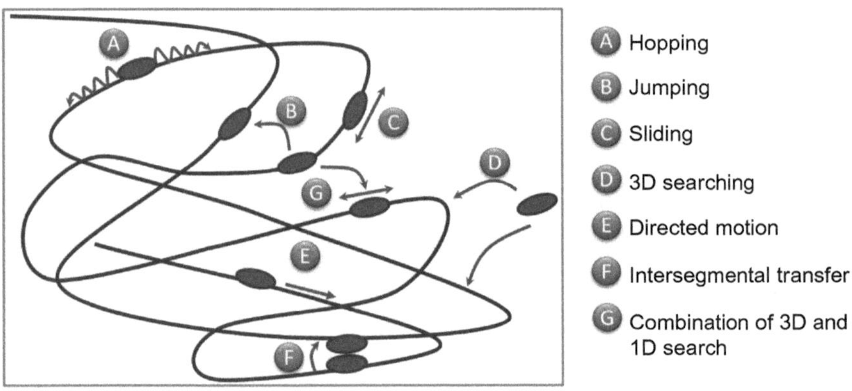

FIG. 2. Potential modes of DNA damage searching by UvrA and UvrAB repair proteins. A number of mechanisms by which proteins can search for lesions are depicted. Hopping is distinguished from jumping by the distance over which the translocation occurs; however, in both cases, the protein remains within close proximity of the DNA. Sliding suggests that the protein remains in constant contact with the DNA, making it difficult to separate from hopping. If the protein dissociates from DNA into bulk solution, then a 3D search is employed to find the target site. Directed motion requires the input of energy in the form of nucleotide. (See Color Insert.)

movement in any one direction. Thus, directed motion in which the protein migrates in only one direction requires an input of energy. It is possible that the ATPase sites in UvrA and UvrB could provide directed motion.

In another example of how $UvrA_2B_2$ could search the genome, it has been noted that this complex can, in principle, bind multiple DNA helices[56] and, therefore, it offers a mechanism for transfer from one DNA region to another without ever dissociating in what is called intersegmental transfer (see Fig. 2). In this mode of searching, two DNA molecules would have to make close contact. Such mechanisms of protein movement on DNA have been proposed for restriction endonucleases and for lac repressor.[57,58]

Clearly, there are a number of proposed mechanisms by which proteins can negotiate DNA to find their target sites; however, definitive evidence for one mechanism versus another is extremely difficult to obtain using bulk methods. Therefore, more recently, there has been an explosion of studies using a variety of single-molecule techniques that permit each individual protein complex to be followed with good time and spatial resolution.[40,43,45,59–69] These studies are likely to provide a clear understanding of which method(s) of protein transfer occurs. Understanding the precise mechanism of how proteins locate their target sites is of crucial importance and is discussed in more detail below.

C. Necessary Experimental Components to Observe Single Molecules in Action

The original presentation of the lac repressor finding its target site faster than the Smoluchowski-defined diffusional limit[70,71] was the first experimental demonstration that the search mechanism may involve a 1D component instead of being purely 3D. This subsequently prompted a number of papers exploring the theoretical aspects of this potential mode of target location (see Ref. 55). In addition, a number of experimental studies have provided evidence for and against these models which are summarized below. Single-molecule techniques offer a direct method of determining how proteins find their targets and have more recently been deployed in a number of experimental formats. A few of these are summarized here:

1. Optical tweezers: By using a focused (normally infrared) beam, it is possible to exert forces on dielectric spheres, thus "trapping" them at the waist of the incident laser. This approach allows for the position of the trapped bead to be manipulated in three dimensions, and by attaching a single strand of DNA to the bead, the DNA is likewise manipulated. In an ingenious experiment where one end of DNA was tethered to a bead and the other to RNAP attached to a surface, it was shown that RNAP scans the groove of DNA in search of its promoter[72] because movement of the surface led to

rotation of the trapped bead. These data have provided strong evidence that sliding is used by RNAP. In another study, Yanagida and coworkers[50] suspended a single extended DNA filament between two trapped beads and found that RNAP performed a random walk along the DNA. Upon release of the DNA tension, the DNA reformed a random coil conformation and it was noted that the proteins bound to the DNA targets with greater efficiency. This could imply either that DNA tension is preventing efficient target location or that the proteins bind and release rapidly and that their redistribution to the promoter sites is favored by the formation of a compact DNA structure that allows the proteins to three-dimensionally hop between sites on the DNA, as suggested by a later study using restriction endonucleases.[73] The use of laser tweezers has permitted a number of elegant studies on the single-molecule properties of DNA–protein interactions,[74–78] including obtaining base pair resolution of RNAP stepping.[79] Despite these advances, optical trapping is not a trivial technique to apply as it requires specialist equipment and expertise. In addition, data collection is not multiplexed and is therefore slow as each investigation typically provides information on a single DNA strand. However, the recent use of multichannel fluidics could overcome some of the latter shortcomings.[47]

2. Magnetic tweezers offer an approach that harnesses much of the power of laser tweezers with the potential for more multiplexing. This approach has been used to investigate the function of a number of enzymes bound to DNA that alter its end-to-end length. Such length variations occur through alterations in single-stranded and double-stranded character[80] or topoisomerase-induced relaxation of DNA,[81] among others. For magnetic tweezers to provide detailed information on the physical basis of target site location, however, further development to combine this with fluorescence detection is required.

3. Direct imaging is currently in a position to provide direct information on the mechanism of target site acquisition for DNA-binding proteins. To date, a number of studies have provided diffusion constants and hopping distances (many summarized in Table I). These studies have invariably required the extension of the DNA substrate from its random coil to facilitate imaging of protein motions. To achieve this, DNA can be laid directly down on a surface by combing.[82,83] This approach allows the highest signal to noise ratio because imaging can be performed in the evanescent field of a total internal reflection microscope (TIRF-equipped microscope). Typically, the surface is activated with a hydrophobic moiety such as polystyrene or polymethylmethacrylate.[82] At low pH, DNA binds to these surfaces at two or more points; indeed, the number of binding points can be visualized by snapping the DNA under intense illumination, at which time the DNA recoils to its bound points on the surface.[83] This

approach is very simple and requires no further flow during data acquisition. Other methods of confining DNA close to the surface have been achieved by tagging one end of the DNA with biotin and conjugating to surface-immobilized streptavidin.[42] Alternatively, lipid-immobilized DNA molecules can be spatially aligned on the coverslip such that under flow they elongate into "curtains".[84] The disadvantages of these methods are, first, that they all require flow (with the exception of combing), which will tend to bias the dataset toward sliders as proteins that dissociate from the DNA are likely to be carried away by the flow. In addition, the effects of the surface are unknown: Are moving complexes stalling due to the binding of the target site or because of interaction with the surface? Finally, as these structures are often located within the same focal plane as the surface, it becomes difficult to distinguish between surface-immobilized beads and those that are immobile but attached to the DNA. Stopping and starting flow using DNA curtains helps to circumvent this latter problem.[30]

4. Elevated DNA platforms or "tightropes"[16] offer perhaps the best compromise between many of the methods described here as they are microns away from the surface and they elongate the DNA, and anything that appears stationary for long enough to see within the visual field must be bound to DNA (acquisitions are performed using high numerical aperture objectives, which shorten the depth of field) (see Fig. 3). These structures can be simply constructed by sequentially flowing the components into a surface-passivated observation flowcell.[16] The simple flow sequence for setting up tightropes is as follows: (i) poly-L-lysine-coated 5 μm sized

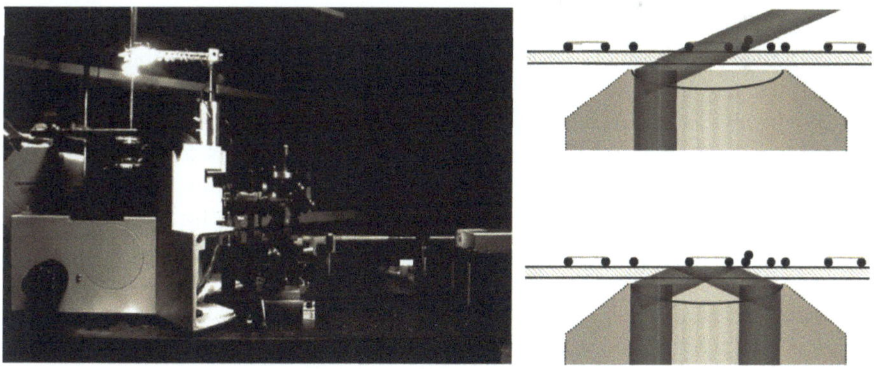

FIG. 3. Oblique angle fluorescence permits a high signal to noise view of DNA tightropes. To generate OAF (top right), a standard TIRF (bottom right) optical path is steered to a subcritical angle, resulting in a far-field illumination beam emerging at a steep angle. Although this was achieved using a custom-built system (left), this is possible using off-the-shelf systems with little difficulty. (See Color Insert.)

beads, (ii) DNA, (iii) wash, and (iv) Qdot-bound proteins. TIRF is adapted to illuminate the proteins through the lens by steering the excitation laser to a subcritical angle; in practice, this simply requires moving the excitation laser toward the center of the objective until the DNA tightropes can be seen. Using this "oblique angle fluorescence" (OAF) configuration provides considerable improvement in the signal to noise ratio over epi-fluorescence. Visualization of specific proteins is mediated through tagging with quantum dots (Qdots) (discussed later). Qdots are suitable for long acquisitions because of their resistance to photobleaching while their extreme brightness permits short integration times. Qdots also have very broad excitation spectra permitting multicolor emissions to be recorded using just a single excitation source. This latter point is extremely important as most DNA enzymes work with partners to perform their job; such is the case for all systems that exploit kinetic proofreading mechanisms. Moreover, as the proteins are elevated far from any surface-bound Qdots, they do not interfere with fluorescent spot discrimination. Finally, the density of DNA can be tuned to high or low values. This permits a single protein on a single DNA strand to be examined in the context of other closely juxtaposed DNA molecules. This is close, but not a perfect equivalent, to random coil DNA, where sequence-distant strands come within close spatial proximity of the DNA-bound proteins, thereby making it possible to study intersegmental transfer or even observe hopping directly[16].

5. To image proteins working on DNA using any of the approaches above first requires the conjugation of a fluorophore. Here we focus on strategies for Qdot conjugation to allow single-molecule imaging during NER. In order to watch UvrA and UvrB work together to search DNA, two different strategies for conjugating Qdots were necessary[16,85] (see Fig. 4). Qdots are commercially available with a number of different ligands attached to the surface and a wide variety of approaches have been used to visualize single molecules both *in vitro* and *in vivo.*[86,87] The two most common are streptavidin and antibodies. In order to conjugate UvrA to Qdots, a biotin ligase site was engineered into the C terminus of UvrA. This 15-amino acid-sequence contains a specific lysine that is covalently attached to biotin by biotin ligase. This modification can be done during protein production within *E. coli* or after purification using purified biotin ligase to levels approaching 95% of the total protein. Mixing the biotinylated UvrA with excess Qdots ensures that only one UvrA monomer is attached per Qdot. As multiple streptavidin moieties are bound to the Qdot surface, atomic force microscopy is routinely used to assure a 1:1 stoichiometry of attachment of protein to DNA.[16,85] Importantly, it is essential to ensure that the attachment of a Qdot does not affect the function of the protein. Therefore, electrophoretic mobility shift assays in agarose gels are used to ensure that the attachment of the Qdot to

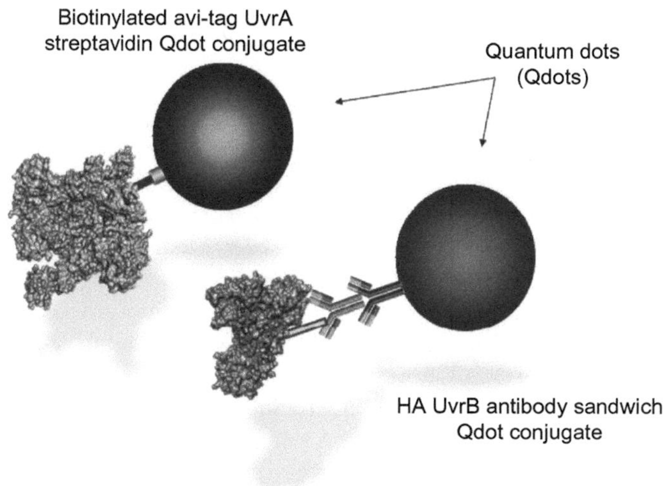

FIG. 4. Quantum dot conjugation strategies. To ensure that there is no cross talk between different quantum dots, a differential conjugation strategy is important. Two approaches that were used for UvrA and UvrB conjugations are highlighted. UvrA (left) was bound to quantum dots using a short peptide sequence extension, which is biotinylated either endogenously or more efficiently using biotin ligase. UvrB could be differentially labeled by using an HA tag, which can be labeled using an antibody sandwich strategy. (See Color Insert.)

the protein does not interfere with its DNA binding or interaction with other proteins.[16] UvrB was conjugated to Qdots by placing antibody recognition sequence of hemagglutinin, HA tag, on its N terminus. Specific high-affinity mouse monoclonal antibodies bind to the HA tag. Qdots coated with goat F (ab′)2 antimouse IgG conjugates are then used to make an "antibody sandwich".[85] Using this dual labeling approach, both proteins could be observed simultaneously as UvrA loads UvrB onto damaged sites.[16] As many researchers use polyhistidine tags and nickel chromatography to purify and tag mammalian proteins, Piehler and coworkers have developed a chemical tag that binds to six or more histidines.[88] They have used this approach to effectively label single molecules both *in vitro* and *in vivo*.[89–91]

Application of these methods allows the physical biochemistry of the DNA search to be examined. This is a problem that is pertinent not only to the mechanism of NER, but also to all protein systems that interact with DNA, and which can be addressed only now with the development of single-molecule methods. Specifically, using direct imaging (see approach 4 above), the first glimpses of how the bacterial NER proteins find lesions have been obtained after over four decades since the discovery of genes.

III. Damage Searching by $UvrA_2$ and $UvrA_2B_2$

The first aspect of any DNA protein interaction is specificity, which often refers to the affinity of a protein for its cognate binding site. However, there is a vast amount of very similar DNA, which leads to the question: How are targets found? This is the central theme of this chapter and by using the single-molecule techniques outlined earlier, we are able to begin to address this question. It is likely that the NER system uses a combination of the many possible mechanisms such as sliding, hopping, and intersegmental transfer. Investigations at the single-molecule level have permitted direct observation of these processes and also allow quantitative characterization of their relative abundance. With more details emerging from such analyses, it may also become clear why one mechanism is favored over another (Fig. 5).

Observation of $UvrA_2$ bound to DNA found it to remain statically bound for ~7 s before detaching and binding elsewhere. This is exactly the equivalent of a 3D search mechanism, where the DNA is searched through random encounters. However, it was noted that $UvrA_2$ appeared to jump to nearby strands, with an average jump distance of 1.2 μm.[16] The significance of this distance has yet to be determined, but it may be a result of the architecture of the assay. A high density of DNA tightropes resembles a sheet of binding surface; as a molecule is released from the DNA, what is the probability that it will pass back onto the sheet versus diffusing into solution? Does the relatively high incidence of rebinding suggest that DNA keeps the protein close to itself until it finds the right orientation to bind, possibly through electrostatics? If it is the latter, then further studies using salt to shield these charges will provide valuable insight. An approximation of the search time for a lesion suggested that the 3D mode of search was very inefficient[16]; therefore, it is unlikely that $UvrA_2$ alone can search the DNA effectively. Surprisingly, however, it was found that the addition of UvrB to form a $UvrA_2B_2$ complex led to a change in the way $UvrA_2B_2$ searched the DNA relative to $UvrA_2$ alone: UvrB collapsed the 3D search into 1D diffusion along the DNA.[16] Recalculation of the 1D search time resulted in a huge increase in the rate at which DNA could be examined, bringing the search time for the $UvrA_2B_2$ molecules within the cells down from hours to minutes. It is interesting to note that not all of the molecules participated in this process; why is this so? One possibility is that the complex composition ($UvrA_2B_2$ vs. $UvrA_2B$) may affect motility. It was found that the UvrAB complexes used three different modes of motion: 1D diffusion, directed, and paused. Directed motion was observed only in the presence of ATP; however, it is possible that these represented under-sampled populations of diffusers. Paused movers appeared to diffuse and then pause, often pausing at the same site many times, and although the DNA was not intentionally damaged, some nicks were detected. Perhaps these "pausers" were sensing damage but not

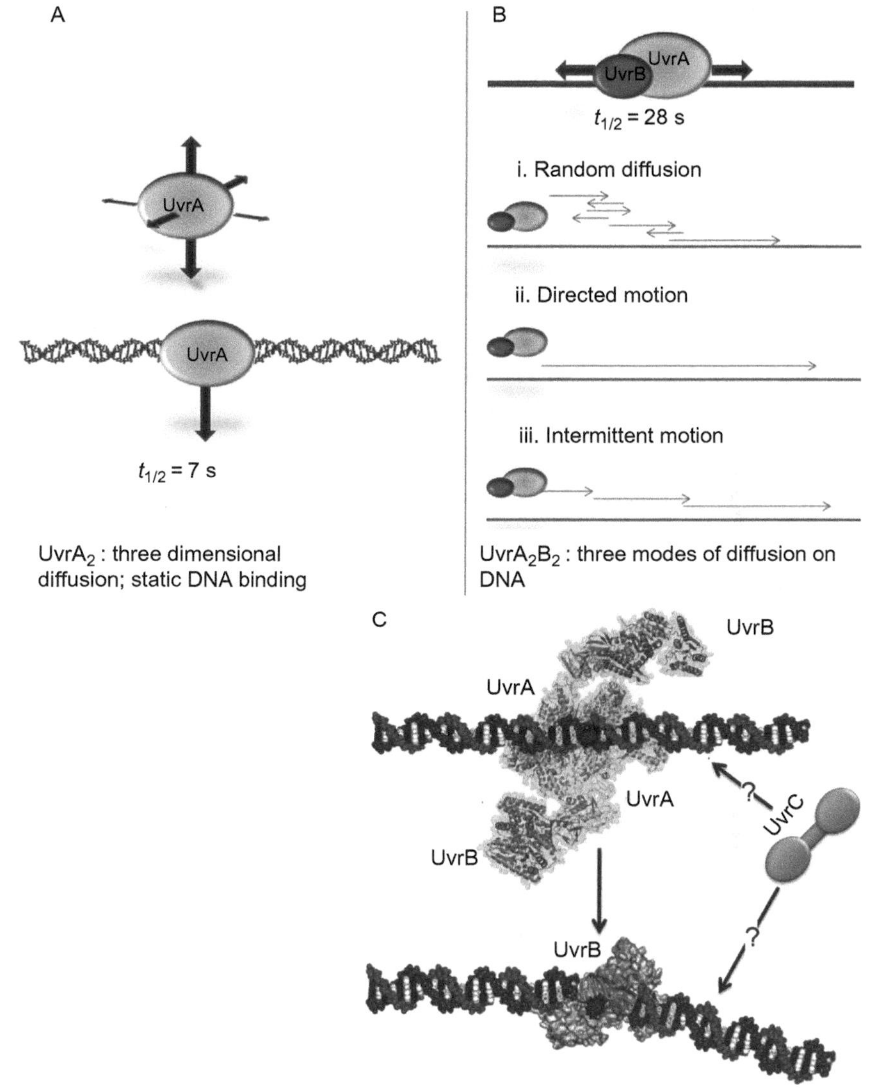

FIG. 5. Summary of motion of $UvrA_2$ and $UvrA_2B_2$ on DNA. (A) $UvrA_2$ has been shown to perform a 3D search with an average bound life time of 7 s on DNA. $UvrA_2$ was also observed jumping from DNA helix to another helix over distances of 1–2 μm (not shown). (B) UvrB collapses UvrA's search mode from 3D to 1D sliding and increases its average lifetime on DNA to 40 s. Three different sliding modes were observed for $UvrA_2B_2$. (C) The nature of how UvrC finds its way to the UvrB–DNA preincision complex is not known. The low concentrations of UvrC in the cell would make a 3D search very inefficient if UvrC remained statically bound for longer than 1 s. (See Color Insert.)

forming a preincision complex[16]. Although the dominant UvrAB species was determined to comprise two each of UvrA and UvrB, there remains the possibility that other complex combinations[18,92,93] may underlie the one-dimensionally diffusion-competent complex.

A comparison of the diffusion constants in Table I shows that the NER proteins are the slowest of all the complexes. This is not a consequence of the DNA tightrope assay as studies of base excision repair have shown much faster diffusion.[37]Therefore, the NER proteins engage in a slow search; is this because the proteins are hunting for lesions as they move? Or has this to do with the multienzyme processing that is occurring? It is tempting to speculate that there are components of both of these scenarios present. Lesions are thought to be found by detecting deformation of the DNA,[9] which makes the proteins sensitive to the conformational state of the DNA; furthermore, as UvrB wraps the DNA around itself,[94] this may lead to substantial local energy barriers if the DNA is held at each end as in the tightrope assay. $UvrA_2$ is thought to pass the lesion to UvrB for checking, one of the kinetic proofreading steps; this process will not be instantaneous and therefore may also slow the search process. Wallace and coworkers[37] showed that there is a correlation between the diffusion constant of protein motion and the "alpha" value. The alpha value is a description of how the molecule moves: if $\alpha=1$, then the protein is diffusing freely; $\alpha=2$ is directed motion; and $\alpha<1$ is subdiffusion or anomalous diffusion. It was found that slower movers have $\alpha<1$. Furthermore, they showed that a critical phenylalanine residue in the tip of an interrogation loop was responsible for this subdiffusive behavior of a DNA glycosylase as it moves along the DNA.[37] Models of subdiffusion[95,96] propose a number of explanations including the possibility that the molecule pauses and then continues. Our own simulations suggest the same, and therefore suggest that perhaps the slow diffusion of the $UvrA_2B_2$ is due to interrogation of the DNA. Interestingly, we have found that pause distributions occur over very long and very short timescales, offering a potential explanation for the observed paused movers. It is interesting to note that UvrA also contains a Phe residue at the tip of the zinc finger, which has been shown to be essential for damage recognition.[4]

Target acquisition is not only the domain of $UvrA_2B_2$ but also the problem that faces the other components of NER, which include UvrC, UvrD, Pol I, and DNA ligase. UvrC needs to efficiently locate a preincision complex with UvrB bound to the DNA. As the very slow release of UvrB is on the order of hours,[92,97] it means that as damage accumulates, UvrB becomes trapped on the DNA. Therefore, efficient location by UvrC frees UvrB to search for more lesions. However, UvrC is present only at 20 molecules/cell, making this search process the likely rate-limiting step of the whole repair reaction.[98] As UvrC is an endonuclease, its cellular concentration reflects a balance between its molecular imperative to cut DNA[99,100] and its preincision complex search efficiency. In addition, as it has an affinity for UvrB and shares a binding site with UvrA on UvrB, it seems that a useful switch has evolved, preventing interaction with a searching complex until a lesion is found. However, UvrB is in excess of UvrA and therefore it is possible that UvrB interacts with UvrC,[101] sequestering it from interacting with DNA until a preincision complex is found. Therefore, perhaps, UvrB forms a complex with UvrC and accompanies it to the preincision complex and at the same time protects DNA from spurious incisions.[102]

Single-molecule approaches offer another advantage beyond the physical description of lesion search: direct observation of repair in action being made possible, revealing both the sequence of events and their kinetics. The canonical mechanism of NER involves a complex of UvrA and UvrB searching for lesions, followed by UvrA's ejection upon encountering damage. By observing this process, it will be possible to confirm this sequence of events and also determine the kinetics of UvrA release once the lesion is encountered. Furthermore, abortive complexes formed stochastically[99,102] that may contribute to kinetic proofreading will also be detected. Together these data will provide a comprehensive understanding of how the preincision complex is formed. The next process of UvrC recruitment can also be studied similarly and, as DNA helicase working together with DNA polymerase I has been proposed to remove UvrC and the damaged oligonucleotide,[103] we will also be able to discover whether UvrB is associated with the damaged strand or bound to the opposite strand, and whether UvrB is present as a dimer or a monomer. This latter question links back to our earlier discussion of hopping versus sliding; one of the key arguments against sliding is that it probes only one strand. Therefore, is the role of the $UvrA_2B_2$ complex to permit UvrB loading onto either strand? All these questions can be studied using multicolor direct imaging of repair proteins labeled with differently colored Qdots.

The physical principles that underlie NER are critical to our understanding of this process. Therefore, it is important that bulk and single-molecule studies go hand in hand. Furthermore, investigations in simple bacterial systems are an ideal platform for understanding these new concepts; diving into more complex mammalian counterpart systems will preclude a thorough understanding. Therefore, it is imperative that we revisit the bacterial system to achieve a robust grounding in the principles that likely underlie mammalian NER.

IV. Future Directions

A. Observing Protein Nanomachines at Work

The end goal of the investigations of protein machines working on DNA is to provide a kinetically correct, ordered mechanism. NER offers an ideal platform for creating such a view of DNA repair; it is a relatively simple multiprotein system that has been studied for many years, providing a body of knowledge from which we can now apply single-molecule methods to address the physical principles that underlie the mechanism. It has only recently become possible to image multiple proteins at the single-molecule level albeit currently there are many limitations, such as protein labeling, excitation of multiple dyes, and cross talk between multiple-emission color channels. However, with the introduction of Qdots that are all excited by a single wavelength, the possibilities for multiplexing are rapidly becoming reinforced. This still leaves key problems such as multiple specific protein labeling, and although imaging multiple colors is challenging, the use of image splitting and recombining technologies has enabled multiple differentially filtered images to be projected onto a single camera chip, permitting real-time multicolor imaging. Furthermore, with the very recent commercialization of complementary metal-oxide semiconductor (CMOS) cameras, the available field of view has increased enormously without the cost of readout speed. Although issues of sensitivity and quantum efficiency still loom before CMOS, the rapid advances in EMCCD tell us that a new generation of highly sensitive rapid-imaging cameras is around the corner. Much larger image chips enable more colors to be imaged at the same time; three and four colors are currently possible with even more being conceivable in the near future. These new cameras combined with multicolor protein tagging and single-molecule tracking may allow real-time imaging of all the steps of DNA repair processes in living cells in the future.

B. Overcoming the Brownian Motion Barrier

What happens as a diffusing molecule moves? We return to the problem that it is not trivial to distinguish hopping from sliding. Evidence for both scenarios exists but the definitive test is to directly observe motion along DNA (see Table I). Currently, the time resolution for such motions is limited by photon capture, that is, an image is formed by the photons collected by the detector; more photons give a more precise estimation of the position of the protein. However, more photons take longer to collect, in which time a diffuser can move, reducing the accuracy of its position determination. This paradox can be addressed by collecting photons more quickly; one way to do this is to use dark-field microscopy. A gold nanoparticle attached to the protein of interest will scatter light in proportion to the rate (intensity) of photons striking it. If the incident light enters from an oblique angle, then it will not be directly detected, thus creating a “dark” background. This approach has been successfully used to measure the motion of the molecular motor myosin V[104] with submillisecond time resolution. By observing slowly diffusing DNA–protein complexes moving with high spatial and temporal resolution, it will be possible to resolve whether proteins slide or hop. Additionally, gold nanorods offer the ability to follow the polarization of the molecule as it negotiates DNA; this is an important measure of whether or not the molecule follows the helical pitch of the DNA.

Recent technical advances have allowed for the development of new single-molecule approaches to study how proteins interact with DNA. A physical description of these interactions is vital as simply knowing the sequence of events only describes half of the picture. Furthermore, bulk methods average behavior across the entire population, whereas single-molecule approaches allow for a complete description of the various heterogeneous pathways and processes that molecules undertake in reality. This is clearly evident from the number of modes of motion that UvrAB employs to find its target site. The next challenge that faces the field is to construct more complex systems, and watch the entire process of NER occur in real time. First, this will be achieved *in vitro* and, ultimately, *in vivo.*[39] Though this is not an easy challenge, as our approaches become more interdisciplinary, it is both a worthy and attainable one.

References

1. Peng Y, Wang H, Santana dos Santos L, Kisker C, Van Houten B. Chapter 13. Nucleotide excision repair from bacteria to humans: Structure-function studies. In: Penning TM, editor. *Chemical carcinogenesis, current cancer research*. New York, NY: Springer Science + Business Media, LLC; 2011.
2. Jaciuk M, Nowak E, Skowronek K, Tanska A, Nowotny M. Structure of UvrA nucleotide excision repair protein in complex with modified DNA. *Nat Struct Mol Biol* 2011;**18**:191–7.
3. Croteau DL, DellaVecchia MJ, Perera L, Van Houten B. Cooperative damage recognition by UvrA and UvrB: identification of UvrA residues that mediate DNA binding. *DNA Repair (Amst)* 2008;**7**:392–404.
4. Croteau DL, DellaVecchia MJ, Wang H, Bienstock RJ, Melton MA, Van Houten B. The C-terminal zinc finger of UvrA does not bind DNA directly but regulates damage-specific DNA binding. *J Biol Chem* 2006;**281**:26370–81.
5. Wagner K, Moolenaar GF, Goosen N. Role of the insertion domain and the zinc-finger motif of Escherichia coli UvrA in damage recognition and ATP hydrolysis. *DNA Repair (Amst)* 2011;**10**:483–96.
6. Wagner K, Moolenaar GF, Goosen N. Role of the two ATPase domains of Escherichia coli UvrA in binding non-bulky DNA lesions and interaction with UvrB. *DNA Repair (Amst)* 2010;**9**:1176–86.
7. Van Houten B, Gamper H, Hearst JE, Sancar A. Analysis of sequential steps of nucleotide excision repair in Escherichia coli using synthetic substrates containing single psoralen adducts. *J Biol Chem* 1988;**263**:16553–60.
8. Truglio JJ, Croteau DL, Van Houten B, Kisker C. Prokaryotic nucleotide excision repair: the UvrABC system. *Chem Rev* 2006;**106**:233–52.
9. Van Houten B, Croteau DL, DellaVecchia MJ, Wang H, Kisker C. "Close-fitting sleeves": DNA damage recognition by the UvrABC nuclease system. *Mutat Res* 2005;**577**:92–117.
10. Truglio JJ, Karakas E, Rhau B, Wang H, DellaVecchia MJ, Van Houten B, et al. Structural basis for DNA recognition and processing by UvrB. *Nat Struct Mol Biol* 2006;**13**:360–4.

11. Skorvaga M, DellaVecchia MJ, Croteau DL, Theis K, Truglio JJ, Mandavilli BS, et al. Identification of residues within UvrB that are important for efficient DNA binding and damage processing. *J Biol Chem* 2004;**279**:51574–80.
12. Skorvaga M, Theis K, Mandavilli BS, Kisker C, Van Houten B. The beta -hairpin motif of UvrB is essential for DNA binding, damage processing, and UvrC-mediated incisions. *J Biol Chem* 2002;**277**:1553–9.
13. Karakas E, Truglio JJ, Croteau D, Rhau B, Wang L, Van Houten B, et al. Structure of the C-terminal half of UvrC reveals an RNase H endonuclease domain with an Argonaute-like catalytic triad. *EMBO J* 2007;**26**:613–22.
14. Truglio JJ, Rhau B, Croteau DL, Wang L, Skorvaga M, Karakas E, et al. Structural insights into the first incision reaction during nucleotide excision repair. *EMBO J* 2005;**24**:885–94.
15. Husain I, Van Houten B, Thomas DC, Abdel-Monem M, Sancar A. Effect of DNA polymerase I and DNA helicase II on the turnover rate of UvrABC excision nuclease. *Proc Natl Acad Sci USA* 1985;**82**:6774–8.
16. Kad NM, Wang H, Kennedy GG, Warshaw DM, Van Houten B. Collaborative dynamic DNA scanning by nucleotide excision repair proteins investigated by single-molecule imaging of quantum-dot-labeled proteins. *Mol Cell* 2010;**37**:702–13.
17. Malta E, Moolenaar GF, Goosen N. Dynamics of the UvrABC nucleotide excision repair proteins analyzed by fluorescence resonance energy transfer. *Biochemistry* 2007;**46**: 9080–8.
18. Pakotiprapha D, Inuzuka Y, Bowman BR, Moolenaar GF, Goosen N, Jeruzalmi D, et al. Crystal structure of Bacillus stearothermophilus UvrA provides insight into ATP-modulated dimerization, UvrB interaction, and DNA binding. *Mol Cell* 2008;**29**:122–33.
19. Pakotiprapha D, Liu Y, Verdine GL, Jeruzalmi D. A structural model for the damage-sensing complex in bacterial nucleotide excision repair. *J Biol Chem* 2009;**284**:12837–44.
20. DellaVecchia MJ, Croteau DL, Skorvaga M, Dezhurov SV, Lavrik OI, Van Houten B. Analyzing the handoff of DNA from UvrA to UvrB utilizing DNA-protein photoaffinity labeling. *J Biol Chem* 2004;**279**:45245–56.
21. Reardon JT, Sancar A. Thermodynamic cooperativity and kinetic proofreading in DNA damage recognition and repair. *Cell cycle (Georgetown, Tex.)* 2004;**3**:141–4.
22. Luijsterburg MS, von Bornstaedt G, Gourdin AM, Politi AZ, Mone MJ, Warmerdam DO, et al. Stochastic and reversible assembly of a multiprotein DNA repair complex ensures accurate target site recognition and efficient repair. *J Cell Biol* 2010;**189**:445–63.
23. Hopfield JJ. Kinetic proofreading: a new mechanism for reducing errors in biosynthetic processes requiring high specificity. *Proc Natl Acad Sci USA* 1974;**71**:4135–9.
24. Mellon I, Hanawalt PC. Induction of the Escherichia coli lactose operon selectively increases repair of its transcribed DNA strand. *Nature* 1989;**342**:95–8.
25. Selby CP, Sancar A. Molecular mechanism of transcription-repair coupling. *Science* 1993;**260**:53–8.
26. Gascon J, Oubina A, Perez-Lezaun A, Urmeneta J. Sensitivity of selected bacterial species to UV radiation. *Curr Microbiol* 1995;**30**:177–82.
27. Chandrasekhar D, Van Houten B. In vivo formation and repair of cyclobutane pyrimidine dimers and 6-4 photoproducts measured at the gene and nucleotide level in Escherichia coli. *Mutat Res* 2000;**450**:19–40.
28. Chandrasekhar D, Van Houten B. High resolution mapping of UV-induced photoproducts in the Escherichia coli lacI gene. Inefficient repair of the non-transcribed strand correlates with high mutation frequency. *J Mol Biol* 1994;**238**:319–32.
29. Gruskin EA, Lloyd RS. Molecular analysis of plasmid DNA repair within ultraviolet-irradiated Escherichia coli.II. UvrABC-initiated excision repair and photolyase-catalyzed dimer monomerization. *J Biol Chem* 1988;**263**:12738–43.

30. Gorman J, Greene EC. Visualizing one-dimensional diffusion of proteins along DNA. *Nat Struct Mol Biol* 2008;**15**:768–74.
31. Elowitz MB, Surette MG, Wolf PE, Stock JB, Leibler S. Protein mobility in the cytoplasm of Escherichia coli. *J Bacteriol* 1999;**181**:197–203.
32. Berg HC. *Random walks in biology*. Princeton, N.J: Princeton University Press; 1983.
33. Gowers DM, Halford SE. Protein motion from non-specific to specific DNA by three-dimensional routes aided by supercoiling. *EMBO J* 2003;**22**:1410–8.
34. Rau DC, Sidorova NY. Diffusion of the restriction nuclease EcoRI along DNA. *J Mol Biol* 2010;**395**:408–16.
35. Bonnet I, Biebricher A, Porte PL, Loverdo C, Benichou O, Voituriez R, et al. Sliding and jumping of single EcoRV restriction enzymes on non-cognate DNA. *Nucleic Acids Res* 2008;**36**:4118–27.
36. Biebricher A, Wende W, Escude C, Pingoud A, Desbiolles P. Tracking of single quantum dot labeled EcoRV sliding along DNA manipulated by double optical tweezers. *Biophys J* 2009;**96**:L50–2.
37. Dunn AR, Kad NM, Nelson SR, Warshaw DM, Wallace SS. Single Qdot-labeled glycosylase molecules use a wedge amino acid to probe for lesions while scanning along DNA. *Nucleic Acids Res* 2011;**39**:7487–98.
38. Wang H, DellaVecchia MJ, Skorvaga M, Croteau DL, Erie DA, Van Houten B. UvrB domain 4, an autoinhibitory gate for regulation of DNA binding and ATPase activity. *J Biol Chem* 2006;**281**:15227–37.
39. Elf J, Li GW, Xie XS. Probing transcription factor dynamics at the single-molecule level in a living cell. *Science* 2007;**316**:1191–4.
40. Gorman J, Plys AJ, Visnapuu ML, Alani E, Greene EC. Visualizing one-dimensional diffusion of eukaryotic DNA repair factors along a chromatin lattice. *Nat Struct Mol Biol* 2010;**17**:932–8.
41. Gorman J, Chowdhury A, Surtees JA, Shimada J, Reichman DR, Alani E, et al. Dynamic basis for one-dimensional DNA scanning by the mismatch repair complex Msh2-Msh6. *Mol Cell* 2007;**28**:359–70.
42. Blainey PC, van Oijen AM, Banerjee A, Verdine GL, Xie XS. A base-excision DNA-repair protein finds intrahelical lesion bases by fast sliding in contact with DNA. *Proc Natl Acad Sci USA* 2006;**103**:5752–7.
43. Jeong C, Cho WK, Song KM, Cook C, Yoon TY, Ban C, et al. MutS switches between two fundamentally distinct clamps during mismatch repair. *Nat Struct Mol Biol* 2011;**18**:379–85.
44. Kochaniak AB, Habuchi S, Loparo JJ, Chang DJ, Cimprich KA, Walter JC, et al. Proliferating cell nuclear antigen uses two distinct modes to move along DNA. *J Biol Chem* 2009;**284**:17700–10.
45. Tafvizi A, Huang F, Fersht AR, Mirny LA, van Oijen AM. A single-molecule characterization of p53 search on DNA. *Proc Natl Acad Sci USA* 2011;**108**:563–8.
46. Graneli A, Yeykal CC, Robertson RB, Greene EC. Long-distance lateral diffusion of human Rad51 on double-stranded DNA. *Proc Natl Acad Sci USA* 2006;**103**:1221–6.
47. Amitani I, Liu B, Dombrowski CC, Baskin RJ, Kowalczykowski SC. Watching individual proteins acting on single molecules of DNA. *Methods Enzymol* 2010;**472**:261–91.
48. Finkelstein IJ, Visnapuu ML, Greene EC. Single-molecule imaging reveals mechanisms of protein disruption by a DNA translocase. *Nature* 2010;**468**:983–7.
49. Kim JH, Larson RG. Single-molecule analysis of 1D diffusion and transcription elongation of T7 RNA polymerase along individual stretched DNA molecules. *Nucleic Acids Res* 2007;**35**: 3848–58.
50. Harada Y, Funatsu T, Murakami K, Nonoyama Y, Ishihama A, Yanagida T. Single-molecule imaging of RNA polymerase-DNA interactions in real time. *Biophys J* 1999;**76**:709–15.

51. Etson CM, Hamdan SM, Richardson CC, van Oijen AM. Thioredoxin suppresses microscopic hopping of T7 DNA polymerase on duplex DNA. *Proc Natl Acad Sci USA* 2010;**107**:1900–5.
52. Porecha RH, Stivers JT. Uracil DNA glycosylase uses DNA hopping and short-range sliding to trap extrahelical uracils. *Proc Natl Acad Sci USA* 2008;**105**:10791–6.
53. Schurr JM. The one-dimensional diffusion coefficient of proteins absorbed on DNA. Hydrodynamic considerations. *Biophys Chem* 1979;**9**:413–4.
54. von Hippel PH, Berg OG. Facilitated target location in biological systems. *J Biol Chem* 1989;**264**:675–8.
55. Berg OG, Winter RB, von Hippel PH. Diffusion-driven mechanisms of protein translocation on nucleic acids. 1. Models and theory. *Biochemistry* 1981;**20**:6929–48.
56. Wagner K, Moolenaar G, van Noort J, Goosen N. Single-molecule analysis reveals two separate DNA-binding domains in the Escherichia coli UvrA dimer. *Nucleic Acids Res* 2009;**37**:1962–72.
57. Halford SE, Welsh AJ, Szczelkun MD. Enzyme-mediated DNA looping. *Annu Rev Biophys Biomol Struct* 2004;**33**:1–24.
58. Fickert R, Muller-Hill B. How Lac repressor finds lac operator in vitro. *J Mol Biol* 1992;**226**:59–68.
59. Wang F, Greene EC. Single-molecule studies of transcription: from one RNA polymerase at a time to the gene expression profile of a cell. *J Mol Biol* 2011;**412**:814–31.
60. van Oijen AM, Loparo JJ. Single-molecule studies of the replisome. *Annu Rev Biophys* 2010;**39**:429–48.
61. Rigler R. Fluorescence and single molecule analysis in cell biology. *Biochem Biophys Res Commun* 2010;**396**:170–5.
62. Rajagopalan S, Huang F, Fersht AR. Single-Molecule characterization of oligomerization kinetics and equilibria of the tumor suppressor p53. *Nucleic Acids Res* 2010;**39**:2294–303.
63. Manley S, Gillette JM, Lippincott-Schwartz J. Single-particle tracking photoactivated localization microscopy for mapping single-molecule dynamics. *Methods Enzymol* 2010;**475**:109–20.
64. Lymperopoulos K, Kiel A, Seefeld A, Stohr K, Herten DP. Fluorescent probes and delivery methods for single-molecule experiments. *Chemphyschem* 2010;**11**:43–53.
65. Pierobon P, Cappello G. Quantum dots to tail single bio-molecules inside living cells. *Adv Drug Deliv Rev* 2012;**64**:167–78.
66. Lavelle C, Praly E, Bensimon D, Le Cam E, Croquette V. Nucleosome remodeling machines and other molecular motors observed at the single molecule level. *FEBS J* 2011;**278**:3596–607.
67. Gill JP, Wang J, Millar DP. DNA polymerase activity at the single-molecule level. *Biochem Soc Trans* 2011;**39**:595–9.
68. English BP, Hauryliuk V, Sanamrad A, Tankov S, Dekker NH, Elf J. Single-molecule investigations of the stringent response machinery in living bacterial cells. *Proc Natl Acad Sci USA* 2011;**108**:E365–73.
69. Friedman JI, Stivers JT. Detection of damaged DNA bases by DNA glycosylase enzymes. *Biochemistry* 2010;**49**:4957–67.
70. Riggs AD, Bourgeois S, Cohn M. The lac repressor-operator interaction. 3. Kinetic studies. *J Mol Biol* 1970;**53**:401–17.
71. Riggs AD, Suzuki H, Bourgeois S. Lac repressor-operator interaction I. Equilibrium studies. *J Mol Biol* 1970;**48**:67–83.
72. Sakata-Sogawa K, Shimamoto N. RNA polymerase can track a DNA groove during promoter search. *Proc Natl Acad Sci USA* 2004;**101**:14731–5.
73. van den Broek B, Lomholt MA, Kalisch SM, Metzler R, Wuite GJ. How DNA coiling enhances target localization by proteins. *Proc Natl Acad Sci USA* 2008;**105**:15738–42.

74. Bianco PR, Kowalczykowski SC. Translocation step size and mechanism of the RecBC DNA helicase. *Nature* 2000;**405**:368–72.
75. Brower-Toland BD, Smith CL, Yeh RC, Lis JT, Peterson CL, Wang MD. Mechanical disruption of individual nucleosomes reveals a reversible multistage release of DNA. *Proc Natl Acad Sci USA* 2002;**99**:1960–5.
76. Perkins TT, Li HW, Dalal RV, Gelles J, Block SM. Forward and reverse motion of single RecBCD molecules on DNA. *Biophys J* 2004;**86**:1640–8.
77. Skinner GM, Baumann CG, Quinn DM, Molloy JE, Hoggett JG. Promoter binding, initiation, and elongation by bacteriophage T7 RNA polymerase. A single-molecule view of the transcription cycle. *J Biol Chem* 2004;**279**:3239–44.
78. van Oijen AM, Blainey PC, Crampton DJ, Richardson CC, Ellenberger T, Xie XS. Single-molecule kinetics of lambda exonuclease reveal base dependence and dynamic disorder. *Science* 2003;**301**:1235–8.
79. Abbondanzieri EA, Greenleaf WJ, Shaevitz JW, Landick R, Block SM. Direct observation of base-pair stepping by RNA polymerase. *Nature* 2005;**438**:460–5.
80. Lee JB, Hite RK, Hamdan SM, Xie XS, Richardson CC, van Oijen AM. DNA primase acts as a molecular brake in DNA replication. *Nature* 2006;**439**:621–4.
81. Strick TR, Croquette V, Bensimon D. Single-molecule analysis of DNA uncoiling by a type II topoisomerase. *Nature* 2000;**404**:901–4.
82. Allemand JF, Bensimon D, Jullien L, Bensimon A, Croquette V. pH-dependent specific binding and combing of DNA. *Biophys J* 1997;**73**:2064–70.
83. Lyon WA, Fang MM, Haskins WE, Nie S. A dual-beam optical microscope for observation and cleavage of single DNA molecules. *Anal Chem* 1998;**70**:1743–8.
84. Graneli A, Yeykal CC, Prasad TK, Greene EC. Organized arrays of individual DNA molecules tethered to supported lipid bilayers. *Langmuir* 2006;**22**:292–9.
85. Wang H, Tessmer I, Croteau DL, Erie DA, Van Houten B. Functional characterization and atomic force microscopy of a DNA repair protein conjugated to a quantum dot. *Nano Lett* 2008;**8**:1631–7.
86. Pinaud F, Clarke S, Sittner A, Dahan M. Probing cellular events, one quantum dot at a time. *Nat Methods* 2010;**7**:275–85.
87. Medintz IL, Uyeda HT, Goldman ER, Mattoussi H. Quantum dot bioconjugates for imaging, labelling and sensing. *Nat Mater* 2005;**4**:435–46.
88. Clarke S, Pinaud F, Beutel O, You C, Piehler J, Dahan M. Covalent monofunctionalization of peptide-coated quantum dots for single-molecule assays. *Nano Lett* 2010;**10**:2147–54.
89. DeRocco V, Anderson T, Piehler J, Erie DA, Weninger K. Four-color single-molecule fluorescence with noncovalent dye labeling to monitor dynamic multimolecular complexes. *Biotechniques* 2010;**49**:807–16.
90. Roullier V, Clarke S, You C, Pinaud F, Gouzer GG, Schaible D, et al. High-affinity labeling and tracking of individual histidine-tagged proteins in live cells using Ni2+ tris-nitrilotriacetic acid quantum dot conjugates. *Nano Lett* 2009;**9**:1228–34.
91. You C, Wilmes S, Beutel O, Lochte S, Podoplelowa Y, Roder F, et al. Self-controlled monofunctionalization of quantum dots for multiplexed protein tracking in live cells. *Angewandte Chemie* 2010;**49**:4108–12 (International ed. in English).
92. Orren DK, Sancar A. The (A)BC excinuclease of Escherichia coli has only the UvrB and UvrC subunits in the incision complex. *Proc Natl Acad Sci USA* 1989;**86**:5237–41.
93. Orren DK, Sancar A. Formation and enzymatic properties of the UvrB.DNA complex. *J Biol Chem* 1990;**265**:15796–803.
94. Wang H, Lu M, Tang MS, Van Houten B, Ross JB, Weinfeld M, et al. DNA wrapping is required for DNA damage recognition in the Escherichia coli DNA nucleotide excision repair pathway. *Proc Natl Acad Sci USA* 2009;**106**:12849–54.

95. Saxton MJ. A biological interpretation of transient anomalous subdiffusion I. Qualitative model. *Biophys J* 2007;**92**:1178–91.
96. Guigas G, Weiss M. Sampling the cell with anomalous diffusion - the discovery of slowness. *Biophys J* 2008;**94**:90–4.
97. Yeung AT, Mattes WB, Grossman L. Protein complexes formed during the incision reaction catalyzed by the Escherichia coli UvrABC endonuclease. *Nucleic Acids Res* 1986;**14**:2567–82.
98. Van Houten B. Nucleotide excision repair in Escherichia coli. *Microbiol Rev* 1990;**54**:18–51.
99. Caron PR, Grossman L. Incision of damaged versus nondamaged DNA by the Escherichia coli UvrABC proteins. *Nucleic Acids Res* 1988;**16**:7855–65.
100. Verhoeven EE, van Kesteren M, Moolenaar GF, Visse R, Goosen N. Catalytic sites for 3′ and 5′incision of Escherichia coli nucleotide excision repair are both located in UvrC. *J Biol Chem* 2000;**275**:5120–3.
101. Moolenaar GF, Franken KL, van de Putte P, Goosen N. Function of the homologous regions of the Escherichia coli DNA excision repair proteins UvrB and UvrC in stabilization of the UvrBC-DNA complex and in 3′-incision. *Mutat Res* 1997;**385**:195–203.
102. Moolenaar GF, Hoglund L, Goosen N. Clue to damage recognition by UvrB: residues in the beta-hairpin structure prevent binding to non-damaged DNA. *EMBO J* 2001;**20**:6140–9.
103. Orren DK, Selby CP, Hearst JE, Sancar A. Post-incision steps of nucleotide excision repair in Escherichia coli. Disassembly of the UvrBC-DNA complex by helicase II and DNA polymerase I. *J Biol Chem* 1992;**267**:780–8.
104. Dunn AR, Spudich JA. Dynamics of the unbound head during myosin V processive translocation. *Nat Struct Mol Biol* 2007;**14**:246–8.

Transcription-Coupled DNA Repair in Prokaryotes

Ann Ganesan, Graciela Spivak, and Philip C. Hanawalt

Department of Biology, Stanford University, Stanford, California, USA

Transcription-coupled repair (TCR) is a subpathway of nucleotide excision repair (NER) that acts specifically on lesions in the transcribed strand of expressed genes. First reported in mammalian cells, TCR was then documented in *Escherichia coli*. In this organism, an RNA polymerase arrested at a lesion is displaced by the transcription repair coupling factor, *Mfd*. This protein recruits the NER lesion-recognition factor *UvrA*, and then dissociates from the DNA. *UvrA* binds UvrB, and the assembled UvrAB* complex initiates repair. In mutants lacking active *Mfd*, TCR is absent. A gene transcribed by the bacteriophage T7 RNA polymerase in *E. coli* also requires *Mfd* for TCR. The CSB protein (missing or defective in cells of patients with Cockayne syndrome, complementation group B) is essential for TCR in humans. CSB and its homologs in higher eukaryotes are likely functional equivalents of *Mfd*.

I. Introduction

Nucleotide excision repair (NER), an intensively studied DNA repair mechanism, was discovered in *Escherichia coli* and is now believed to occur in essentially all living organisms. The major features of the genetic control of NER and the biochemical pathways involved have been well documented;

Progress in Molecular Biology
and Translational Science, Vol. 110
DOI: 10.1016/B978-0-12-387665-2.00002-X

however, there is still much to be learned about the details of this process and its relation to other DNA transactions, especially replication and transcription.[1]

In *E. coli*, the first step in NER depends upon a complex (UvrAB*) composed of two molecules of the *UvrA* protein and one or two molecules of the UvrB protein finding and recognizing DNA damage. This is challenging because of the enormous tangle of undamaged DNA compared to the low frequency of lesions and the small number of molecules of the UvrAB* complex in each bacterium, especially if the SOS stress response system, which can regulate the levels of *UvrA* and UvrB, has not been induced (see Chapter 1 in this volume, "Dynamics of Lesion Processing by Bacterial Nucleotide Excision Repair Proteins" for a more complete discussion of this subject). Another strategic problem is the large variety and abundance of proteins associated with DNA as part of normal metabolism. After the UvrAB* complex recognizes and binds to a lesion, *UvrA* dissociates from the complex, leaving UvrB bound to the DNA. The UvrC endonuclease then associates with the UvrB–DNA complex and nicks the DNA on either side of the lesion. The resulting oligonucleotide is removed by the UvrD helicase and DNA polymerase I. The polymerase synthesizes a replacement patch using the undamaged complementary DNA strand as a template, and finally, ligase (LigA) joins the newly synthesized patch to the adjacent preexisting strand.

Two forms of NER have been described (Fig. 1): transcription-coupled repair (TCR) and global genomic repair (GGR). The former removes lesions in the transcribed strand of active genes while the latter removes lesions in the genome overall. In the case of GGR, the UvrAB* complex recognizes DNA lesions directly. In contrast, in TCR, the RNA polymerase elongation complex (EC) first encounters a lesion in the transcribed strand of a gene. If the RNA polymerase remains arrested at the lesion, it prevents repair; however, the *Mfd* protein, originally known as the transcription repair coupling factor (TRCF), can dissociate the RNA polymerase and the nascent transcript from the site and recruit *UvrA*, which attracts UvrB. As a result, the transcribed strand of the gene is repaired more rapidly than the nontranscribed strand or the genome overall. One of the remaining questions about TCR is exactly how the rate of repair is enhanced, considering that more enzymatic steps are involved in TCR than in GGR.

II. Background: Genomic Heterogeneity in NER and the Discovery of TCR

Based on her studies of mutation and survival in *E. coli*, Evelyn Witkin concluded that UV-induced damage might be repaired more efficiently in some parts of the genome than in others.[2] Subsequent studies of mammalian cells in

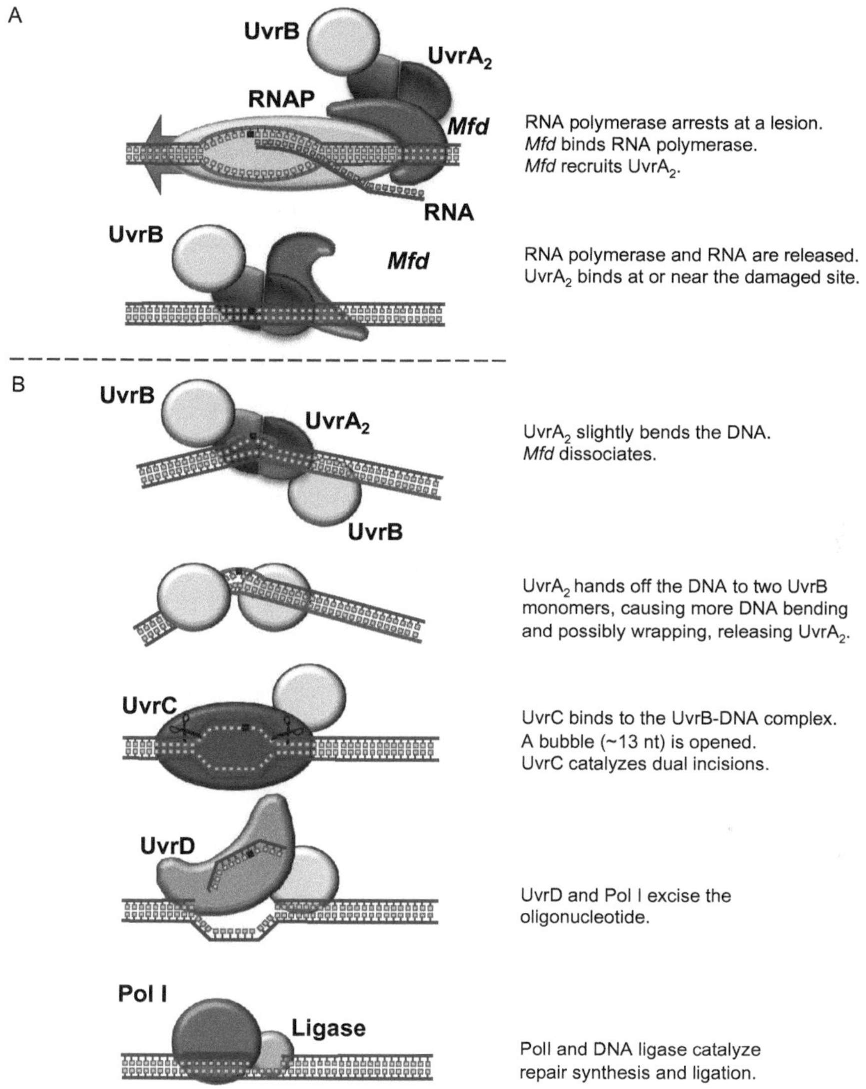

FIG. 1. Transcription-coupled and global genomic repair of DNA lesions in *Escherichia coli* (adapted from Ref. 1). (A) These steps are specific to transcription-coupled repair. (B) These steps are common to transcription-coupled repair and global genomic repair. Although the figure depicts the dissociation of RNA polymerase after the UvrAB* complex has been recruited, the sequence of these events is not known. The relative sizes of the nucleic acids and proteins depicted may not reflect their actual sizes and conformations. (See Color Insert.)

culture produced direct evidence that NER was not completely random with respect to the DNA sequences repaired. For example, Zolan *et al.*[3] observed that furocoumarin adducts were repaired more slowly in the alpha DNA sequences of African green monkey cells than in the bulk of the DNA. Similar results were obtained for aflatoxin B1 adducts.[4] Shortly thereafter, Mansbridge and Hanawalt[5] found that in cells derived from a patient with xeroderma pigmentosum complementation group C (XPC), some domains of the DNA were repaired rapidly after ultraviolet (UV) irradiation while the rest were repaired slowly, if at all. Then Bohr *et al.*[6] observed that in UV-irradiated Chinese hamster ovary (CHO) cells, the active *DHFR* gene was repaired more rapidly than the genome overall, and Mellon *et al.*[7] showed that the preferential repair of the *DHFR* gene in CHO and in human cells in culture was due to the rapid repair of the transcribed strand of the expressed gene. The nontranscribed strand was repaired at the same rate as the genome overall. Extending those studies, the authors found that the transcribed strand of the *lac* operon of *E. coli* was repaired more rapidly than the nontranscribed strand when the operon was induced, but not when it was repressed.[8] The studies showing preferential repair of UV damage in transcriptionally active genes in cultured CHO and human cells had led to the idea that chromatin might affect the rate of repair, and that DNA damage in active genes might be more accessible to repair complexes than damage in inactive regions of the genome.[6,9] The observation that the rapid repair was specific to the transcribed strand of a gene made this idea unlikely, however,[7] especially because of the results obtained with *E. coli*, which lacks the chromatin structure of eukaryotic DNA. It then was proposed that RNA polymerase blocked at a lesion in the transcribed strand might facilitate recognition of the damage by the NER incision complex.[8]

III. The Role of RNA Polymerase in TCR

Extending the analysis of the mechanism(s) underlying TCR, Selby and Sancar used purified proteins *in vitro* to study the role of the *E. coli* RNA polymerase in repair. They observed that a cyclobutane pyrimidine dimer (CPD) in the transcribed strand of a gene blocked the RNA polymerase EC, which then prevented incision of the DNA by the repair complex.[10] Thus, the blocked RNA polymerase inhibited repair and could not directly account for TCR. In further studies, Selby and Sancar isolated a protein that they called "transcription repair coupling factor" (TRCF). This protein was able to dissociate the blocked RNA polymerase and the nascent RNA transcript from the DNA and facilitate excision *in vitro*.[11] They later established that the *mfd* gene of *E. coli* coded for this protein, also known as *Mfd*.[12] Curiously, subsequent work with *E. coli* implicated the mismatch repair proteins MutL and MutS in TCR and in mutation frequency decline (MFD),[13,14] although the reactions reconstituted *in vitro* by Selby and Sancar did not require these proteins.

Although the RNA polymerase blocked at a lesion inhibits repair, it is an essential participant in TCR. It plays an important role in scanning for damage as it translocates, and having encountered a lesion, if blocked, it becomes a target of *Mfd*. The interaction with the arrested RNA polymerase activates *Mfd*, which then recruits *UvrA*. The β subunit of the polymerase, coded by the *rpoB* gene, interacts with *Mfd*, and certain mutations affecting the N-terminal region of the RpoB protein prevent this interaction and abolish TCR.[15–18]

An unexpected observation was that in *E. coli*, a gene engineered to be transcribed by the RNA polymerase of bacteriophage T7 showed TCR when it was actively transcribed but not when it was downregulated.[19] Even more surprising, the TCR promoted by the T7 polymerase *in vivo* was not observed in an *mfd*$^-$ strain, implicating *Mfd* in the response although Selby and Sancar had shown that *Mfd* did not interact with the T7 polymerase *in vitro*.[12] The biochemical details of this phenomenon remain to be determined.

IV. The Role of *Mfd* in TCR

In 1956, Witkin[20] documented a phenomenon in UV-irradiated *E. coli* that was later called "mutation frequency decline",[21] and she identified a gene (*mfd*) associated with it.[2,22] The phenomenon was characterized by a decrease in certain suppressor mutations when protein synthesis was inhibited in UV-irradiated *E. coli*, and it did not occur in excision-deficient mutants or in *mfd*$^-$ mutants. Subsequently, Bockrath and Palmer[23] presented evidence that MFD reflected the repair of premutational lesions (probably T<>C CPDs[24]) in tRNA genes and that the repair occurred more rapidly in the transcribed strand than in the nontranscribed strand. They suggested that the nascent RNA transcript might provide the template for efficient repair of the transcribed strand, a plausible hypothesis at the time.

Based on the observations and ideas of Bockrath and Palmer, Selby and Sancar cloned and sequenced the *mfd* gene, purified the *Mfd* protein, and characterized it.[11,12,15,25,26] They ascertained that *Mfd* and TRCF were identical, and they established that when the RNA polymerase EC was blocked at a lesion, *Mfd* could dissociate the polymerase and the nascent RNA transcript from the DNA. Because part of the *Mfd* protein sequence was homologous to a portion of the UvrB protein, and because they found that *Mfd* could bind *UvrA* (but not UvrB), they proposed that that *Mfd* could recruit *UvrA*, facilitating repair. Subsequent studies have supported and extended these ideas, corroborating the idea that the initiation of NER is different for TCR than for GGR.

The *E. coli Mfd* is a monomeric 130 kDa protein containing eight structural domains[27] (Fig. 2). The N-terminal portion, consisting of D1a, D2, D1b, and D3, is joined to the C-terminal portion (D4, D5, D6, and D7) by a flexible linker of 23 amino acids. Although probably not normally associated with the EC,[29] *Mfd* can interact with the β subunit of the RNA polymerase when the EC is arrested by a DNA lesion, a DNA-bound protein, or by nucleotide depletion.[16] The *Mfd* protein appears to approach the EC from the upstream direction, the interaction being inhibited by a σ factor bound to the β subunit of the polymerase but not by a downstream protein bound to the DNA. If the EC encounters a pause site, backtracks a few nucleotides, and becomes transiently arrested, *Mfd* can reactivate it and allow it to continue transcribing if the needed nucleoside triphosphates are present. It apparently does this by pushing the polymerase forward to realign the RNA transcript in the active site. Similarly, if the EC has been arrested by nucleotide depletion *Mfd* can facilitate its restart if nucleotides become available.[16,28,30–32] In contrast, if nucleotides are not available or if the EC is blocked by a DNA lesion such as a CPD, or a protein bound to the DNA, *Mfd* can displace the RNA polymerase and the nascent transcript from the DNA. It has been reported that *Mfd* can also displace an EC that encounters a head-on replisome complex,[33,34] although an alternative result has also been described.[35]

Studies of the structure and function of *Mfd* have shown that the enzymatic activities of the purified wild-type protein in solution are inhibited by an interaction between two of the protein domains, the N-terminal D2 and the C-terminal D7.[36] This interaction interferes with the RNA polymerase displacement, ATPase, and DNA translocase activities. Changes in the protein that prevent the interaction of D2 and D7 not only enhance these enzymatic activities but also result in a protein that is more sensitive to proteases, indicating significant changes in conformation.[37–40] Presumably, the conformation of the protein is regulated *in vivo* so that *Mfd* only displaces RNA polymerase from the DNA when the polymerase is arrested.

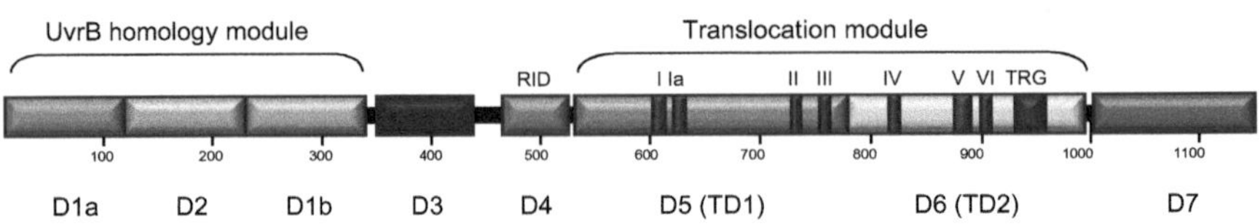

FIG. 2. Schematic linear representation of the *Mfd* protein (adapted from Ref. 28): The UvrA-interacting domain (Domain 2) is homologous to the N-terminal UvrA-binding domain of UvrB. The RNA polymerase interacting domain (RID) is shown in pale blue. The translocation domain of *Mfd*, shown in green, contains seven SF2 helicase motifs, indicated with Roman numerals. The TRG motif, shown in purple, is homologous to the translocase domain of RecG. Numbers below the bars refer to amino acids. (See Color Insert.)

Interaction between *Mfd* and RNA polymerase is crucial to EC displacement, and it involves the RNA polymerase interaction domain (RID), D4, of *Mfd* and the β subunit of the polymerase. Studies of the effects of mutations in these regions, together with structural studies of the individual proteins and of the *Mfd*–RID complexed with the β1 portion of the RNA polymerase β subunit, have provided information about the nature of the interaction and the amino acids involved. For example, a mutation in D4, L499R, abolishes the displacement of RNA polymerase from DNA but not the DNA binding, ATP binding, or ATP hydrolysis by *Mfd*.[28] In many of the structural studies, the *Mfd* protein analyzed has been in the closed conformation in which the enzymatic activities, including RNA polymerase displacement, are inhibited.[17,28,32,36] What determines the conformation of the protein *in vivo*, and exactly how the *Mfd* protein initiates a productive interaction with the EC that results in the forward movement or the dissociation of the RNA polymerase from DNA remains puzzling.

In contrast to *Mfd*, when an EC encounters another EC blocked by a protein bound to the DNA, the upstream EC may help the downstream EC overcome the block. Two upstream ECs to "push" the downstream EC past the block are more efficient than one. Unlike *Mfd*, the upstream EC does not displace the downstream EC but allows it to continue transcription.[41,42] What determines the outcome of such encounters is not understood.

In addition to the RID, the superfamily 2 translocation module (D5 and D6) of *Mfd* is required for the displacement of a stalled or blocked EC. Some sequences of this "motor domain" are homologous to sequences in the RecG protein, and like RecG, *Mfd* is an ATP-dependent double-strand DNA translocase. Certain substitutions in the TRG domain (e.g., R953A and Q963A) abrogate RNA polymerase displacement *in vitro* and *in vivo* without interfering with DNA binding and nucleotide hydrolysis.[28,37]

In addition to displacing the EC blocked at a lesion, *Mfd* recruits *UvrA*. The UvrB homology module (D1a, D2, D1b) of *Mfd* is absolutely required for this,[39] but the details of the mechanism involved have not been elucidated. In the case of GGR, Orren and Sancar[43] and Lin and Sancar[44] proposed that a heterotrimer composed of two *UvrA* molecules and one UvrB molecule initiated the identification of a lesion. Subsequently, Malta *et al.*[45] presented evidence that the identification involved a heterotetramer of two *UvrA* molecules and two UvrB molecules, with the two UvrB molecules then being loaded onto the DNA by *UvrA*. More recently, using Q-dot technology, Kad *et al.*[46] confirmed that the $UvrA_2UvrB_2$ heterotetramer could form even in the absence of DNA, but found that it was not necessary for loading UvrB on DNA. Although UvrB did not bind to DNA in the absence of *UvrA*, the *UvrA* homodimer did bind to DNA in the absence of UvrB, and UvrB could then associate with the bound *UvrA*. Under the experimental conditions used, UvrB was mostly present as a monomer, and the *UvrA* homodimer was able to bind two-independent UvrB molecules. Data presented by Pakotiprapha *et al.*,[47] however, favored the heterotrimer. Considering this, one might imagine that the *UvrA* dimer binds an *Mfd* molecule at a lesion and then attracts a UvrB molecule, which causes *Mfd* to dissociate; however, it seems more likely that *Mfd* recruits the *UvrA* dimer and transfers it to DNA, then dissociates before UvrB binds to *UvrA*. This idea is based on the observation that *Mfd* can displace UvrB from a UvrAB* complex *in vitro*.[12]

The D7 is not required for TCR, but its autoregulatory function may prevent unwanted interactions with *UvrA* that could reduce GGR capacity, especially if there are more *Mfd* than *UvrA* molecules in a cell.[39]

Mfd may represent a special case of proteins that facilitate recognition of DNA lesions by UvrAB*. In *in vitro* reactions, photolyase from *E. coli*, but not from yeast, increases the rate of incision of DNA containing CPDs,[48,49] and it enhances the survival of UV-irradiated *uvrA* mutants in which the insertion domain of *UvrA* has been deleted.[50] Similarly, the YbaZ protein, an alkyltransferase-like protein that binds to O(6)-alkylguanine lesions but does not repair them, facilitates NER of O(6)-ethylguanine and O(6)-propylguanine by interacting with *UvrA*.[51] In contrast to photolyase and YbaZ, *Mfd* recognizes RNA polymerase blocked at a DNA lesion rather than a specific type of damage, and it can therefore recruit *UvrA* to a variety of lesions. Unlike *Mfd*, neither photolyase nor YbaZ is known to have significant UvrB homology, and the basis for their interaction with *UvrA* is not known.

V. The Role of *UvrA* in TCR

The *E. coli UvrA* is a 103-kD protein in which 11 domains have been identified[52] (Fig. 3). This protein, which is required for both TCR and GGR, may be the least prevalent of the proteins involved in NER, although estimates vary from approximately 20 molecules per cell to approximately 200 before SOS induction.[43,53] The active form of *UvrA* is a homodimer, and the low ratio of *UvrA* dimers to lesions in irradiated cells suggests that *UvrA* should turn over relatively rapidly to attain the rates of repair observed. The amount of *UvrA* may be kept at a low level under normal conditions to minimize "gratuitous" NER in undamaged DNA.

Certain mutations in *uvrA* caused reduced levels of GGR relative to TCR. This result was interpreted to mean that these mutants are deficient in recognizing DNA damage and initiating GGR but are able to catalyze TCR because RNA polymerase and *Mfd* contribute to damage recognition.[39] None of the TCR-proficient *uvrA* mutants tested was completely deficient in GGR, however, and none of the GGR-proficient *uvrA* mutants lacked TCR.[39] In contrast,

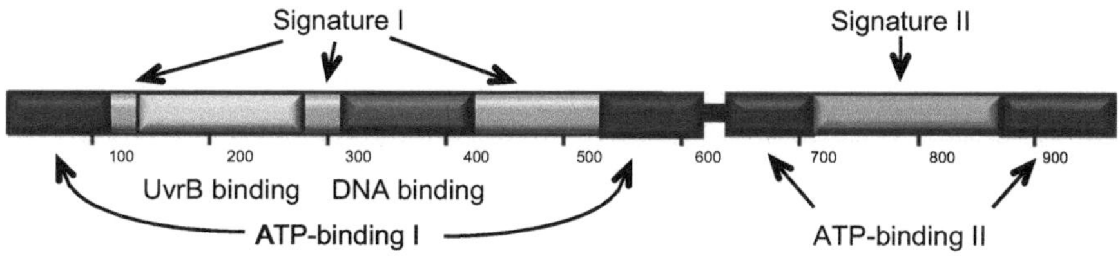

FIG. 3. Schematic linear representation of the *UvrA* protein (adapted from Ref. 52). Its domains are colored: ATP-binding I, red; signature I, pink; ATP-binding II, dark blue; signature II, green; UvrB-binding, yellow; insertion domain (labeled DNA binding), purple; and linker, black. Inactivation of either one of the two ATP-binding domains causes the loss of ATPase activity *in vitro* and UV sensitivity *in vivo*. Numbers below the bars refer to amino acids. (See Color Insert.)

in eukaryotes, including human cells, mutants lacking TCR but not GGR or GGR but not TCR have been identified. For example, cells from XPC patients show TCR but not GGR, while cells from patients with Cockayne syndrome (CSA or CSB) lack TCR but retain the capacity for GGR. Studies of cells deficient in the CSB (ERCC6) protein have led to the idea that this protein and *Mfd* play similar roles in TCR; however, the interactions of CSB with the other components of TCR in mammalian cells are more complex and less clearly understood than those of *Mfd*.

The ability of *UvrA* to interact with UvrB is required for TCR as well as for GGR, and the interaction has been analyzed in detail. Specific mutations in *UvrA* (e.g., R176, R206, and D219 but not E185) or in UvrB (e.g., E215 and R223 but not E222) prevent the interaction and thus inhibit repair.[47,52,54]

VI. The Role of UvrB in TCR

Like *UvrA*, UvrB (Fig. 4) is required for both GGR and TCR. Because the principal difference between TCR and GGR is in the initial steps of damage recognition, once *UvrA* has been recruited and *Mfd* released from the vicinity of a lesion, the transfer of UvrB to DNA and the subsequent incision and resynthesis are thought to involve the same biochemical steps as GGR. The UvrAB* complex recognizes damage in GGR, and when the *UvrA* protein dissociates, UvrB must be positioned to allow the UvrC endonuclease to bind and incise the damaged strand, not the undamaged strand. Assuming the same is true for TCR, one might expect that, if there were a particular kind of damage that only arrested the RNA polymerase when the lesion was in the nontranscribed strand, the repair of the nontranscribed strand would be more rapid than the repair of the transcribed strand. Although *Mfd* would recruit

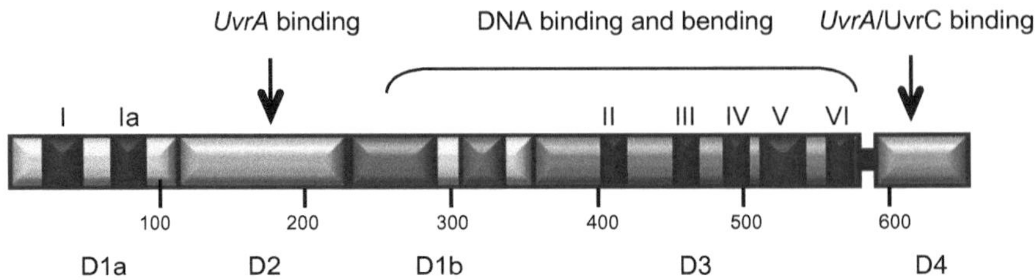

FIG. 4. Schematic linear representation of the UvrB protein (adapted from Refs. 55,56). The UvrA-binding region in the amino-terminal domain is homologous to the UvrA-binding region of *Mfd*. The carboxyl-terminal domain, D4 (gray), is an autoinhibitory domain.[57] The UvrA-binding region in D4 overlaps the UvrC-binding site. The helicase motifs, shown in red in D1a and D3, participate in DNA binding. Numbers below the bars refer to amino acids. (See Color Insert.)

UvrA to the DNA at the site of the arrested polymerase, after *Mfd* dissociated, UvrB should recognize the damage on the nontranscribed strand, bind to it, and attract UvrC. To our knowledge, this situation has not been observed.

VII. Other Examples of Transcription-Related DNA Damage Processing in Bacteria

Several reports have described phenomena associated with transcription and termed TCR; however, in some cases, the studies do not include direct measurements of DNA damage removal or results showing that the transcribed strand of an active gene is repaired more rapidly than the nontranscribed strand. We have used the term "transcription-related DNA damage processing" for these responses until more information becomes available.

A. NusA

The relatively neat picture of TCR involving RNA polymerase, *Mfd*, and the UvrAB* complex has become more complicated with the observation that the transcription terminator/antiterminator NusA may also be implicated. In fact, NusA may be involved in two forms of transcription-related DNA damage processing, one being *Mfd* dependent and the other being *Mfd* independent. Support for this idea comes from studies of the effect of *nusA* mutants on the survival of several derivatives of *E. coli* K-12 treated with nitrofurazone, NQO, MMS, or UV, together with the evidence that NusA associates with the β and β' subunits of the bacterial RNA polymerase and with *UvrA*[58–61]; however, no direct evaluation of its effect on DNA damage removal has been reported. To evaluate the role of NusA, in repair it will be essential to measure lesion removal in *nusA* mutants and to ascertain whether the mutations have differential effects on the repair of the transcribed and nontranscribed strands of an active gene. A clearer understanding of the interactions of *Mfd* and NusA, both of which can associate with the β subunit of RNA polymerase, would also be helpful.

Several other proteins that associate with RNA polymerase (e.g., NusG) have yet to be assessed for their relationship to DNA repair.

B. Base Excision Repair

While NER primarily removes "bulky" or distorting DNA lesions, base excision repair (BER) usually removes oxidative base damage and abnormal bases. Although the evidence for TCR of "bulky" lesions that arrest RNA polymerase, including CPDs, cisplatin, cholesterol, and menthol, is convincing, the results of studies of oxidative lesions have been more controversial, and the reports of TCR of thymine glycols in human cells based on an antibody assay have been retracted.[62–65] In contrast, recent analyses of the repair of 5-hydroxy-uracil in human cells and cell extracts have shown that the repair of this lesion is coupled to transcription.[66]

Studies of transcriptional mutagenesis (TM) in nondividing *E. coli* have implicated NER and TCR in the repair of uracil and 8-oxoguanine at a specific site in the transcribed strand of a luciferase gene.[67,68] The evidence is based on the effects of *uvrA* and *mfd* mutations on the frequency of repair compared to TM, the measured end point being luciferase activity resulting from TM. Although neither uracil nor 8-oxoguanine appears to be a significant block to the *E. coli* RNA polymerase *in vitro*,[69] *Mfd* appears to contribute to the repair of these lesions. The experimental system does not allow measurements of repair of the nontranscribed strand of the gene, however, so the relative repair rates of the transcribed and the nontranscribed strands under these conditions is not known. Because the *mfd* mutation was not completely epistatic to the *uvrA* mutation, it was proposed that BER components are involved in some of the *Mfd*-dependent repair. Further studies are needed to identify these components and to validate the model proposed.[68]

VIII. Conclusions

The study of *E. coli* continues to provide important mechanistic insights into DNA damage-processing pathways such as TCR, but with new questions being generated as previous ones are answered. For example, direct measurements of lesion removal have shown that *Mfd* is absolutely required for TCR as it was originally described, but inactivating *Mfd* has little effect on the survival of *E. coli* exposed to DNA-damaging agents such as UV. In contrast, defects in the CSB protein, thought to be the functional homolog of *Mfd* in mammals including humans, can result in serious developmental and neurological problems (reviewed in Ref. 1). It has been suggested that NusA may be involved in a form of TCR that is independent of *Mfd* and can account for the mild effect of *mfd* mutations;[58,59] however, there is as yet no direct biochemical evidence for TCR in the absence of *Mfd*. Furthermore, the biological response (MFD) that reflects TCR and was the basis for the subsequent identification of *Mfd* is absent when the protein is not active. In *Bacillus subtilis*, *Mfd* has been implicated in recombination,[70] and in both *B. subtilis*[71] and *Campylobacter jejuni*,[72] it has been reported to have a mutagenic effect. In addition, *Mfd* has been reported to release ECs arrested by the bacteriophage HK022 Nun protein.[73] Until the biochemical bases for these observations are known, interpreting an *Mfd*-dependent biological response as TCR without direct measurements of repair is of questionable validity.

Note added in proof: In contrast to data implicating TCR in the repair of oxidative lesions, Schalow *et al.* (J. Bacteriol. 2012;194: 2637-2645) have reported that neither *mfd* nor uvrA mutations affect the survival, the rate of recovery of transcription, or the rate of removal of formamidopyrimidine DNA glycosylase sensitive sites in the DNA of actively growing *E. coli* treated with H_2O_2, consistent with the idea that neither TCR nor GGR contribute significantly to the repair of oxidative lesions produced by this compound.

In addition, Deaconescue *et al.* (*Proc Nat Acad Sci USA* 2012;**109**:3353–3358) recently published studies of the structure of *Mfd*, providing more details about the interaction of *Mfd* with *UvrA* and the conformational changes the protein undergoes during its interactions with other components of TCR. Their data, like those of Manelyte *et al.*,[39] indicate that *Mfd* can dissociate a stalled EC without interacting with *UvrA*; they propose that the likely sequence of events in TCR is the interaction of *Mfd* with a stalled EC by protein–protein contacts and the dissociation of the EC prior to the recruitment of *UvrA*.

Acknowledgments

Support for our research on TCR for many years was provided by grants from the National Cancer Institute, NIH 5R35 CA44349, RO1 CA91456, and RO1 CA90915. Current support is provided from grant 1RO1 ES0188344 from the National Institute of Environmental Health Sciences, NIH.

References

1. Hanawalt PC, Spivak G. Transcription-coupled DNA repair: two decades of progress and surprises. *Nat Rev Mol Cell Biol* 2008;**9**:958–70.
2. Witkin EM. Radiation-induced mutations and their repair. *Science* 1966;**152**:1345–52.
3. Zolan ME, Cortopassi GA, Smith CA, Hanawalt PC. Deficient repair of chemical adducts in alpha DNA of monkey cells. *Cell* 1982;**28**:613–9.

4. Leadon SA, Zolan ME, Hanawalt PC. Restricted repair of aflatoxin B1 induced damage in alpha DNA of monkey cells. *Nucleic Acids Res* 1983;**11**:5675–89.
5. Mansbridge JN, Hanawalt PC. Domain-limited repair of DNA in ultraviolet irradiated fibroblasts from xeroderma pigmentosum complemetation group C. In: Friedberg EC, Bridges BR, editors. *Cellular responses to DNA damage, UCLA symposium on molecular and cellular biology, new series*, vol. 11. New York: Alan R.Liss; 1983. pp. 195–207.
6. Bohr VA, Smith CA, Okumoto DS, Hanawalt PC. DNA repair in an active gene: removal of pyrimidine dimers from the DHFR gene of CHO cells is much more efficient than in the genome overall. *Cell* 1985;**40**:359–69.
7. Mellon I, Spivak G, Hanawalt PC. Selective removal of transcription-blocking DNA damage from the transcribed strand of the mammalian DHFR gene. *Cell* 1987;**51**:241–9.
8. Mellon I, Hanawalt PC. Induction of the *Escherichia coli* lactose operon selectively increases repair of its transcribed DNA strand. *Nature* 1989;**342**:95–8.
9. Mellon I, Bohr VA, Smith CA, Hanawalt PC. Preferential DNA repair of an active gene in human cells. *Proc Natl Acad Sci USA* 1986;**83**:8878–82.
10. Selby CP, Sancar A. Transcription preferentially inhibits nucleotide excision repair of the template DNA strand in vitro. *J Biol Chem* 1990;**265**:21330–6.
11. Selby CP, Witkin EM, Sancar A. *Escherichia coli mfd* mutant deficient in "mutation frequency decline" lacks strand-specific repair: in vitro complementation with purified coupling factor. *Proc Natl Acad Sci USA* 1991;**88**:11574–8.
12. Selby CP, Sancar A. Molecular mechanism of transcription-repair coupling. *Science* 1993;**260**:53–8.
13. Mellon I, Champe GN. Products of DNA mismatch repair genes *mutS* and *mutL* are required for transcription-coupled nucleotide-excision repair of the lactose operon in *Escherichia coli*. *Proc Natl Acad Sci USA* 1996;**93**:1292–7.
14. Li BH, Bockrath R. Mutation frequency decline in *Escherichia coli*.I. Effects of defects in mismatch repair. *Mol Gen Genet* 1995;**249**:585–90.
15. Selby CP, Sancar A. Structure and function of transcription-repair coupling factor I. Structural domains and binding properties. *J Biol Chem* 1995;**270**:4882–9.
16. Park JS, Marr MT, Roberts JW. *E. coli* transcription repair coupling factor (Mfd protein) rescues arrested complexes by promoting forward translocation. *Cell* 2002;**109**:757–67.
17. Smith AJ, Savery NJ. RNA polymerase mutants defective in the initiation of transcription-coupled DNA repair. *Nucleic Acids Res* 2005;**33**:755–64.
18. Ganesan AK, Smith AJ, Savery NJ, Zamos P, Hanawalt PC. Transcription coupled nucleotide excision repair in *Escherichia coli* can be affected by changing the arginine at position 529 of the beta subunit of RNA polymerase. *DNA Repair (Amst)* 2007;**6**:1434–40.
19. Ganesan AK, Hanawalt PC. Transcription-coupled nucleotide excision repair of a gene transcribed by bacteriophage T7 RNA polymerase in *Escherichia coli*. *DNA Repair (Amst)* 2010;**9**:958–63.
20. Witkin EM. Time, temperature, and protein synthesis: a study of ultraviolet-induced mutation in bacteria. *Cold Spring Harb Symp Quant Biol* 1956;**21**:123–40.
21. Doudney CO, Haas FL. Modification of ultra violet-induced mutation frequency and survival in bacteria by post-irradiation treatment. *Proc Nat Acad Sci USA* 1958;**44**:390–401.
22. Witkin EM. Mutation frequency decline revisited. *Bioessays* 1994;**16**:437–44.
23. Bockrath RC, Palmer JE. Differential repair of premutational UV-lesions at tRNA genes in *E. coli*. *Mol Gen Genet* 1977;**156**:133–40.
24. Bockrath R, Li BH. Mutation frequency decline in *Escherichia coli* II. Kinetics support the involvement of transcription-coupled excision repair. *Mol Gen Genet* 1995;**249**:591–9.
25. Selby CP, Sancar A. Mechanisms of transcription-repair coupling and mutation frequency decline. *Microbiol Rev* 1994;**58**:317–29.

26. Selby CP, Sancar A. Structure and function of transcription-repair coupling factor II. Catalytic properties. *J Biol Chem* 1995;**270**:4890–5.
27. Savery NJ. The molecular mechanism of transcription-coupled DNA repair. *Trends Microbiol* 2007;**15**:326–33.
28. Deaconescu AM, Chambers AL, Smith AJ, Nickels BE, Hochschild A, Savery NJ, et al. Structural basis for bacterial transcription-coupled DNA repair. *Cell* 2006;**124**:507–20.
29. Butland G, Peregrin-Alvarez JM, Li J, Yang W, Yang X, Canadien V, et al. Interaction network containing conserved and essential protein complexes in *Escherichia coli*. *Nature* 2005;**433**:531–7.
30. Roberts J, Park J-SS. Mfd, the bacterial transcription repair coupling factor: translocation, repair and termination. *Curr Opin Microbiol* 2004;**7**:120–5.
31. Park J-SS, Roberts JW. Role of DNA bubble rewinding in enzymatic transcription termination. *Proc Natl Acad Sci USA* 2006;**103**:4870–5.
32. Westblade LF, Campbell EA, Pukhrambam C, Padovan JC, Nickels BE, Lamour V, et al. Structural basis for the bacterial transcription-repair coupling factor/RNA polymerase interaction. *Nucleic Acids Res* 2010;**38**:8357–69.
33. Trautinger BW, Jaktaji RP, Rusakova E, Lloyd RG. RNA polymerase modulators and DNA repair activities resolve conflicts between DNA replication and transcription. *Mol Cell* 2005;**19**:247–58.
34. Pomerantz RT, O'Donnell M. Direct restart of a replication fork stalled by a head-on RNA polymerase. *Science* 2010;**327**:590–2.
35. Dutta D, Shatalin K, Epshtein V, Gottesman ME, Nudler E. Linking RNA polymerase backtracking to genome instability in *E. coli*. *DNA Repair (Amst)* 2011;**146**:533–43.
36. Murphy MN, Gong P, Ralto K, Manelyte L, Savery NJ, Theis K. An N-terminal clamp restrains the motor domains of the bacterial transcription-repair coupling factor Mfd. *Nucleic Acids Res* 2009;**37**:6042–53.
37. Chambers AL, Smith AJ, Savery NJ. A DNA translocation motif in the bacterial transcription-repair coupling factor, Mfd. *Nucleic Acids Res* 2003;**31**:6409–18.
38. Smith AJ, Szczelkun MD, Savery NJ. Controlling the motor activity of a transcription-repair coupling factor: autoinhibition and the role of RNA polymerase. *Nucleic Acids Res* 2007;**35**: 1802–11.
39. Manelyte L, Kim Y-IT, Smith AJ, Smith RM, Savery NJ. Regulation and rate enhancement during transcription-coupled DNA repair. *Mol Cell* 2010;**40**:714–24.
40. Srivastava DB, Darst SA. Derepression of bacterial transcription-repair coupling factor is associated with a profound conformational change. *J Mol Biol* 2011;**406**:275–84.
41. Epshtein V, Nudler E. Cooperation between RNA polymerase molecules in transcription elongation. *Science* 2003;**300**:801–5.
42. Epshtein V, Toulme F, Rahmouni AR, Borukhov S, Nudler E. Transcription through the roadblocks: the role of RNA polymerase cooperation. *EMBO J* 2003;**22**:4719–27.
43. Orren DK, Sancar A. The (A)BC excinuclease of *Escherichia coli* has only the UvrB and UvrC subunits in the incision complex. *Proc Natl Acad Sci USA* 1989;**86**:5237–41.
44. Lin JJ, Sancar A. (A)BC excinuclease: the *Escherichia coli* nucleotide excision repair enzyme. *Mol Microbiol* 1992;**6**:2219–24.
45. Malta E, Moolenaar GF, Goosen N. Dynamics of the UvrABC nucleotide excision repair proteins analyzed by fluorescence resonance energy transfer. *Biochemistry* 2007;**46**:9080–8.
46. Kad NM, Wang H, Kennedy GG, Warshaw DM, Van Houten B. Collaborative Dynamic DNA Scanning by Nucleotide Excision Repair Proteins Investigated by Single-Molecule Imaging of Quantum-Dot-Labeled Proteins. *Mol Cell* 2010;**37**:702–13.
47. Pakotiprapha D, Liu Y, Verdine GL, Jeruzalmi D. A structural model for the damage-sensing complex in bacterial nucleotide excision repair. *J Biol Chem* 2009;**284**:12837–44.

48. Sancar A, Franklin KA, Sancar GB. *Escherichia coli* DNA photolyase stimulates uvrABC excision nuclease in vitro. *Proc Natl Acad Sci USA* 1984;**81**:7397–401.
49. Sancar GB, Smith FW. Interactions between yeast photolyase and nucleotide excision repair proteins in *Saccharomyces cerevisiae* and *Escherichia coli*. *Mol Cell Biol* 1989;**9**:4767–76.
50. Wagner K, Moolenaar GF, Goosen N. Role of the insertion domain and the zinc-finger motif of *Escherichia coli* UvrA in damage recognition and ATP hydrolysis. *DNA Repair (Amst)* 2011;**10**:483–96.
51. Mazon G, Philippin G, Cadet J, Gasparutto D, Fuchs RP. The alkyltransferase-like *ybaZ* gene product enhances nucleotide excision repair of O(6)-alkylguanine adducts in *E. coli*. *DNA Repair (Amst)* 2009;**8**:697–703.
52. Pakotiprapha D, Inuzuka Y, Bowman BR, Moolenaar GF, Goosen N, Jeruzalmi D, et al. Crystal structure of *Bacillus stearothermophilus* UvrA provides insight into ATP-modulated dimerization, UvrB interaction, and DNA binding. *Mol Cell* 2008;**29**:122–33.
53. Crowley DJ, Hanawalt PC. Induction of the SOS response increases the efficiency of global nucleotide excision repair of cyclobutane pyrimidine dimers, but not 6-4 photoproducts, in UV-irradiated *Escherichia coli*. *J Bacteriol* 1998;**180**:3345–52.
54. Wagner K, Moolenaar GF, Goosen N. Role of the two ATPase domains of *Escherichia coli* UvrA in binding non-bulky DNA lesions and interaction with UvrB. *DNA Repair (Amst)* 2010;**9**:1176–86.
55. Theis K, Chen PJ, Skorvaga M, Van Houten B, Kisker C. Crystal structure of UvrB, a DNA helicase adapted for nucleotide excision repair. *EMBO J* 1999;**18**:6899–907.
56. Manelyte L, Guy CP, Smith RM, Dillingham MS, McGlynn P, Savery NJ. The unstructured C-terminal extension of UvrD interacts with UvrB, but is dispensable for nucleotide excision repair. *DNA Repair (Amst)* 2009;**8**:1300–10.
57. Wang H, Della VMJ, Skorvaga M, Croteau DL, Erie DA, Van Houten B. UvrB domain 4: an autoinhibitory gate for regulation of DNA binding and ATPase activity. *J Biol Chem* 2006;**281**:15227–37.
58. Cohen SE, Lewis CA, Mooney RA, Kohanski MA, Collins JJ, Landick R, et al. Roles for the transcription elongation factor NusA in both DNA repair and damage tolerance pathways in *Escherichia coli*. *Proc Natl Acad Sci USA* 2010;**107**:15517–22.
59. Cohen SE, Walker GC. New discoveries linking transcription to DNA repair and damage tolerance pathways. *Transcription* 2011;**2**:37–40.
60. McGlynn P. Linking transcription with DNA repair, damage tolerance, and genome duplication. *Proc Natl Acad Sci USA* 2010;**107**:15314–5.
61. Bylund GO, Nord S, Lovgren JM, Wikstrom PM. Alterations in the β flap and β' dock domains of the RNA polymerase abolish NusA-mediated feedback regulation of the metY-nusA-infB operon. *J Bacteriol* 2011;**193**:4113–22.
62. Leadon SA. Retraction. *DNA Repair (Amst)* 2003;**2**:361.
63. Gowen LC, Avrutskaya AV, Latour AM, Koller BH, Leadon SA. Retraction. *Science* 2003;**300**:1657.
64. Cooper PK, Nouspikel T, Clarkson SG. Retraction. *Science* 2005;**308**:1740.
65. Le Page F, Kwoh EE, Avrutskaya A, Gentil A, Leadon SA, Sarasin A, et al. Transcription-coupled repair of 8-oxoguanine: requirement for XPG, TFIIH, and CSB and implications for Cockayne syndrome. *Cell* 2005;**123**:711.
66. Banerjee D, Mandal SM, Das A, Hegde ML, Das S, Bhakat KK, et al. Preferential repair of oxidized base damage in the transcribed genes of mammalian cells. *J Biol Chem* 2011;**286**:6006–16.
67. Bregeon D, Doddridge ZA, You HJ, Weiss B, Doetsch PW. Transcriptional mutagenesis induced by uracil and 8-oxoguanine in *Escherichia coli*. *Mol Cell* 2003;**12**:959–70.

68. Clauson CL, Saxowsky TT, Doetsch PW. Dynamic flexibility of DNA repair pathways in growth arrested *Escherichia coli*. *DNA Repair (Amst)* 2010;**9**:842–7.
69. Viswanathan A, Doetsch PW. Effects of nonbulky DNA base damages on *Escherichia coli* RNA polymerase-mediated elongation and promoter clearance. *J Biol Chem* 1998;**273**:21276–81.
70. Ayora S, Rojo F, Ogasawara N, Nakai S, Alonso JC. The Mfd protein of Bacillus subtilis 168 is involved in both transcription-coupled DNA repair and DNA recombination. *J Mol Biol* 1996;**256**:301–18.
71. Ross C, Pybus C, Pedraza-Reyes M, Sung HM, Yasbin RE, Robleto E. Novel role of *mfd*: effects on stationary-phase mutagenesis in *Bacillus subtilis*. *J Bacteriol* 2006;**188**:7512–20.
72. Han J, Sahin O, Barton YW, Zhang Q. Key role of Mfd in the development of fluoroquinolone resistance in *Campylobacter jejuni*. *PLoS Pathog* 2008;**4**:e1000083.
73. Washburn RS, Wang Y, Gottesman ME. Role of *E. coli* transcription-repair coupling factor Mfd in Nun-mediated transcription termination. *J Mol Biol* 2003;**329**:655–62.

The Functions of MutL in Mismatch Repair: The Power of Multitasking

Alba Guarné

Department of Biochemistry and Biomedical Sciences, McMaster University, Hamilton, Ontario, Canada

DNA mismatch repair enhances genomic stability by correcting errors that have escaped polymerase proofreading. One of the critical steps in DNA mismatch repair is discriminating the new from the parental DNA strand as only the former needs repair. In *Escherichia coli*, the latent endonuclease MutH carries out this function. However, most prokaryotes and all eukaryotes lack a *mutH* gene. MutL is a key component of this system that mediates protein–protein interactions during mismatch recognition, strand discrimination, and strand removal. Hence, it had long been thought that the primary function of MutL was coordinating sequential mismatch repair steps. However, recent studies have revealed that most MutL homologs from organisms lacking MutH encode a conserved metal-binding motif associated with a weak endonuclease activity. As MutL homologs bearing this activity are found only in organisms relying on MutH-independent DNA mismatch repair, this finding unveils yet another crucial function of the MutL protein at the strand discrimination step. In this chapter, we review recent functional and structural work

Progress in Molecular Biology
and Translational Science, Vol. 110
DOI: 10.1016/B978-0-12-387665-2.00003-1

aimed at characterizing the multiple functions of MutL and discuss how the endonuclease activity of MutL is regulated by other repair factors.

I. Overview of DNA Mismatch Repair

DNA mismatch repair corrects errors that have escaped the proofreading activity of replicative polymerases, in turn increasing replication fidelity by two to three orders of magnitude (reviewed in Refs. 1–4). Defects in the mismatch repair genes cause increased mutation rates and genetic instability,[5] the hallmark of Lynch syndrome. Families with Lynch syndrome, commonly known as human non-polyposis colorectal cancer, have an increased risk of cancers of the colon, stomach, small intestine, liver, gallbladder ducts, upper urinary tract, brain, skin, and prostate—women with this disorder also have a high risk of cancer of the endometrium and ovaries.[6,7]

Beyond correcting replication errors, DNA mismatch repair proteins recognize a variety of DNA lesions and coordinate transcription-coupled nucleotide excision repair, meiotic recombination, cell cycle checkpoint control, and apoptosis.[8,9] Recognition of alkylation lesions and intrastrand cross-links by mismatch repair proteins results in cell cycle arrest and apoptosis when damage is too extensive. In turn, mismatch repair-deficient tumor cells are highly resistant to treatment with alkylating agents due to their inability to induce apoptosis.[10,11] Underscoring the importance of these additional functions, about a fourth of all sporadic cancers have associated mismatch repair defects.[9] These roles of the mismatch repair proteins are not considered further here, but the reader is directed to excellent reviews on the topic.[2,4,12–14]

A. DNA Mismatch Repair in *Escherichia coli*

The components of the mismatch repair system were originally identified and characterized in *Escherichia* coli.[15] In this paradigmatic system, three proteins—MutS, MutL, and MutH—are required to recognize and initiate the mismatch repair response (Table I and Fig. 1).

MutS is a dimeric ATPase from the ABC (adenosine triphosphate (ATP)-*b*inding *c*assette) superfamily.[16] It recognizes single base–base mismatches and insertion/deletion loops (indels) containing up to 4 unpaired bases with 10- to 1500-fold higher affinity than perfectly matched DNA.[17–20] The DNA- and ATP-binding domains of MutS reside in opposite ends of the polypeptide chain; however, the two regions are connected by a long transmitter helix that mediates the nucleotide-dependent proofreading of the mismatch.[21–24] MutS is normally bound to adenosine diphosphate (ADP), but nucleotide-exchange is favored once a mismatch is encountered and a concomitant

TABLE I

Proteins Involved in DNA Mismatch Repair Correction

γ-Proteobacteria (*E. coli*)	Other eubacteria (*B. subtilis*)	Eukaryotes (Humans)	Function
MutS	MutS	MutSα (MSH2–MSH6)	Recognition of base mismatches and 1–2 base indels
		MutSβ (MSH2–MSH3)	Recognition of ≥ 2 base indels
MutL	MutL*	MutLα (MLH1–PMS2*)	Molecular matchmaker in MMR (*endonuclease)
		MutLβ (MLH1–PMS1)	?
		MutLγ (MLH1–MLH3*)	Molecular matchmaker in meiosis (*endonuclease)
MutH	?	?	Nicks hemimethylated DNA (strand discrimination)
RecJ, ExoVII	RecJ, ExoVII	Exo1	$5' \rightarrow 3'$ exonuclease
ExoVII, ExoI, ExoX	–	–	$3' \rightarrow 5'$ exonuclease
Helicase II (UvrD)	Helicase II (UvrD)	–	$3' \rightarrow 5'$ helicase
Polymerase III	Polymerase III	Polymerase δ	DNA resynthesis
β	β	PCNA	Processivity subunit (sliding clamp)
γ-Complex	γ-Complex	RFC	Clamp loader complex
SSB	SSB	RPA	Protects ssDNA during excision and resynthesis
Ligase A	Ligase	Ligase 1	Seals nick after DNA resynthesis

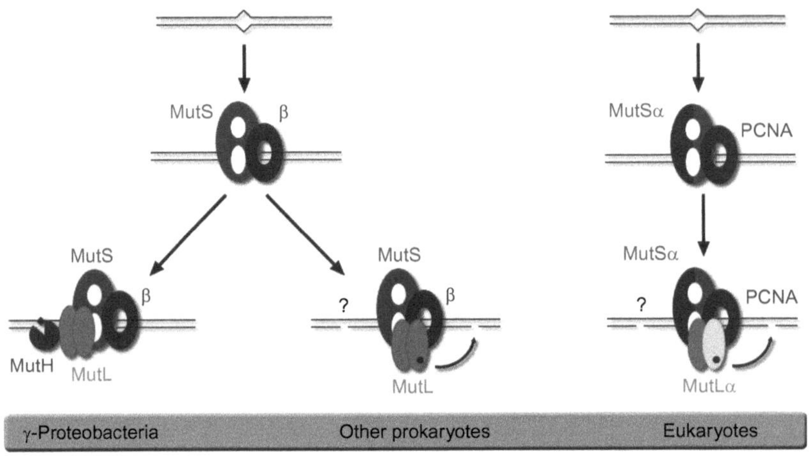

FIG. 1. Mechanisms of mismatch recognition and strand discrimination. In bacteria, homodimeric MutS recognizes mismatches or small insertion/deletion loops (indels) assisted by the β subunit of polymerase III. In an ATP-dependent manner, MutS recruits MutL to the complex. In eukaryotes, heterodimeric MutSα is primarily responsible for correcting mismatches or small indels assisted by PCNA, the processivity subunit of polymerase δ. Similarly, MutSα recruits MutLα to this complex in an ATP-dependent manner. In γ-proteobacteria (left), formation of the MutS–MutL–heteroduplex complex activates the endonuclease activity of MutH that selectively nicks the newly synthesized strand. However, most prokaryotes (center) and all eukaryotes (right) lack a *mutH* gene. MutL homologs from these organisms encompass a conserved endonuclease motif (depicted as a red dot) that selectively cuts the nicked strand of the DNA. However, the origin of the primary nick remains unclear. (See Color Insert.)

conformational change allows the MutS•ATP complex to become a sliding clamp.[22,25–27] This conformational change favors the recruitment of additional mismatch repair factors to the site of damage and, in turn, signals repair.

MutL is one of the factors recruited to mismatches in a MutS- and ATP-dependent manner (Fig. 1).[28] MutL is a dimeric, weak ATPase from the GHL (Gyrase B–Hsp90–MutL) family of ATPases.[29,30] As its primary role is to mediate the protein–protein interactions during mismatch recognition and strand removal, MutL has been historically classified as a molecular matchmaker.[31] Formation of a ternary complex between MutS, MutL, and mismatched DNA (MutS–MutL–heteroduplex) is sufficient to activate the latent endonuclease MutH.

MutH is a member of the type II family of restriction endonucleases that nicks the unmethylated strand of hemimethylated guanine, adenine, thymine, cytosine (GATC) sequences.[32,33] In *E. coli*, all adenines within GATC sequences are methylated by dam methylase; however, after the passage of the replication fork, the newly replicated strand is transiently unmethylated. By nicking the unmethylated strand at these hemimethylated GATC sites, MutH provides an entry point for MutL-dependent loading of helicase II (also known as UvrD) and single-strand binding (SSB) protein. Therefore, the primary role of MutH is to ensure that repair is targeted exclusively to the newly synthesized strand.

Mismatch repair is bidirectional; therefore, mismatches and indels are repaired regardless of whether MutH nicks the newly synthesized strand 5′ or 3′ to the lesion.[34–36] If the nick occurs 5′ to the mismatch, MutL recruits the exonucleases RecJ or ExoVII to remove the newly synthesized strand.[37,38] Conversely, ExoI, ExoVII, or ExoX are recruited when MutH nicks 3′ to the mismatch.[37–41] In both cases, the uvrD helicase, with 3′→5′ polarity is recruited, and likely oriented, by MutL to unwind DNA toward the mismatch.[42–46]

Once the erroneous strand has been removed, the SSB protein stabilizes the gap so that polymerase III and DNA ligase can resynthesize the strand correctly.[40] Importantly, the β subunit of polymerase III—commonly referred to as β, sliding, or processivity clamp, owing to its role in enhancing the processivity of polymerase III—has binding sites for MutS and MutL, and it has been suggested that these interactions may be important to sense DNA lesions and recruit mismatch repair proteins to the sites of damage.[47–49]

B. Strand Discrimination in Mismatch Repair in Organisms Lacking MutH

While homologs of MutS and MutL are found in all domains of life, the *dam methylase* and *mutH* genes are only found in a subset of γ-proteobacteria (Table I and Fig. 1). A strand discrimination endonuclease has not been identified in other organisms, but the excision step can be recreated *in vitro* using purified proteins, as far as a nicked DNA is supplied to the reaction.[50]

In humans, five paralogs of MutS and four paralogs of MutL have been identified (Table I). MutS paralogs associate to form heterodimers with specialized functions. Studies in yeast, mice, and humans indicated that MutSα (formed by the association of the *M*ut*S* *h*omologs MSH2 and MSH6) is primarily responsible for correcting single-base mismatches or small indels (Table I),[51–53] whereas MutSβ (formed by the association of MSH2 and MSH3) is primarily responsible for correcting larger indels (up to 16 nucleotides).[54] The other homologs of MutS are involved in mitochondrial transactions and meiotic recombination.[3] Likewise, three MutL heterodimers formed by the association of hMLH1 with hPMS2 (hMutLα), hPMS1 (hMutLβ), and hMLH3 (hMutLγ) have been identified (Table I).[55–58] Of these, MutLα is the only MutL heterodimer required for mismatch repair.[56] MutLγ has a defined role in meiotic recombination[59] and, while no function has been ascribed to human MutLβ, the interactome of hPMS1 includes a number of proteins from the ubiquitylation pathway, suggesting that the function of MutLβ may be modulated by ubiquitylation.[60]

While strand discontinuities and gaps can direct mismatch repair *in vitro*, the mechanism of strand discrimination remains obscure (Fig. 1). Interestingly, when the excision step of the mismatch repair process is reconstituted *in vitro*, different proteins are required depending on whether the nick is positioned 5′ or 3′ to the mismatch. MutSα, the SSB protein replication protein A (RPA), and the exonuclease Exo1 are required when the nick is located 5′ to the mismatch.[50,61,62] In addition to these factors, MutLα, the proliferating cell nuclear antigen PCNA (the processivity subunit of DNA polymerase δ) and RFC (the *r*eplication *f*actor *C*, responsible for loading PCNA onto DNA) are also necessary when the nick is placed 3′ to the mismatch.[63]

The requirement of Exo1 when the nick is found 3′ to the mismatch came as a surprise because Exo1 is a 5′→3′ exonuclease.[64–66] It was initially proposed that Exo1 had cryptic 3′→5′ exonuclease activity,[63] but no evidence was found. The apparent inconsistency was resolved when Modrich and coworkers found that human PMS2, one of the protomers that form the MutLα heterodimer, has a latent endonuclease activity dependent on PCNA.[67] MutLα can cut DNA 5′ or 3′ to the mismatch, but it preferentially cleaves between the mismatch and a preexisting nick. Therefore, the endonuclease activity of MutLα provides entry sites for Exo1,[67] explaining why MutLα is strictly necessary to repair mismatches from a 3′ nick, but dispensable when the nick is found 5′ to the mismatch.[63]

C. The Multiple Faces of MutL

Eukaryotic MutL proteins coordinate not only DNA mismatch repair, but also transcription-coupled nucleotide excision repair, DNA damage surveillance, meiotic recombination, cell cycle checkpoint control, and apoptosis.[8] The physical interactions of eukaryotic MutL homologs with MutS homologs PCNA, RFC, and Exo1 are well documented and have been shown to be critical for mismatch repair.[28,68–76] Additionally, MutLα interacts with MRE11 during the processing of double-strand breaks[77–79] and proteins from the ATR-Chk1 pathway.[10,11,80] MutLα also interacts with BRCA1 and the BRCA1-associated helicase BRIP1/BACH1,[60,81] which have critical roles in homologous recombination, double-strand break, and intrastrand cross-link repair.[82,83]

In a similar manner, it has been proposed that by recruiting different protein effectors, *E. coli* MutL determines the biological outcome of a lesion recognized by MutS.[84,85] During mismatch repair, MutL interacts with MutS and mismatched DNA to activate the endonuclease activity of MutH.[32,33,86] It interacts with helicase II (also known as uvrD) and activates its helicase activity.[42,43] MutL also interacts with several single-strand exonucleases[87] and it has been shown to interact with the processivity clamp (β) and its clamp loader.[47,88] Beyond mismatch repair, MutL activates the very short patch repair pathway through its interaction with the endonuclease vsr, the base excision repair pathway through its interaction with the MutY glycosylase, and nucleotide excision repair through its interaction with the UvrB nuclease.[85,89–91]

Given the extent of their interactomes, it is not surprising that MutL proteins are classified as molecular matchmakers.[31] However, the discovery of the endonuclease activity of MutLα and later on in bacterial MutL homologs from organisms lacking the *mutH* gene[67,92–96] has brought MutL from behind the scenes to center stage.

II. MutL is a Multidomain Protein

MutL is composed of two structured domains connected by a flexible linker. The N-terminal region of the protein (~330 residues) encompasses an ATPase domain that is highly conserved from *E. coli* to humans.[97,98] The C-terminal region (~200 residues) mediates either homo- (in prokaryotes) or hetero- (in eukaryotes) dimerization. Despite the limited sequence conservation of this domain, secondary structure prediction and structural studies have revealed that this region of the protein is structurally conserved for most MutL homologs.[42,96,99] The metal-binding motif associated with the endonuclease activity found in some MutL proteins is located in this region of the protein.[67,93–96] The linker connecting these two conserved regions (100–150 residues) is variable in sequence and susceptible to protease degradation, reinforcing the idea of its flexibility and lack of defined structure.[42,97]

ATP binding at the N-terminal region of the protein triggers the engagement of the ATPase domains and a substantial conformational change that has been elegantly monitored by size-exclusion chromatography using purified *E. coli* MutL (Fig. 2).[42,97] Similar conformational changes have also been observed for eukaryotic MutLα.[100,102–104] Importantly, atomic force microscopy has demonstrated that ATP binding induces asymmetric conformational changes in human and yeast MutLα that yield four major conformations of the heterodimer.[100]

Comparison of the ATPase activities of yeast MLH1 and PMS1 (the yeast homolog of human PMS2) has revealed important differences between the two protomers of MutLα. Yeast MLH1 binds ATP with higher affinity than yeast PMS1.[102] Additionally, the effect of replacing key residues in the ATPase site of yMLH1 is more deleterious than generating the equivalent mutations in the ATPase site of yPMS1—an effect that is also seen for human MLH1 and PMS2.[102–104] Importantly, mutations in the ATPase site of hMLH1 disrupt

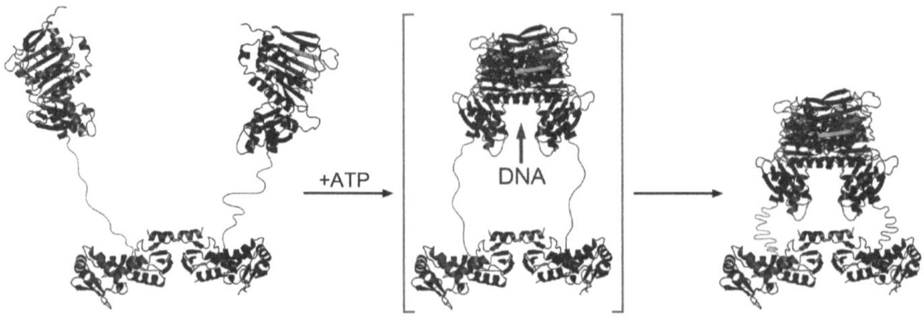

FIG. 2. Architecture of MutL. Ribbon diagrams of the N- and C-terminal domains of *E. coli* MutL (PDB IDs: 1B63 and 1X9Z, respectively), joined by a flexible linker represented by a color-coded line. The protomers of the dimer are shown in purple and teal with the AMPPnP molecules bound to the N-terminal domain shown in yellow. The conformational changes imposed by nucleotide binding are modeled based on the conformational changes indentified by size-exclusion chromatography and atomic force microscopy data.[97,100,101] The potential semicondensed state is proposed by analogy to the semicondensed conformations observed in yeast and human MutLα.[100] The DNA-binding site is indicated with a gray arrow. (See Color Insert.)

DNA mismatch repair *in vivo*, while the equivalent mutations in hPMS2 have minimal effect.[105] Collectively, these results suggest that the ATPase activity of MLH1 is more important than that of PMS2 for normal mismatch repair. However, it is still unclear how each protomer contributes to the specific functions of MutL.

Despite the differences between the two protomers, ATP binding stabilizes the association of the ATPase domains of the two protomers of MutLα, in turn defining the ring-like conformation of the heterodimer.[100,104] A recent study has shown that yeast MutLα diffuses along DNA via a hopping/stepping mechanism and proposes that yMutLα adopts a ring-like architecture to travel along DNA.[106] The work also suggests that yMutLα bypasses obstacles in an open-ring configuration, underscoring the importance of the nucleotide-induced conformational changes and the functional asymmetry of the MutL heterodimer.

A. The ATPase Domain

The structures of the N-terminal regions of *E. coli*, MutL, hMLH1, hPMS2, and yPMS1, have been determined.[97,98,101,107] This region of MutL is made up of two α/β subdomains. The first one, encompassing the N-terminal 200 residues of the protein, is composed of an eight-stranded mixed β-sheet and five α-helices, and contains the four conserved motifs characteristic of the GHL family of ATPases.[29,30] The second α/β subdomain, encompassing the following ~100 residues of the protein, is made up of five mixed β-strands and two α-helices.[97,101] It is in this second subdomain that residues that have been deemed important for DNA binding reside (Fig. 3).

The structures of the apo- and nucleotide-bound forms of *E. coli* MutL and hPMS2 are available and they reveal that the relative orientation between α/β subdomains changes upon nucleotide binding.[97,98,101] Comparison of the structure of the N-terminal domain of *E. coli* MutL bound to AMPPnP with those of the nucleotide-bound forms of hMLH1, hPMS2, and yPMS1 unveils an identical orientation between α/β subdomains, supporting the idea that nucleotide binding determines the relative orientation between subdomains. In *E. coli* MutL, nucleotide binding triggers the association of this domain and causes the organization of several loops surrounding the nucleotide-binding site, including the loop connecting helices αB and αD often referred to as the ATP lid (Fig. 3).[101]

Human PMS2 (and its yeast homolog, yPMS1) have a markedly lower affinity for ATP than other MutL homologs.[98,102,103] While the cause of this reduced affinity is unclear, the structures of the ATPase domains of hPMS2 and yPMS1 reveal that these two MutL homologs include unique insertions between strands β9 and β10 in the second α/β subdomain (Fig. 3).[98,107] These insertions could affect the cooperativity between the two α/β subdomains and, consequently, alter the ATP- or DNA-binding affinities of these MutL homologs, though this idea remains to be tested.

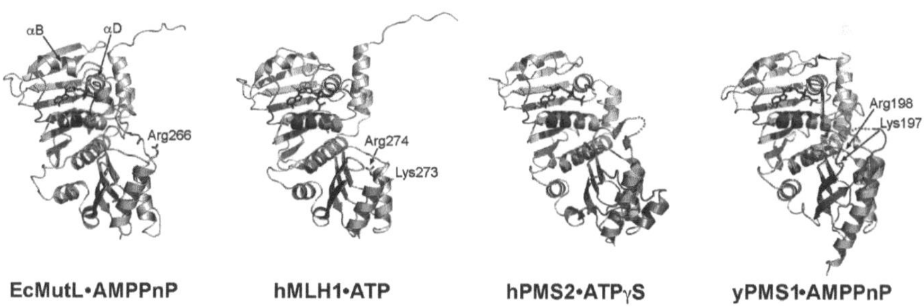

FIG. 3. Crystal structures of the ATPase domain of MutL. From left to right, ribbon diagrams of the structures of *E. coli* MutL•AMPPnP, human MLH1•ATP, human PMS2•ATPγS, and *Saccharomyces cerevisiae* PMS1•AMPPnP (PDB IDs: 1B63, 3NA3, 1H7U, and 3H4L, respectively). The nucleotides are shown as blue sticks and the nucleotide-binding motifs characteristic of the GHL family of ATPases are colored in green. The insertions characteristic to the hPMS2 paralog (PMS1 in yeast) are colored in orange. Residues important for DNA binding are shown as color-coded sticks and labeled. (See Color Insert.)

B. DNA Binding

MutL homologs bind both single- and double-stranded DNA in a sequence- and mismatch-independent manner.[46,108,109] High-affinity binding depends on DNA length, suggesting its cooperative nature.[110] While the N-terminal domain of MutL is sufficient to bind DNA, the presence of the dimerization domain enhances binding.[42] Similarly, studies using purified yMutLα and the N-terminal domains of yMLH1 and yPMS1 reveal that the two ATPase domains of the heterodimer bind DNA independently, but suggest that the C-terminal region of the protein enhances binding.[110,111]

The structure of the N-terminal domain of *E. coli* MutL bound to AMPPnP resembles a saddle (Fig. 2).[101] Mutation of Arg266, which resides in the inner face of the saddle, abrogates DNA binding, and hence it has been proposed that this groove is responsible for DNA binding (Figs. 2 and 3).[112] Accordingly, mutation of Arg273/Arg274 in yMLH1 or Lys197/Arg198 in yPMS1, which also reside on the inner face of the saddle, also reduces DNA-binding affinity and causes loss of mismatch repair activity.[107,111] Binding of both ATP and DNA protects the N-terminal domains of human and yeast MutL proteins from proteolysis, reinforcing the idea that the two α/β subdomains at the N-terminal region of MutL function cooperatively.[103,110,113] DNA binding is enhanced by ATP binding and, reciprocally, the ATPase activity of MutL is stimulated by single- and double-stranded DNA, though binding to single-stranded DNA probably does not occur in physiological conditions.[25,98,101,106] However, the specific effect of DNA binding on the ATPase activity of the protein is still poorly understood.

C. The Dimerization Domain

The structures of the dimerization domains of the *E. coli*, *Bacillus subtilis*, and *Neisseria gonorrhoeae* MutL (referred to as MutL-CTD) have been determined.[42,96,114] The N and C termini of the domain mediate dimerization, while the intervening region defines an external subdomain that protrudes to the solvent (Fig. 4). The dimerization subdomain consists of a four-stranded antiparallel β-sheet. One side of this β-sheet defines the dimerization interface, while the other is packed against helix αE. The C terminus defines a short α-helix that crosses the dimerization interface and is packed against the dimerization subdomain of the neighboring protomer (Fig. 4). The external subdomain is formed by three tightly packed α-helices and a three-stranded antiparallel β-sheet that has a sharp bend in the middle and is completely exposed to the solvent. The two subdomains are connected via helix αA and the loop joining helices αD and αE. In organisms lacking the latent endonuclease MutH, MutL homologs contain a metal-binding motif with the $DQHAX_2EX_4E$ consensus sequence.[67,99] This motif is located in helix αA at the interface between the two subdomains.[96,114]

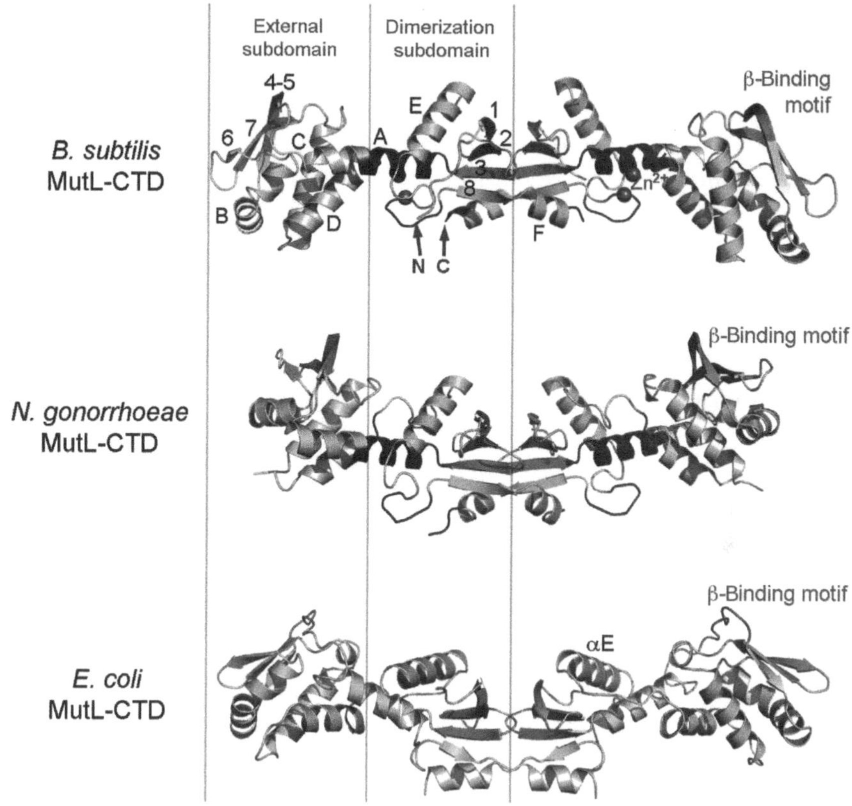

FIG. 4. Crystal structures of the dimerization domain of MutL. The structures of *B. subtilis* MutL (residues 433–627), *N. gonorrhoeae* MutL (residues 463–656), and *E. coli* MutL (residues 432–615) are shown as ribbon diagrams with the secondary structure motifs labeled. The conserved motifs defining the endonuclease site are highlighted in purple and the β-binding motifs are colored orange and labeled. The Zn^{2+} metal ions found in the structure of *B. subtilis* MutL are shown as teal spheres. Helices are labeled with letters (A–F) and strands with consecutive numbers from the N to the C terminus. (See Color Insert.)

Despite the low sequence conservation among the dimerization regions of MutL proteins from different organisms, the structures available reveal a common topology (Fig. 4). Even helix αA, which only in certain MutL homologs encompasses the endonuclease motif, is structurally conserved. However, the orientation of this helix is different depending on whether the protein contains the conserved endonuclease motif (like *B. subtilis* and *N. gonorrhoeae* MutL) or not (like *E. coli* MutL). In MutL proteins containing the conserved endonuclease motif, helix αA undergoes a rigid-body movement that changes the relative orientation between the dimerization and external subdomains (Fig. 4). In turn, this conformational change alters the orientation of helix αE and the αE-β8 loop that in the structure of *E. coli* MutL-CTD includes an additional helix, while in the structures of *B. subtilis* and *N. gonorrhoeae*, MutL-CTD adopts an extended conformation. Collectively, the conformational changes around helix αA shape the endonuclease site of the protein (see below).

Recently, the structure of the C-terminal domain of hMLH1 has also been determined (PDB ID: 3RBN). Human MLH1 is the common subunit in the hMutLα, hMutLβ, and hMutLγ heterodimers; however, when produced in the absence of its functional partners, it can also form stable homodimers (J. Wong and A. Guarné, unpublished results). The structure of the C-terminal domain of MLH1 is significantly different from those of other MutL homologs

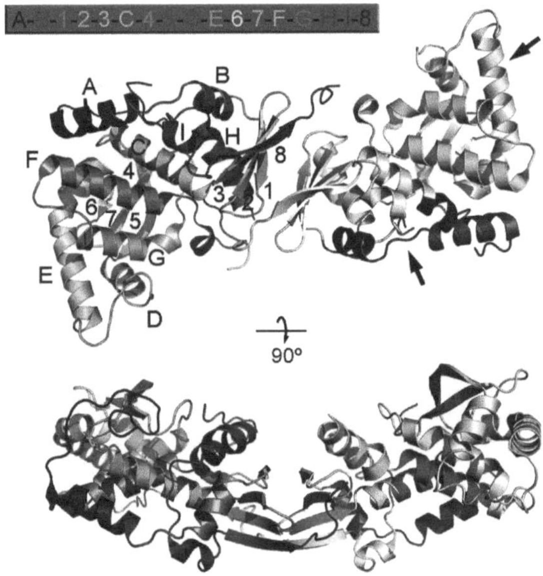

FIG. 5. Crystal structure of the dimerization domain of human MLH1. Orthogonal views of the MLH1 dimer (residues 479–756) with one protomer colored as a rainbow from the N (blue) to the C terminus (red) with the secondary structure elements labeled. The other protomer is shown as a white ribbon with the N-terminal extension—found in MLH1, but not in other MutL homologs—shown in blue. The location of the insertions setting apart MLH1 from other MutL homologs are marked with black arrows and the putative PCNA-binding motif is highlighted in purple. The view on the bottom panel resembles the orientation shown in Fig. 4. (See Color Insert.)

(Figs. 4 and 5). This is not surprising considering that the C-terminal region of hMLH1 (residues 475–756) cannot be aligned with any other MutL paralog.[42,96,98,99] As for other MutL homologs, the dimerization interface of hMLH1 is still defined by a four-stranded antiparallel β-sheet. However, the domain adopts a compact and predominantly helical structure that cannot be easily divided into subdomains. This compact structure is primarily caused by the N terminus of the domain wrapping around helix αC—the equivalent of helix αA on the structures of *B. subtilis* and *E. coli* MutL, the insertion of helix αE, and the extension of helix αF (Fig. 5). Additionally, the relative orientation of helices αC, αE, and αF (αA, αC, and αD, respectively, in other MutL-CTD structures) is markedly different. It is conceivable that these differences are important to form functional MutLα, MutLβ, and MutLγ heterodimers. Indeed, trypsin protection assays suggest that formation of a stable hMutLα heterodimer requires a longer fragment of PMS2 than the region predicted from sequence and structural homology (J. W. and A. G., unpublished results).[96]

III. Architecture of the Endonuclease Domain

In addition to the endonuclease motif, five highly conserved motifs have been identified within the C-terminal domain of MutL.[99] Four of them (GQ, [A/S]C[K/R], C[P/N]HGRP, and FXR) are found exclusively in MutL homologs that contain the conserved endonuclease motif (DQHAX_2EX_4E), whereas the fifth (QXXL[L/I]XP) is found in MutL homologs that have, and also in those that do not have, endonuclease activity.[99] It was predicted that the four additional motifs identified in MutL proteins carrying the conserved endonuclease motif would define a single active site,,[99,115] and the structures of the C-terminal domains of *B. subtilis* and *N. gonorrhoeae* MutL have recently confirmed this prediction.[96,114]

A. The Endonuclease Site

The endonuclease site of MutL is a composite active site where the DQHAX_2EX_4E, [A/S]C[K/R], and C[P/N]HGRP motifs are provided by one MutL protomer, while the FXR motif is contributed by the other protomer of the dimer (Fig. 6).[96] Accordingly, this motif is absent in eukaryotic MutL homologs bearing endonuclease activity, but present in yeast and human MLH1.[99] Except for the [A/S]C[K/R] motif, contributed by the external subdomain, all the motifs that define the endonuclease site reside in the dimerization subdomain of the protein (Fig. 6).

Aquifex aeolicus MutL, which lacks the external subdomain, is a proficient endonuclease, suggesting that this subdomain of the protein has minimal effect on the endonuclease activity of MutL (Fig. 4).[93,116] *A. aeolicus* MutL has the

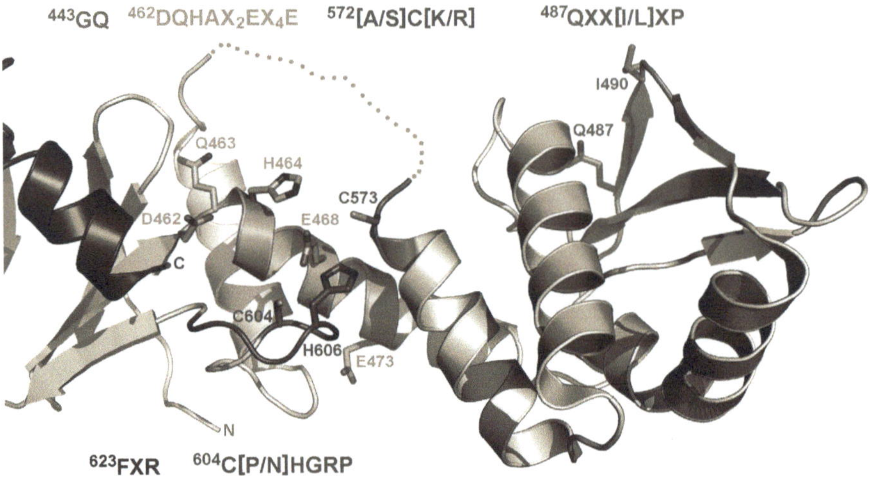

FIG. 6. Organization of the endonuclease site of *B. subtilis* MutL. Ribbon diagram of the C-terminal domain of *B. subtilis* MutL-CTD (PDB ID: 3KDK) shown in light gray with the C-terminal helix of the neighboring protomer of the dimer shown in dark gray. The conserved motifs related to the endonuclease activity of the protein are shown in pink (^{443}GQ), yellow (^{462}DQHAX$_2$-EX$_4$E), cyan (^{572}SCK), green (^{604}CPHRGP), and red (^{623}FXR), with the strictly conserved residues shown as color-coded sticks. The disordered loop connecting the external and dimerization domains is shown as a dotted line. The N- and C-terminal ends of the domain are labeled **N** and **C**. The conserved motif responsible for the interaction with the β-sliding clamp (^{487}QXXIXP) is shown in orange with the key residues shown as color-coded sticks and labeled. (See Color Insert.)

conserved [A/S]C[K/R] motif at the linker connecting the αA and αE helices; however, it is unlikely that it will adopt a similar conformation to that seen in the structures of *B. subtilis* and *N. gonorrhoeae* MutL. Therefore, the [A/S]C [K/R] motif presumably affects the endonuclease activity of MutL only marginally. The role of the conserved GQ motif is also unclear. In contrast to the [A/S]C[K/R], C[P/N]HGRP, and FXR motifs that cluster around the endonuclease motif (DQHAX$_2$EX$_4$E), the GQ motif resides at the dimerization interface (Figs. 4 and 6). However, it is close to the αD–αE loop and hence may indirectly contribute to the overall stability of the endonuclease site.

Not all the conserved residues in the endonuclease motif are equally important either (Fig. 6). For instance, mutation of the Gln463 or Glu473 (D*Q*HAX$_2$EX$_4$*E*) does not disrupt mismatch repair activity in *B.* subtilis.[96] Similarly, mutation of the equivalent glutamic acid in hPMS2 (Glu710, DQHAX$_2$EX$_4$*E*) does not affect mismatch repair activity.[99] Conversely, mutation of Asp462, His464, or Glu468 in *B. subtilis* MutL (*D*Q*H*AX$_2$*E*X$_4$E) or the corresponding residues in hPMS2 (Asp699, His701, and Glu705) completely abrogates mismatch repair function.[67,96,99]

B. MutL is an Mn^{2+}-Dependent Endonuclease

The endonuclease activity of MutLα is strictly dependent on manganese, and the importance of Asp699 and Glu705 (*D*QHAX$_2$*E*X$_4$E) for metal binding has been demonstrated.[67] However, the spatial arrangement of these residues in the structure *B. subtilis* MutL suggests that they are unlikely to form a single metal ion-binding site (Fig. 6).[96] Mutation of Asp462 or His464 (*D*Q*H*AX$_2$EX$_4$E) in *B. subtilis* MutL, or the equivalent residues in human PMS2 or yeast PMS1, abrogates endonuclease activity and mismatch repair function.[67,94,96,99,117] Similarly, mutation of Asp523 (*D*QHAX$_2$EX$_4$E) in yeast MLH3 disrupts the activities of the protein in mismatch repair and meiotic crossing over.[118] Collectively, these results reveal the importance of these two conserved residues and suggest that their side chains, and potentially the main and/or side chains within the β1–β2 loop, could define the binding site of the catalytic metal.

Interestingly, while the endonuclease activity is strictly dependent on manganese,[67,96] human PMS2 and *B. subtilis* MutL bind zinc.[99] Accordingly, two Zn^{2+} metal ions are found in the structure of *B. subtilis* MutL (Fig. 4).[96] The Zn^{2+} ion occupying "site A" is defined by the side chain of Glu468 from the DQHAX$_2$*E*X$_4$E motif, as well as those of Cys604 and His606 from the *C*P*H*GRP motif (Fig. 7A). A well-ordered water molecule completes the tetrahedral coordination of this site. The Zn^{2+} ion bound at "site B" is coordinated by the side chains of His464 and Glu468 from the DQ*H*AX$_2$*E*X$_4$E motif and loosely contacts the side chain of Cys573 from the S*C*K motif (Fig. 7A). The usual coordination distances for Zn^{2+} metal ions range from 2.11 to 2.39 Å; however, the thiol group of Cys573 is located 2.67–2.93 Å away, indicating that this residue does not belong to the first coordination shell of the Zn^{2+} ion bound to "site B."[120,121] Studies in solution show that mutation of Glu468, Cys604, or His606 abrogates Zn^{2+} binding, whereas mutation of His464 does not affect metal binding, suggesting that "site A," but not "site B," is a bona fide Zn^{2+}-binding site.[96]

Numerous studies have mined protein structure databases to analyze the coordination numbers, geometry, and composition of metal-binding sites found in proteins.[120–127] Based on these studies, the coordination number and residue composition of the Zn^{2+}-binding site of MutL ("site A") would preclude Mn^{2+} binding, suggesting that the catalytic metal ion binds elsewhere. However, if Zn^{2+} is not mimicking the binding of the catalytic metal ion, what is its role?

C. The Regulatory Zn^{2+}-Binding Site

In addition to the structure of *B. subtilis* MutL bound to Zn^{2+}, two other independent structures of its apo-form are available.[96] Superimposition of these structures reveals that the relative orientation between the two

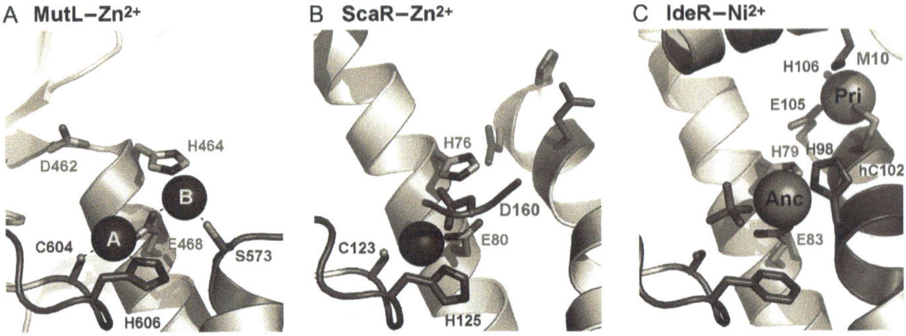

FIG. 7. Organization of the Zn^{2+}-binding site of *B. subtilis* MutL. (A) Detailed view of the endonuclease site of *B. subtilis* MutL (PDB ID: 3KDK) with the conserved motifs shown in yellow ($DQHAX_2EX_4E$), cyan (SCK), and green (CPHRGP). The side chains of the conserved residues defining the metal-binding sites are shown as color-coded sticks and labeled. The zinc metal ions occupying sites A and B are shown as purple spheres and labeled. The two structured water molecules completing the tetrahedral coordination are omitted for clarity. (B) Detailed view of the Zn^{2+}-binding site of *S. gordonii* ScaR (PDB ID: 3HRU) shown on the same orientation and color coding as in (A). Notice that in this case the metal ion is fully coordinated by protein side chains, as opposed to MutL where the fourth ligand is provided by a water molecule. (C) Detailed view of the primary (Pri) and ancillary (Anc) metal-binding sites of a two-domain variant of *M. tuberculosis* IdeR (PDB ID: 2ISY) shown on the same orientation and color coding as in (A). The Ni^{2+} metal ions are shown as green spheres. The ancillary metal ion (Ni^{2+}) is coordinated by the side chains of His79, Glu83, and His98 and the octahedral coordination of this metal ion is completed by one of the oxygens of a well-ordered phosphate group (shown as color-coded sticks) and two water molecules (omitted for clarity). These water molecules are absent on the structure of full-length IdeR,[119] where the side chains of Glu172 and Gln175 complete the coordination shell of the ancillary metal ion-binding site. The Ni^{2+} at the primary metal-binding site is coordinated by the side chains of Met10, oxidized Cys102 (labeled hCys102), Glu105, and His106, as well as the main chain of Cys102 and two water molecules (also omitted for clarity). (See Color Insert.)

subdomains of the protein is highly variable, except when Zn^{2+} is bound to MutL. Upon metal binding, both protomers of the dimer adopt exactly the same conformation,[96] suggesting that Zn^{2+} binding restricts the flexibility between the external and dimerization subdomains. Indeed, "site A" has a coordination number characteristic of structural Zn^{2+} metal ions—as opposed to penta- or hexacoordination that is preferred for catalytic Zn^{2+} metal ions[128,129]—supporting the hypothesis that the metal bound to this site has a structural role.

The Zn^{2+} metal-binding site found in MutL closely resembles the secondary metal-binding site found in *Streptococcus gordonii* ScaR, a Mn^{2+}-dependent transcriptional regulator from the DtxR/MntR family (Fig. 7B).[130] Members of this family maintain manganese and iron homeostasis by down-regulating metal uptake systems.[131–133] Direct binding of Mn^{2+} or Fe^{2+}—and to a lesser extent other transition metal ions such as Ni^{2+}, Zn^{2+}, Co^{2+}, and

Cd^{2+}—allosterically activates binding to DNA and, therefore, the metal works as an obligate corepressor.[132,134] ScaR senses Mn^{2+} like DtxR, but is, however, structurally more similar to the Fe^{2+}-dependent regulators DtxR and IdeR.[130] Two metal ions bind to *Mycobacterium tuberculosis* IdeR.[135] Binding to an ancillary site promotes protein dimerization, but activation of DNA binding requires both sites occupied by metal ions (Fig. 7C). The primary site is located at the domain interface between the central dimerization and DNA-binding domains of the protein, while the ancillary site is located between the dimerization domain and a C-terminal domain that regulates the oligomerization state of the protein and stabilizes the active conformation of the protein (Fig. 7).[119,136–138] The primary metal-binding site, which is important for DNA binding, is strictly conserved between ScaR and IdeR; however, this site is empty in all the crystal structures of ScaR.[130] The ancillary site found in IdeR is not conserved in ScaR, but a secondary metal-binding site about 5 Å away is found in the structures of ScaR–Zn^{2+} and ScaR–Cd^{2+}. This secondary site is novel in the DtxR/MntR family and is virtually identical to the Zn^{2+}-binding site identified in MutL, with the exception that the Zn^{2+} metal ion is fully coordinated by protein side chains (Fig. 7).

While the role and metal specificity of this novel metal-binding site found in ScaR remain unclear, it has been postulated that His76 (or His79 in IdeR) may play a role in coordinating the interactions between the two metal-binding sites of the protein.[130] Therefore, it is plausible that His464 (the equivalent residue in *B. subtilis* MutL) also plays a role in coordinating interactions between the structural Zn^{2+} and the catalytic metal of the protein. Accordingly, the integrity of this histidine is critical for proficient DNA mismatch repair.[96,99]

D. The Endonuclease Activity of MutL

Modrich and coworkers demonstrated that the endonuclease activity of eukaryotic MutLα can be directed toward the discontinuous strand of a nicked duplex in the presence of PCNA, RFC, and MutSα.[67,94] The primary role of RFC is loading PCNA and hence it has only an indirect effect on the endonuclease activity of MutLα.[139] Conversely, PCNA and MutSα stimulate and impose the mismatch dependency of the reaction, respectively, through their direct interaction with MutLα.[139] In the absence of these protein factors, MutLα retains a weak activity that nicks DNA in a nonspecific manner.[67] The overall activity detected in this one-protein experiment is weak, but it has been instrumental in characterizing the endonuclease activity of several prokaryotic MutL proteins.[67,93–96]

The endonuclease activity of MutLα is also stimulated by ATP, but not by ADP, indicating that the conformational changes imposed by ATP binding are important to activate the endonuclease activity of MutL (Fig. 2).[97,100,101] Similar to eukaryotic MutLα, the DNA-nicking activity of *B. subtilis* MutL is exclusively detected in the presence of manganese and its endonuclease activity is also strongly stimulated by ATP, but not other adenosine compounds.[96] Interestingly, the addition of Zn^{2+} and Co^{2+} to reactions containing Mn^{2+} ions stimulates the endonuclease activity of *B. subtilis* MutL significantly, reinforcing the idea that two different metal ions with distinct roles are required for the endonuclease activity of MutL.[96] However, in the presence of metal ion mixtures, cutting of the two strands of the duplex is favored over nicking, presumably due to the presence of the two active sites in the *B. subtilis* MutL homodimer.[96]

The endonuclease activity of *N. gonorrhoeae* MutL is more promiscuous than that of *B. subtilis* or eukaryotic MutL homologs and can be activated by both Mn^{2+} and Mg^{2+}, and to a lesser extent by Ca^{2+}.[92] In contrast to what has been reported for human, yeast, and *B. subtilis* MutL, ATP inhibits the endonuclease activity of full-length *N. gonorrhoeae* MutL.[67,94,96] ATP had also been shown to inhibit the endonuclease activity of *Thermus thermophilus* and *Aquifex aeolicus* MutL,[93] though inhibition was later ascribed to a concentration effect.[95] It is cumbersome that mutation of key residues in the endonuclease motif of *N. gonorrhoeae* MutL (D491N or E502Q, $DQHAX_2EX_4E$) barely affects the endonuclease activity of the protein.[92] However, the double mutation of Asp491 and Glu502 radically reduces the endonuclease activity of the protein, indicating that this motif is indeed responsible for this activity.[92] While some of the apparently conflicting results obtained with *N. gonorrhoeae* MutL have been partly resolved in studies using *A. aeolicus* and *T. thermophilus* MutL,[93,95,116] detailed mechanistic studies are still required to understand the catalysis of this reaction.

IV. Regulation of the Endonuclease Activity of MutL

In vivo, the nicking activity of MutL must be tightly regulated to ensure its mismatch dependency. Modrich and coworkers found that PCNA and RFC stimulate the activity of MutLα in both homo- and heteroduplexes.[67] This situation is reminiscent of the MutH activation by MutL, where MutH stimulates ATP hydrolysis by MutL and this, in turn, can activate the endonuclease activity of MutH even in the absence of a mismatch.[97,140] However, it is the formation of a MutS–MutL–heteroduplex complex that fully activates the endonuclease activity of MutH.[25] Similarly, it is the presence of MutSα that preferentially stimulates the endonuclease activity of MutL on heteroduplex DNA.[67] In this section, we discuss how DNA and ATP binding, as well as the interaction with other repair factors, regulate the endonuclease activity of MutL.

A. DNA Binding

One key difference between the endonuclease activities of *B. subtilis* and *N. gonorrhoeae* MutL is the requirement for the ATPase domain of the protein. *B. subtilis* MutL-CTD does not bind DNA on its own and, consequently, its endonuclease activity is undetectable in the absence of the ATPase domain of the protein.[96] Conversely, both the full-length and the C-terminal domain of *N. gonorrhoeae* MutL bind DNA and have measurable endonuclease activity.[92] Accordingly, the dimerization domain of *N. gonorrhoeae* MutL can bind DNA even in the absence of the ATPase domain of the protein,[92] reinforcing the idea that the lack of activity in *B. subtilis* MutL-CTD is due to the low affinity of this domain for DNA.

The N-terminal domains of *E. coli*, human, and yeast homologs of MutL bind DNA even in the absence of the dimerization domain of the protein.[98,101,107] However, the dimerization domains of *E. coli* and human MutLα do not bind DNA on their own (J. W. and A. G., unpublished results).[42] The endonuclease activity of eukaryotic MutLα has not been assayed in the absence of the ATPase domain and, hence, it is unclear whether this domain is strictly required for the activity of MutL. The conformational change imposed by the association of the two ATPase domains upon nucleotide binding should bring DNA bound at the inner groove of the N-terminal domain of the protein in close proximity to the endonuclease domain of the protein (Fig. 2).[100] Indeed, a recent study where ATPase and endonuclease domains of *A. aeolicus* MutL were obtained separately reveals that the ATPase domain stimulates the endonuclease activity in a Zn^{2+}-dependent manner,[116] suggesting the direct interaction between the two domains of the protein. This implicitly suggests that ATP binding may stimulate the endonuclease activity of the protein by providing a mechanism to bypass the DNA binding defect of the endonuclease domain.[96] Therefore, the lack of DNA binding at the C-terminal domains of human MutLα and *B. subtilis* MutL may entail a powerful mechanism to avoid indiscriminate nicking of DNA.

B. Stimulatory Effect of the Processivity Clamp

PCNA and RFC (the clamp loader for PCNA) stimulate the endonuclease activity of human and yeast MutLα.[67,94] This stimulation has been attributed to the direct interaction between MutLα and PCNA.[139] The interaction between these two proteins has been demonstrated and a conserved motif located in the C terminus of yMLH1 has been deemed important for PCNA binding.[73,76] However, while the consensus sequence of this motif (*Q*IX[*L*/*I*]XX*F*A) loosely resembles a PCNA-interacting motif (*Q*XX[*M*/*L*/*I*]XX*F*[Y/F]), its conformation is markedly different from other bona fide PCNA-interacting motifs. The N terminus of PCNA-interacting motifs is usually extended and typically makes a short antiparallel β-zipper with the C-terminal β-strand of the clamp, while the

C terminus of the motif forms a short 3_{10} helix that inserts the two aromatic residues of the motif in hydrophobic pockets at the outer surface of the clamp.[141–144] On the contrary, the *Q*IG*L*TD*F* motif found in hMLH1 is located at the C terminus of helix αC and it is partially occluded from the solvent by helix αA (Fig. 5). Given its conformation and the fact that the disruption of this motif weakens, but does not abolish, the interaction with PCNA,[73] it is likely that additional PCNA-interacting motifs are present in MutLα.

The direct interaction between human PMS2 (or yeast PMS1) with PCNA has not been studied, but the specific interaction between the endonuclease domain of *B. subtilis* MutL and β, the bacterial counterpart of PCNA, has been demonstrated.[145] This interaction is mediated by the conserved ^{487}*Q*EM*I*VP motif, which resides in the external subdomain of the protein and is conspicuously exposed to the solvent (Figs. 4 and 6). The consensus sequence of this motif resembles a β-binding motif[146] and its structure is virtually identical to that seen in the structures of other β-binding peptides bound to β.[96] Mutations within this motif abolish the interaction between the C-terminal domain of *B. subtilis* MutL and β and cause mismatch repair defects that are comparable to those of null *mutL* strains.[96,145] As this motif is conserved in human PMS2 and other eukaryotic MutL proteins encompassing the endonuclease motif,[99] it is presumed that it also mediates the interaction with PCNA.

The structure of the β sliding clamp from *B. subtilis* is unknown, but superimposition of the endonuclease domain of *B. subtilis* MutL onto the structure of the β clamp from another Gram-positive organism (*Streptococcus pyogenes*) unveils two interesting facts. First, it suggests that the interaction of one MutL protomer with the clamp would preclude the interaction of the second protomer due to steric hindrance (Fig. 8). Consequently, only one endonuclease site can be activated at a time even in homodimeric MutL homologs, in turn preventing indiscriminate nicking. Second, the relative orientation between β and MutL is such that the endonuclease site of MutL aligns with the central chamber of the clamp, suggesting that β may stimulate the endonuclease activity of MutL by either threading DNA to the active site or tethering the endonuclease domain to DNA. The latter is the mechanism by which PCNA stimulates the activity of the flap endonuclease Fen1.[147]

The presence of PCNA also biases the nicking activity of human MutLα toward the discontinuous strand, an effect that is attributed to the directionality of the loading.[139] PCNA and β are loaded at 3′ double-strand–single-strand junctions with a specific orientation.[142,148,149] Most proteins that bind to PCNA or β, including MutL, do so through a hydrophobic cleft at the oligomerization interface of the protein. The groove resides on one face of the clamp, often referred to as the proximal face due to its proximity to the 3′ end upon loading and, therefore, the intrinsic asymmetry of the β–DNA complex is likely imposed onto the β–DNA–MutL complex.[96,139,145]

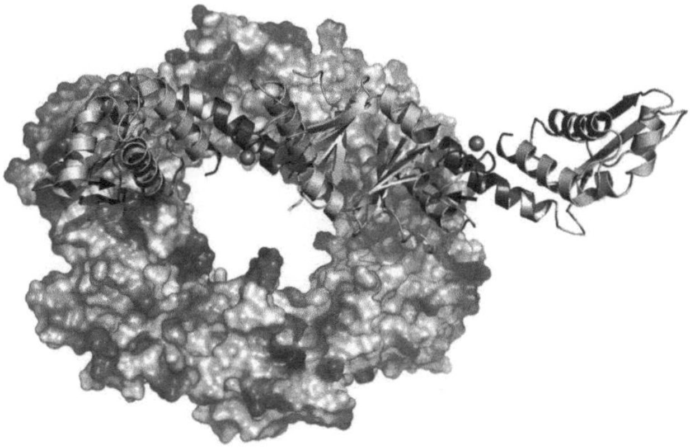

FIG. 8. Docking of MutL onto the structure of β. The structure of *B. subtilis* MutL-CTD (PDB ID: 3KDK) is shown as a ribbon diagram superimposed onto the structure of *Streptococcus pyogenes* β (PDB ID: 2AVT), depicted as an electrostatic potential surface. The endonuclease and the β-binding motifs of BsMutL-CTD are colored dark green and orange, respectively, and the regulatory zinc metal ion is depicted as a pink sphere. The two protein-binding grooves on the surface of the sliding clamp are indicated with black arrows. (For interpretation of the references to color in this figure legend, the reader is referred to the online version of this chapter.)

Interestingly, the conserved QXXL[L/I]XP motif is also found in *E. coli* MutL and other MutL proteins that do not have endonuclease activity. However, disruption of this motif in *E. coli* MutL causes only moderate DNA mismatch repair defects.[145] This is not surprising because it has been previously shown that β is dispensable for the MutH activation and strand excision steps, despite being essential for overall mismatch repair in *E.* coli.[49] Therefore, it has been proposed that the interaction between the C-terminal domain of MutL and processivity clamp is only critical in organisms that lack the MutH endonuclease.[96] Collectively, these findings strengthen the idea that in the absence of MutH, PCNA and β likely coordinate the transfer of information between the early steps in mismatch repair.[47,72,74,76,150]

C. Mismatch Dependency

While we are starting to understand how the conformational changes induced by ATP binding and the interaction with the processivity clamp stimulate the endonuclease activity of MutL, the molecular mechanisms ensuring mismatch dependency remain obscure. The interaction between MutSα and PCNA, or MutS and β in bacteria, increases the mismatch-binding specificity of MutS and recruits this mismatch-binding protein to sites of damage and replication foci.[150–152] However, the interaction between PCNA and MutSα is not required to activate mismatch excision, and mutation of the PCNA-binding site confers only a modest increase in mutation rates.[139,153]

Moreover, PCNA apparently disrupts MutS–MutL–heteroduplex complexes,[154] reinforcing the idea that the role of the processivity clamp is likely transferring information between sequential mismatch repair steps. Therefore, it has been proposed that MutSα also exerts its effect on the endonuclease activity of MutLα through the direct interaction between the two proteins.[139]

Despite being critical for mismatch repair, the interaction between MutS and MutL is poorly characterized.[69,71,155,156] MutS–MutL complexes form only transiently in the presence of heteroduplex DNA and ATP.[25,69,155] However, the two subunits of the MutS and MutL dimers have distinct nucleotide affinities and intrinsic ATPase activities, in turn allowing for a significant number of nucleotide-bound states and conformations that have obscured the detailed characterization of the MutS–MutL–heteroduplex complex.[1,3,52] The formation of these complexes has been studied in the context of three different models—translocation, molecular switch, and stationary—that are at odds as to whether MutS–MutL complexes stay bound to the mismatch or move away from it to signal nicking of the newly replicated strand.[69,157–162] Whether moving or stationary, all models agree that ATP binding and hydrolysis regulate the formation of the MutS–MutL complex and, in turn, the activation of downstream effectors in the repair response. ATP binding, but not hydrolysis, by yMutSα is required to form a ternary MutS–MutL–DNA complex.[69] Conversely, nucleotide binding by MutL does not seem to be critical for the interaction with MutS,[28,156] even though the ATPase domain of MutL mediates the interaction.[70,71,86] It is plausible that the interaction with MutS affects ATP or DNA binding to MutL, indirectly regulating its endonuclease activity. However, probing this idea necessitates a detailed analysis of the MutS–MutL–heteroduplex complex.

V. Concluding Remarks

Six years ago, Modrich and coworkers discovered that eukaryotic MutLα is a latent endonuclease that preferentially nicks the discontinuous strand of a mismatched DNA.[67] The subsequent characterization of the endonuclease activity of other MutL homologs has provided important clues as to how organisms lacking the *mutH* gene—both eukaryotes and prokayotes—target the newly replicated strand during mismatch repair. We now understand how MutL nicks DNA, how this activity is regulated by ATP binding, and how it is stimulated by the interaction with the processivity clamp. These findings have profoundly advanced our understanding of MutH-independent mismatch repair, but an important question remains unanswered: How does MutS affect the endonuclease activity of MutL and, in turn, impose the mismatch dependency on the reaction? As we move forward, it is imperative to understand how the formation of the ternary MutS–MutL–heteroduplex complex alters the conformation of MutL. This, in turn, will reveal how conformational changes in MutL modulate the multiple functions of these key mismatch repair proteins.

References

1. Iyer RR, Pluciennik A, Burdett V, Modrich PL. DNA mismatch repair: functions and mechanisms. *Chem Rev* 2006;**106**:302–23.
2. Jiricny J. The multifaceted mismatch-repair system. *Nat Rev Mol Cell Biol* 2006;**7**:335–46.
3. Kunkel TA, Erie DA. DNA mismatch repair. *Annu Rev Biochem* 2005;**74**:681–710.
4. Li GM. Mechanisms and functions of DNA mismatch repair. *Cell Res* 2008;**18**:85–98.
5. Peltomaki P. Lynch syndrome genes. *Fam Cancer* 2005;**4**:227–32.
6. Vasen HF, Mecklin JP, Khan PM, Lynch HT. The international collaborative group on hereditary non-polyposis colorectal cancer (ICG-HNPCC). *Dis Colon Rectum* 1991;**34**:424–5.
7. Vasen HF, Watson P, Mecklin JP, Lynch HT. New clinical criteria for hereditary nonpolyposis colorectal cancer (HNPCC, Lynch syndrome) proposed by the International Collaborative group on HNPCC. *Gastroenterology* 1999;**116**:1453–6.
8. Buermeyer AB, Deschenes SM, Baker SM, Liskay RM. Mammalian DNA mismatch repair. *Annu Rev Genet* 1999;**33**:533–64.
9. Peltomaki P. Role of DNA mismatch repair defects in the pathogenesis of human cancer. *J Clin Oncol* 2003;**21**:1174–9.
10. Fink D, Aebi S, Howell SB. The role of DNA mismatch repair in drug resistance. *Clin Cancer Res* 1998;**4**:1–6.
11. Li GM. The role of mismatch repair in DNA damage-induced apoptosis. *Oncol Res* 1999;**11**:393–400.
12. Stojic L, Brun R, Jiricny J. Mismatch repair and DNA damage signalling. *DNA Repair (Amst)* 2004;**3**:1091–101.
13. Damia G, D'Incalci M. Genetic instability influences drug response in cancer cells. *Curr Drug Targets* 2010;**11**:1317–24.
14. Kovtun IV, McMurray CT. Features of trinucleotide repeat instability in vivo. *Cell Res* 2008;**18**:198–213.
15. Modrich P. Methyl-directed DNA mismatch correction. *J Biol Chem* 1989;**264**:6597–600.
16. Gorbalenya AE, Koonin EV. Superfamily of UvrA-related NTP-binding proteins. Implications for rational classification of recombination/repair systems. *J Mol Biol* 1990;**213**:583–91.
17. Carraway M, Marinus MG. Repair of heteroduplex DNA molecules with multibase loops in Escherichia coli. *J Bacteriol* 1993;**175**:3972–80.
18. Dohet C, Wagner R, Radman M. Methyl-directed repair of frameshift mutations in heteroduplex DNA. *Proc Natl Acad Sci USA* 1986;**83**:3395–7.
19. Learn BA, Grafstrom RH. Methyl-directed repair of frameshift heteroduplexes in cell extracts from Escherichia coli. *J Bacteriol* 1989;**171**:6473–81.
20. Parker BO, Marinus MG. Repair of DNA heteroduplexes containing small heterologous sequences in Escherichia coli. *Proc Natl Acad Sci USA* 1992;**89**:1730–4.
21. Lamers MH, Perrakis A, Enzlin JH, Winterwerp HH, de Wind N, Sixma TK. The crystal structure of DNA mismatch repair protein MutS binding to a G x T mismatch. *Nature* 2000;**407**:711–7.

22. Lebbink JH, Fish A, Reumer A, Natrajan G, Winterwerp HH, Sixma TK. Magnesium coordination controls the molecular switch function of DNA mismatch repair protein MutS. *J Biol Chem* 2010;**285**:13131–41.
23. Lebbink JH, Georgijevic D, Natrajan G, Fish A, Winterwerp HH, Sixma TK, et al. Dual role of MutS glutamate 38 in DNA mismatch discrimination and in the authorization of repair. *EMBO J* 2006;**25**:409–19.
24. Obmolova G, Ban C, Hsieh P, Yang W. Crystal structures of mismatch repair protein MutS and its complex with a substrate DNA. *Nature* 2000;**407**:703–10.
25. Acharya S, Foster PL, Brooks P, Fishel R. The coordinated functions of the E. coli MutS and MutL proteins in mismatch repair. *Mol Cell* 2003;**12**:233–46.
26. Gorman J, Chowdhury A, Surtees JA, Shimada J, Reichman DR, Alani E, et al. Dynamic basis for one-dimensional DNA scanning by the mismatch repair complex Msh2-Msh6. *Mol Cell* 2007;**28**:359–70.
27. Jeong C, Cho WK, Song KM, Cook C, Yoon TY, Ban C, et al. MutS switches between two fundamentally distinct clamps during mismatch repair. *Nat Struct Mol Biol* 2011;**18**: 379–85.
28. Mendillo ML, Hargreaves VV, Jamison JW, Mo AO, Li S, Putnam CD, et al. A conserved MutS homolog connector domain interface interacts with MutL homologs. *Proc Natl Acad Sci USA* 2009;**106**:22223–8.
29. Bergerat A, de Massy B, Gadelle D, Varoutas PC, Nicolas A, Forterre P. An atypical topoisomerase II from Archaea with implications for meiotic recombination. *Nature* 1997;**386**:414–7.
30. Mushegian AR, Bassett Jr. DE, Boguski MS, Bork P, Koonin EV. Positionally cloned human disease genes: patterns of evolutionary conservation and functional motifs. *Proc Natl Acad Sci USA* 1997;**94**:5831–6.
31. Sancar A, Hearst JE. Molecular matchmakers. *Science* 1993;**259**:1415–20.
32. Au KG, Welsh K, Modrich P. Initiation of methyl-directed mismatch repair. *J Biol Chem* 1992;**267**:12142–8.
33. Welsh KM, Lu AL, Clark S, Modrich P. Isolation and characterization of the Escherichia coli mutH gene product. *J Biol Chem* 1987;**262**:15624–9.
34. Lahue RS, Su SS, Modrich P. Requirement for d(GATC) sequences in Escherichia coli mutHLS mismatch correction. *Proc Natl Acad Sci USA* 1987;**84**:1482–6.
35. Lu AL. Influence of GATC sequences on Escherichia coli DNA mismatch repair in vitro. *J Bacteriol* 1987;**169**:1254–9.
36. Bruni R, Martin D, Jiricny J. d(GATC) sequences influence Escherichia coli mismatch repair in a distance-dependent manner from positions both upstream and downstream of the mismatch. *Nucleic Acids Res* 1988;**16**:4875–90.
37. Cooper DL, Lahue RS, Modrich P. Methyl-directed mismatch repair is bidirectional. *J Biol Chem* 1993;**268**:11823–9.
38. Grilley M, Griffith J, Modrich P. Bidirectional excision in methyl-directed mismatch repair. *J Biol Chem* 1993;**268**:11830–7.
39. Burdett V, Baitinger C, Viswanathan M, Lovett ST, Modrich P. In vivo requirement for RecJ, ExoVII, ExoI, and ExoX in methyl-directed mismatch repair. *Proc Natl Acad Sci USA* 2001;**98**:6765–70.
40. Lahue RS, Au KG, Modrich P. DNA mismatch correction in a defined system. *Science* 1989;**245**:160–4.
41. Viswanathan M, Burdett V, Baitinger C, Modrich P, Lovett ST. Redundant exonuclease involvement in Escherichia coli methyl-directed mismatch repair. *J Biol Chem* 2001;**276**: 31053–8.

42. Guarné A, Ramon-Maiques S, Wolff EM, Ghirlando R, Hu X, Miller JH, et al. Structure of the MutL C-terminal domain: a model of intact MutL and its roles in mismatch repair. *EMBO J* 2004;**23**:4134–45.
43. Hall MC, Jordan JR, Matson SW. Evidence for a physical interaction between the Escherichia coli methyl-directed mismatch repair proteins MutL and UvrD. *EMBO J* 1998;**17**:1535–41.
44. Hall MC, Ozsoy AZ, Matson SW. Site-directed mutations in motif VI of Escherichia coli DNA helicase II result in multiple biochemical defects: evidence for the involvement of motif VI in the coupling of ATPase and DNA binding activities via conformational changes. *J Mol Biol* 1998;**277**:257–71.
45. Lee JY, Yang W. UvrD helicase unwinds DNA one base pair at a time by a two-part power stroke. *Cell* 2006;**127**:1349–60.
46. Mechanic LE, Frankel BA, Matson SW. Escherichia coli MutL loads DNA helicase II onto DNA. *J Biol Chem* 2000;**275**:38337–46.
47. Lopez de Saro FJ, Marinus MG, Modrich P, O'Donnell M. The beta sliding clamp binds to multiple sites within MutL and MutS. *J Biol Chem* 2006;**281**:14340–9.
48. Lopez de Saro FJ, O'Donnell M. Interaction of the beta sliding clamp with MutS, ligase, and DNA polymerase I. *Proc Natl Acad Sci USA* 2001;**98**:8376–80.
49. Pluciennik A, Burdett V, Lukianova O, O'Donnell M, Modrich P. Involvement of the beta clamp in methyl-directed mismatch repair in vitro. *J Biol Chem* 2009;**284**:32782–91.
50. Zhang Y, Yuan F, Presnell SR, Tian K, Gao Y, Tomkinson AE, et al. Reconstitution of 5'-directed human mismatch repair in a purified system. *Cell* 2005;**122**:693–705.
51. Harfe BD, Jinks-Robertson S. DNA mismatch repair and genetic instability. *Annu Rev Genet* 2000;**34**:359–99.
52. Schofield MJ, Hsieh P. DNA mismatch repair: molecular mechanisms and biological function. *Annu Rev Microbiol* 2003;**57**:579–608.
53. Li GM. DNA mismatch repair and cancer. *Front Biosci* 2003;**8**:d997–d1017.
54. McCulloch SD, Gu L, Li GM. Bi-directional processing of DNA loops by mismatch repair-dependent and -independent pathways in human cells. *J Biol Chem* 2003;**278**:3891–6.
55. Korhonen MK, Raevaara TE, Lohi H, Nystrom M. Conditional nuclear localization of hMLH3 suggests a minor activity in mismatch repair and supports its role as a low-risk gene in HNPCC. *Oncol Rep* 2007;**17**:351–4.
56. Li GM, Modrich P. Restoration of mismatch repair to nuclear extracts of H6 colorectal tumor cells by a heterodimer of human MutL homologs. *Proc Natl Acad Sci USA* 1995;**92**: 1950–4.
57. Lipkin SM, Wang V, Jacoby R, Banerjee-Basu S, Baxevanis AD, Lynch HT, et al. MLH3: a DNA mismatch repair gene associated with mammalian microsatellite instability. *Nat Genet* 2000;**24**:27–35.
58. Raschle M, Marra G, Nystrom-Lahti M, Schar P, Jiricny J. Identification of hMutLbeta, a heterodimer of hMLH1 and hPMS1. *J Biol Chem* 1999;**274**:32368–75.
59. Santucci-Darmanin S, Neyton S, Lespinasse F, Saunieres A, Gaudray P, Paquis-Flucklinger V. The DNA mismatch-repair MLH3 protein interacts with MSH4 in meiotic cells, supporting a role for this MutL homolog in mammalian meiotic recombination. *Hum Mol Genet* 2002;**11**:1697–706.
60. Cannavo E, Gerrits B, Marra G, Schlapbach R, Jiricny J. Characterization of the interactome of the human MutL homologues MLH1, PMS1, and PMS2. *J Biol Chem* 2007;**282**:2976–86.
61. Genschel J, Bazemore LR, Modrich P. Human exonuclease I is required for 5' and 3' mismatch repair. *J Biol Chem* 2002;**277**:13302–11.
62. Genschel J, Modrich P. Mechanism of 5′-directed excision in human mismatch repair. *Mol Cell* 2003;**12**:1077–86.

63. Dzantiev L, Constantin N, Genschel J, Iyer RR, Burgers PM, Modrich P. A defined human system that supports bidirectional mismatch-provoked excision. *Mol Cell* 2004;**15**:31–41.
64. Lee BI, Shannon M, Stubbs L, Wilson 3rd. DM. Expression specificity of the mouse exonuclease 1 (mExo1) gene. *Nucleic Acids Res* 1999;**27**:4114–20.
65. Lee Bi BI, Nguyen LH, Barsky D, Fernandes M, Wilson 3rd. DM. Molecular interactions of human Exo1 with DNA. *Nucleic Acids Res* 2002;**30**:942–9.
66. Wilson 3rd. DM, Carney JP, Coleman MA, Adamson AW, Christensen M, Lamerdin JE. Hex1: a new human Rad2 nuclease family member with homology to yeast exonuclease 1. *Nucleic Acids Res* 1998;**26**:3762–8.
67. Kadyrov FA, Dzantiev L, Constantin N, Modrich P. Endonucleolytic function of MutLalpha in human mismatch repair. *Cell* 2006;**126**:297–308.
68. Iyer RR, Pluciennik A, Genschel J, Tsai MS, Beese LS, Modrich P. MutLalpha and proliferating cell nuclear antigen share binding sites on MutSbeta. *J Biol Chem* 2010;**285**:11730–9.
69. Mendillo ML, Mazur DJ, Kolodner RD. Analysis of the interaction between the Saccharomyces cerevisiae MSH2-MSH6 and MLH1-PMS1 complexes with DNA using a reversible DNA end-blocking system. *J Biol Chem* 2005;**280**:22245–57.
70. Plotz G, Raedle J, Brieger A, Trojan J, Zeuzem S. N-terminus of hMLH1 confers interaction of hMutLalpha and hMutLbeta with hMutSalpha. *Nucleic Acids Res* 2003;**31**:3217–26.
71. Plotz G, Welsch C, Giron-Monzon L, Friedhoff P, Albrecht M, Piiper A, et al. Mutations in the MutSalpha interaction interface of MLH1 can abolish DNA mismatch repair. *Nucleic Acids Res* 2006;**34**:6574–86.
72. Gu L, Hong Y, McCulloch S, Watanabe H, Li GM. ATP-dependent interaction of human mismatch repair proteins and dual role of PCNA in mismatch repair. *Nucleic Acids Res* 1998;**26**:1173–8.
73. Lee SD, Alani E. Analysis of interactions between mismatch repair initiation factors and the replication processivity factor PCNA. *J Mol Biol* 2006;**355**:175–84.
74. Nielsen FC, Jager AC, Lutzen A, Bundgaard JR, Rasmussen LJ. Characterization of human exonuclease 1 in complex with mismatch repair proteins, subcellular localization and association with PCNA. *Oncogene* 2004;**23**:1457–68.
75. Schmutte C, Sadoff MM, Shim KS, Acharya S, Fishel R. The interaction of DNA mismatch repair proteins with human exonuclease I. *J Biol Chem* 2001;**276**:33011–8.
76. Umar A, Buermeyer AB, Simon JA, Thomas DC, Clark AB, Liskay RM, et al. Requirement for PCNA in DNA mismatch repair at a step preceding DNA resynthesis. *Cell* 1996;**87**:65–73.
77. Dherin C, Gueneau E, Francin M, Nunez M, Miron S, Liberti SE, et al. Characterization of a highly conserved binding site of Mlh1 required for exonuclease I-dependent mismatch repair. *Mol Cell Biol* 2009;**29**:907–18.
78. Her C, Vo AT, Wu X. Evidence for a direct association of hMRE11 with the human mismatch repair protein hMLH1. *DNA Repair (Amst)* 2002;**1**:719–29.
79. Jager AC, Rasmussen M, Bisgaard HC, Singh KK, Nielsen FC, Rasmussen LJ. HNPCC mutations in the human DNA mismatch repair gene hMLH1 influence assembly of hMutLalpha and hMLH1-hEXO1 complexes. *Oncogene* 2001;**20**:3590–5.
80. Liu Y, Fang Y, Shao H, Lindsey-Boltz L, Sancar A, Modrich P. Interactions of human mismatch repair proteins MutSalpha and MutLalpha with proteins of the ATR-Chk1 pathway. *J Biol Chem* 2010;**285**:5974–82.
81. Wang Y, Cortez D, Yazdi P, Neff N, Elledge SJ, Qin J. BASC, a super complex of BRCA1-associated proteins involved in the recognition and repair of aberrant DNA structures. *Genes Dev* 2000;**14**:927–39.
82. Cantor SB, Bell DW, Ganesan S, Kass EM, Drapkin R, Grossman S, et al. BACH1, a novel helicase-like protein, interacts directly with BRCA1 and contributes to its DNA repair function. *Cell* 2001;**105**:149–60.

83. Litman R, Peng M, Jin Z, Zhang F, Zhang J, Powell S, et al. BACH1 is critical for homologous recombination and appears to be the Fanconi anemia gene product FANCJ. *Cancer Cell* 2005;**8**:255–65.
84. Polosina YY, Cupples CG. Wot the 'L-Does MutL do? *Mutat Res* 2010;**705**:228–38.
85. Polosina YY, Cupples CG. MutL: conducting the cell's response to mismatched and misaligned DNA. *Bioessays* 2010;**32**:51–9.
86. Winkler I, Marx AD, Lariviere D, Heinze RJ, Cristovao M, Reumer A, et al. Chemical Trapping of the Dynamic MutS-MutL Complex Formed in DNA Mismatch Repair in Escherichia coli. *J Biol Chem* 2011;**286**:17326–37.
87. Yamaguchi M, Dao V, Modrich P. MutS and MutL activate DNA helicase II in a mismatch-dependent manner. *J Biol Chem* 1998;**273**:9197–201.
88. Li F, Liu Q, Chen YY, Yu ZN, Zhang ZP, Zhou YF, et al. Escherichia coli mismatch repair protein MutL interacts with the clamp loader subunits of DNA polymerase III. *Mutat Res* 2008;**637**:101–10.
89. Heinze RJ, Giron-Monzon L, Solovyova A, Elliot SL, Geisler S, Cupples CG, et al. Physical and functional interactions between Escherichia coli MutL and the Vsr repair endonuclease. *Nucleic Acids Res* 2009;**37**:4453–63.
90. Pitsikas P, Polosina YY, Cupples CG. Interaction between the mismatch repair and nucleotide excision repair pathways in the prevention of 5-azacytidine-induced CG-to-GC mutations in Escherichia coli. *DNA Repair (Amst)* 2009;**8**:354–9.
91. Polosina YY, Mui J, Pitsikas P, Cupples CG. The Escherichia coli mismatch repair protein MutL recruits the Vsr and MutH endonucleases in response to DNA damage. *J Bacteriol* 2009;**191**:4041–3.
92. Duppatla V, Bodda C, Urbanke C, Friedhoff P, Rao DN. The carboxy-terminal domain is sufficient for endonuclease activity of Neisseria gonorrhoeae MutL. *Biochem J* 2009;**423**(2): 265–77.
93. Fukui K, Nishida M, Nakagawa N, Masui R, Kuramitsu S. Bound nucleotide controls the endonuclease activity of mismatch repair enzyme MutL. *J Biol Chem* 2008;**283**:12136–45.
94. Kadyrov FA, Holmes SF, Arana ME, Lukianova OA, O'Donnell M, Kunkel TA, et al. Saccharomyces cerevisiae MutLa is a mismatch repair endonuclease. *J Biol Chem* 2007;**282**(51): 37181–90.
95. Mauris J, Evans TC. Adenosine triphosphate stimulates Aquifex aeolicus MutL endonuclease activity. *PLoS One* 2009;**4**:e7175.
96. Pillon MC, Lorenowicz JJ, Uckelmann M, Klocko AD, Mitchell RR, Chung YS, et al. Structure of the endonuclease domain of MutL: unlicensed to cut. *Mol Cell* 2010;**39**:145–51.
97. Ban C, Yang W. Crystal structure and ATPase activity of MutL: implications for DNA repair and mutagenesis. *Cell* 1998;**95**:541–52.
98. Guarné A, Junop MS, Yang W. Structure and function of the N-terminal 40 kDa fragment of human PMS2: a monomeric GHL ATPase. *EMBO J* 2001;**20**:5521–31.
99. Kosinski J, Plotz G, Guarné A, Bujnicki JM, Friedhoff P. The PMS2 subunit of human MutLalpha contains a metal ion binding domain of the iron-dependent repressor protein family. *J Mol Biol* 2008;**382**:610–27.
100. Sacho EJ, Kadyrov FA, Modrich P, Kunkel TA, Erie DA. Direct visualization of asymmetric adenine-nucleotide-induced conformational changes in MutL alpha. *Mol Cell* 2008;**29**: 112–21.
101. Ban C, Junop M, Yang W. Transformation of MutL by ATP binding and hydrolysis: a switch in DNA mismatch repair. *Cell* 1999;**97**:85–97.
102. Hall MC, Shcherbakova PV, Kunkel TA. Differential ATP binding and intrinsic ATP hydrolysis by amino-terminal domains of the yeast Mlh1 and Pms1 proteins. *J Biol Chem* 2002;**277**: 3673–9.

103. Raschle M, Dufner P, Marra G, Jiricny J. Mutations within the hMLH1 and hPMS2 subunits of the human MutLalpha mismatch repair factor affect its ATPase activity, but not its ability to interact with hMutSalpha. *J Biol Chem* 2002;**277**:21810–20.
104. Tran PT, Liskay RM. Functional studies on the candidate ATPase domains of Saccharomyces cerevisiae MutLalpha. *Mol Cell Biol* 2000;**20**:6390–8.
105. Johnson JR, Erdeniz N, Nguyen M, Dudley S, Liskay RM. Conservation of functional asymmetry in the mammalian MutLalpha ATPase. *DNA Repair (Amst)* 2010;**9**:1209–13.
106. Gorman J, Plys AJ, Visnapuu ML, Alani E, Greene EC. Visualizing one-dimensional diffusion of eukaryotic DNA repair factors along a chromatin lattice. *Nat Struct Mol Biol* 2010;**17**:932–8.
107. Arana ME, Holmes SF, Fortune JM, Moon AF, Pedersen LC, Kunkel TA. Functional residues on the surface of the N-terminal domain of yeast Pms1. *DNA Repair (Amst)* 2010;**9**:448–57.
108. Bende SM, Grafstrom RH. The DNA binding properties of the MutL protein isolated from Escherichia coli. *Nucleic Acids Res* 1991;**19**:1549–55.
109. Drotschmann K, Hall MC, Shcherbakova PV, Wang H, Erie DA, Brownewell FR, et al. DNA binding properties of the yeast Msh2-Msh6 and Mlh1-Pms1 heterodimers. *Biol Chem* 2002;**383**:969–75.
110. Hall MC, Wang H, Erie DA, Kunkel TA. High affinity cooperative DNA binding by the yeast Mlh1-Pms1 heterodimer. *J Mol Biol* 2001;**312**:637–47.
111. Hall MC, Shcherbakova PV, Fortune JM, Borchers CH, Dial JM, Tomer KB, et al. DNA binding by yeast Mlh1 and Pms1: implications for DNA mismatch repair. *Nucleic Acids Res* 2003;**31**:2025–34.
112. Junop MS, Yang W, Funchain P, Clendenin W, Miller JH. In vitro and in vivo studies of MutS, MutL and MutH mutants: correlation of mismatch repair and DNA recombination. *DNA Repair (Amst)* 2003;**2**:387–405.
113. Schorzman AN, Perera L, Cutalo-Patterson JM, Pedersen LC, Pedersen LG, Kunkel TA, et al. Modeling of the DNA-binding site of yeast Pms1 by mass spectrometry. *DNA Repair (Amst)* 2011;**10**:454–65.
114. Namadurai S, Jain D, Kulkarni DS, Tabib CR, Friedhoff P, Rao DN, et al. The C-terminal domain of the MutL homolog from Neisseria gonorrhoeae forms an inverted homodimer. *PLoS One* 2010;**5**:e13726.
115. Yang W. Human MutLalpha: the jack of all trades in MMR is also an endonuclease. *DNA Repair (Amst)* 2007;**6**:135–9.
116. Iino H, Kim K, Shimada A, Masui R, Kuramitsu S, Fukui K. Characterization of C- and N-terminal domains of Aquifex aeolicus MutL endonuclease: N-terminal domain stimulates the endonuclease activity of C-terminal domain in a zinc-dependent manner. *Biosci Rep* 2011;**31**:309–22.
117. Erdeniz N, Nguyen M, Deschenes SM, Liskay RM. Mutations affecting a putative MutLalpha endonuclease motif impact multiple mismatch repair functions. *DNA Repair (Amst)* 2007;**6**:1463–70.
118. Nishant KT, Plys AJ, Alani E. A mutation in the putative MLH3 endonuclease domain confers a defect in both mismatch repair and meiosis in Saccharomyces cerevisiae. *Genetics* 2008;**179**:747–55.
119. Feese MD, Ingason BP, Goranson-Siekierke J, Holmes RK, Hol WG. Crystal structure of the iron-dependent regulator from Mycobacterium tuberculosis at 2.0-A resolution reveals the Src homology domain 3-like fold and metal binding function of the third domain. *J Biol Chem* 2001;**276**:5959–66.
120. Dudev T, Lim C. Principles governing Mg, Ca, and Zn binding and selectivity in proteins. *Chem Rev* 2003;**103**:773–88.

121. Dudev T, Lin YL, Dudev M, Lim C. First-second shell interactions in metal binding sites in proteins: a PDB survey and DFT/CDM calculations. *J Am Chem Soc* 2003;**125**:3168–80.
122. Babu CS, Dudev T, Casareno R, Cowan JA, Lim C. A combined experimental and theoretical study of divalent metal ion selectivity and function in proteins: application to E. coli ribonuclease H1. *J Am Chem Soc* 2003;**125**:9318–28.
123. Dudev M, Wang J, Dudev T, Lim C. Factors governing the metal coordination number in metal complexes from Cambridge Structural Database analyses. *J Phys Chem B* 2006;**110**: 1889–95.
124. Harding MM. The geometry of metal-ligand interactions relevant to proteins. *Acta Crystallogr D Biol Crystallogr* 1999;**55**:1432–43.
125. Harding MM. The architecture of metal coordination groups in proteins. *Acta Crystallogr D Biol Crystallogr* 2004;**60**:849–59.
126. Harding MM. Small revisions to predicted distances around metal sites in proteins. *Acta Crystallogr D Biol Crystallogr* 2006;**62**:678–82.
127. Dokmanic I, Sikic M, Tomic S. Metals in proteins: correlation between the metal-ion type, coordination number and the amino-acid residues involved in the coordination. *Acta Crystallogr D Biol Crystallogr* 2008;**64**:257–63.
128. Ryde U. Carboxylate binding modes in zinc proteins: a theoretical study. *Biophys J* 1999;**77**:2777–87.
129. Karlin S, Zhu ZY. Classification of mononuclear zinc metal sites in protein structures. *Proc Natl Acad Sci USA* 1997;**94**:14231–6.
130. Stoll KE, Draper WE, Kliegman JI, Golynskiy MV, Brew-Appiah RA, Phillips RK, et al. Characterization and structure of the manganese-responsive transcriptional regulator ScaR. *Biochemistry* 2009;**48**:10308–20.
131. Guedon E, Helmann JD. Origins of metal ion selectivity in the DtxR/MntR family of metalloregulators. *Mol Microbiol* 2003;**48**:495–506.
132. Jakubovics NS, Smith AW, Jenkinson HF. Expression of the virulence-related Sca (Mn2+) permease in Streptococcus gordonii is regulated by a diphtheria toxin metallorepressor-like protein ScaR. *Mol Microbiol* 2000;**38**:140–53.
133. Que Q, Helmann JD. Manganese homeostasis in Bacillus subtilis is regulated by MntR, a bifunctional regulator related to the diphtheria toxin repressor family of proteins. *Mol Microbiol* 2000;**35**:1454–68.
134. Schmitt MP, Predich M, Doukhan L, Smith I, Holmes RK. Characterization of an iron-dependent regulatory protein (IdeR) of Mycobacterium tuberculosis as a functional homolog of the diphtheria toxin repressor (DtxR) from Corynebacterium diphtheriae. *Infect Immun* 1995;**63**:4284–9.
135. Pohl E, Holmes RK, Hol WG. Crystal structure of the iron-dependent regulator (IdeR) from Mycobacterium tuberculosis shows both metal binding sites fully occupied. *J Mol Biol* 1999;**285**:1145–56.
136. Wang G, Wylie GP, Twigg PD, Caspar DL, Murphy JR, Logan TM. Solution structure and peptide binding studies of the C-terminal src homology 3-like domain of the diphtheria toxin repressor protein. *Proc Natl Acad Sci USA* 1999;**96**:6119–24.
137. Wisedchaisri G, Chou CJ, Wu M, Roach C, Rice AE, Holmes RK, et al. Crystal structures, metal activation, and DNA-binding properties of two-domain IdeR from Mycobacterium tuberculosis. *Biochemistry* 2007;**46**:436–47.
138. Wylie GP, Rangachari V, Bienkiewicz EA, Marin V, Bhattacharya N, Love JF, et al. Prolylpeptide binding by the prokaryotic SH3-like domain of the diphtheria toxin repressor: a regulatory switch. *Biochemistry* 2005;**44**:40–51.
139. Pluciennik A, Dzantiev L, Iyer RR, Constantin N, Kadyrov FA, Modrich P. PCNA function in the activation and strand direction of MutLalpha endonuclease in mismatch repair. *Proc Natl Acad Sci USA* 2010;**107**:16066–71.

140. Hall MC, Matson SW. The Escherichia coli MutL protein physically interacts with MutH and stimulates the MutH-associated endonuclease activity. *J Biol Chem* 1999;**274**:1306–12.
141. Bunting KA, Roe SM, Pearl LH. Structural basis for recruitment of translesion DNA polymerase Pol IV/DinB to the beta-clamp. *EMBO J* 2003;**22**:5883–92.
142. Gulbis JM, Kelman Z, Hurwitz J, O'Donnell M, Kuriyan J. Structure of the C-terminal region of p21(WAF1/CIP1) complexed with human PCNA. *Cell* 1996;**87**:297–306.
143. Matsumiya S, Ishino S, Ishino Y, Morikawa K. Physical interaction between proliferating cell nuclear antigen and replication factor C from Pyrococcus furiosus. *Genes Cells* 2002;**7**:911–22.
144. Shamoo Y, Steitz TA. Building a replisome from interacting pieces: sliding clamp complexed to a peptide from DNA polymerase and a polymerase editing complex. *Cell* 1999;**99**:155–66.
145. Pillon MC, Miller JH, Guarne A. The endonuclease domain of MutL interacts with the beta sliding clamp. *DNA Repair (Amst)* 2011;**10**:87–93.
146. Dalrymple BP, Kongsuwan K, Wijffels G, Dixon NE, Jennings PA. A universal protein-protein interaction motif in the eubacterial DNA replication and repair systems. *Proc Natl Acad Sci USA* 2001;**98**:11627–32.
147. Tom S, Henricksen LA, Bambara RA. Mechanism whereby proliferating cell nuclear antigen stimulates flap endonuclease 1. *J Biol Chem* 2000;**275**:10498–505.
148. Kong XP, Onrust R, O'Donnell M, Kuriyan J. Three-dimensional structure of the beta subunit of E. coli DNA polymerase III holoenzyme: a sliding DNA clamp. *Cell* 1992;**69**:425–37.
149. Yao N, Hurwitz J, O'Donnell M. Dynamics of beta and proliferating cell nuclear antigen sliding clamps in traversing DNA secondary structure. *J Biol Chem* 2000;**275**:1421–32.
150. Simmons LA, Davies BW, Grossman AD, Walker GC. Beta clamp directs localization of mismatch repair in Bacillus subtilis. *Mol Cell* 2008;**29**:291–301.
151. Flores-Rozas H, Clark D, Kolodner RD. Proliferating cell nuclear antigen and Msh2p-Msh6p interact to form an active mispair recognition complex. *Nat Genet* 2000;**26**:375–8.
152. Kleczkowska HE, Marra G, Lettieri T, Jiricny J. hMSH3 and hMSH6 interact with PCNA and colocalize with it to replication foci. *Genes Dev* 2001;**15**:724–36.
153. Iyer RR, Pohlhaus TJ, Chen S, Hura GL, Dzantiev L, Beese LS, et al. The MutSalpha-proliferating cell nuclear antigen interaction in human DNA mismatch repair. *J Biol Chem* 2008;**283**:13310–9.
154. Bowers J, Tran PT, Joshi A, Liskay RM, Alani E. MSH-MLH complexes formed at a DNA mismatch are disrupted by the PCNA sliding clamp. *J Mol Biol* 2001;**306**:957–68.
155. Blackwell LJ, Wang S, Modrich P. DNA chain length dependence of formation and dynamics of hMutSalpha.hMutLalpha.heteroduplex complexes. *J Biol Chem* 2001;**276**:33233–40.
156. Selmane T, Schofield MJ, Nayak S, Du C, Hsieh P. Formation of a DNA mismatch repair complex mediated by ATP. *J Mol Biol* 2003;**334**:949–65.
157. Allen DJ, Makhov A, Grilley M, Taylor J, Thresher R, Modrich P, et al. MutS mediates heteroduplex loop formation by a translocation mechanism. *EMBO J* 1997;**16**:4467–76.
158. Fishel R. Mismatch repair, molecular switches, and signal transduction. *Genes Dev* 1998;**12**:2096–101.
159. Gradia S, Acharya S, Fishel R. The human mismatch recognition complex hMSH2-hMSH6 functions as a novel molecular switch. *Cell* 1997;**91**:995–1005.
160. Jiang J, Bai L, Surtees JA, Gemici Z, Wang MD, Alani E. Detection of high-affinity and sliding clamp modes for MSH2-MSH6 by single-molecule unzipping force analysis. *Mol Cell* 2005;**20**:771–81.
161. Junop MS, Obmolova G, Rausch K, Hsieh P, Yang W. Composite active site of an ABC ATPase: MutS uses ATP to verify mismatch recognition and authorize DNA repair. *Mol Cell* 2001;**7**:1–12.
162. Wang H, Hays JB. Signaling from DNA mispairs to mismatch-repair excision sites despite intervening blockades. *EMBO J* 2004;**23**:2126–33.

The Fpg/Nei Family of DNA Glycosylases: Substrates, Structures, and Search for Damage

Aishwarya Prakash, Sylvie Doublié, and Susan S. Wallace

Department of Microbiology and Molecular Genetics, The Markey Center for Molecular Genetics, University of Vermont, Burlington, Vermont, USA

During the initial stages of the base excision DNA repair pathway, DNA glycosylases are responsible for locating and removing the majority of endogenous oxidative base lesions. The bifunctional formamidopyrimidine DNA glycosylase (Fpg) and endonuclease VIII (Nei) are members of the Fpg/Nei family, one of the two families of glycosylases that recognize oxidized DNA bases, the other being the HhH/GPD (or Nth) superfamily. Structural and biochemical developments over the past decades have led to novel insights into the mechanism of damage recognition by the Fpg/Nei family of enzymes. Despite the overall structural similarity among members of this family, these enzymes exhibit distinct features that make them unique. This review summarizes the current structural knowledge of the Fpg/Nei family members, emphasizes their substrate specificities, and describes how these enzymes search for lesions.

I. Introduction

Escherichia coli formamidopyrimidine (Fapy) DNA glycosylase (Fpg) was originally discovered in Tomas Lindahl's laboratory as a DNA glycosylase that removes methylFapyG from alkylated DNA.[1] The *E. coli* gene for Fpg was

Progress in Molecular Biology
and Translational Science, Vol. 110
DOI: 10.1016/B978-0-12-387665-2.00004-3

subsequently cloned[2] and the protein further characterized in a number of laboratories.[3–6] *E. coli mutM* mutants were identified in Jeffrey Miller's laboratory as mutators that gave rise to G→T transversions.[7] When the MutM protein was subsequently purified, it was found to be identical to Fpg.[8] Following these initial findings, there were a number of biochemical studies showing that 8-oxoguanine (8-oxoG) was also a substrate for Fpg and that Fpg preferred 8-oxoG over methylFapyG.[9,10] Because of this substrate preference and because guanine is the most readily oxidized DNA base, the conclusion was drawn that 8-oxoguanine was the biologically relevant substrate for Fpg. These studies led to the formulation of the GO model for 8-oxoG repair,[11] which proposed that when guanine is oxidized to 8-oxoguanine, it is removed by Fpg. If 8-oxoG is not removed prior to replication, A is often inserted opposite the 8-oxoG by DNA polymerases.[12–15] If this occurs, the A can be removed by another glycosylase called MutY.[16] The GO model also included MutT, which removes 8-oxoguanine nucleoside triphosphates from the nucleotide pool by hydrolyzing them to 8-oxodGMP.[17] Taken together, these data supported the idea that 8-oxoguanine is a biologically important, potentially mutagenic oxidative DNA lesion. However, recent studies have shown that unmethylated FapyG is also a good substrate for Fpg[18,19] and like 8-oxoG, A can also be incorporated opposite FapyG[20,21] and the incorporated A can be removed by MutY.[22] FapyG, which is formed from the same adduct radical as 8-oxoG,[23] appears to be responsible for a substantial number of mutations originally attributed to 8-oxoG and thus is also a biologically relevant substrate.[24]

E. coli nei (endonuclease VIII) was originally discovered in the Wallace laboratory as an activity that recognizes oxidized pyrimidines.[25,26] The gene was cloned and the protein sequence was shown to be very similar to that of Fpg.[27] *nei* mutants had little or no phenotype, but, when coupled with an *nth* mutation, they were mutators leading to C→T transitions.[27] The *nth* gene encodes endonuclease III, which also recognizes oxidized pyrimidines with a substrate specificity that substantially overlaps that of Nei (for reviews see Refs. 28,29).

It was not until the twenty-first century and the sequencing of the human genome that *in silico* analysis allowed the Wallace, Mitra, and Seeberg laboratories to identify, clone, and characterize three Fpg/Nei homologs in mammalian cells, the so-called Neil1 (nei-like), Neil2, and Neil3 proteins.[30–34] Mouse Neil1 and Neil3 were also found in mice nullizygous for *nth*.[35] The substrate specificities of human NEIL1 and NEIL2 have been well characterized.[30–34,36–40] In addition, NEIL1 forms specific interactions with a number of replication proteins and is cell cycle-regulated.[41–44] Thus, it has been proposed that NEIL1 acts as a cow catcher ahead of the replication fork, eliminating potentially mutagenic lesions.[42–44] NEIL2 prefers lesions in single-stranded DNA over those in duplex DNA and interacts with a number of transcription factors including RNA polymerase II, and has been suggested to act in transcription-coupled repair.[45] Although attempts had been made to determine the activity of NEIL3,[33,46,47] it is only recently that NEIL3 has been purified and characterized,[24,48] and its glycosylase activity shown to be similar to that of NEIL2.[24] In mice, Neil3 is present during embryonic development[49] and was found in brain stem cells.[49,50] In humans, expression of NEIL3 has only been observed in thymus.[51]

II. Fpg/Nei Phylogeny

Sequence alignments of members of the Fpg/Nei family of glycosylases indicate that they share many structural and biochemical features.[34] Some of the hallmark motifs of this family include conserved residues in the helix-two-turns-helix motif (H2TH), a zinc finger motif, and a common catalytic mechanism involving either an N-terminal proline (e.g., in NEIL1 and NEIL2) or a valine residue (as in human NEIL3 and the giant mimivirus Nei2 (MvNei2)) as the active site nucleophile. Despite these commonalities, each glycosylase prefers a different spectrum of oxidative lesions. Moreover, some of these subfamilies have changed significantly in sequence from their common ancestor, making it difficult to infer the evolution of these enzymes.

Phylogenetic analysis and functional studies of the Fpg/Nei family indicate that in *Actinobacteria* alone, six gene clades occur, two within the Nei proteins and four within the Fpg clade.[52] The plant and fungi clade is clearly part of the Fpg family while within metazoans, Neil2 and Neil3 form their own clade separate from Neil1. The Neil1 protein, like members of the plant and fungi Fpg/Nei proteins, does not have the canonical zinc finger, but possesses a "zincless finger" motif, which lacks the four characteristic cysteine residues that coordinate a zinc ion. This motif superimposes well with the zinc finger domains of EcoNei and EcoFpg, despite the absence of sequence homology. In contrast to Neil1, both Neil2 and Neil3 possess a zinc finger domain; the former contains a C–H–C–C-type zinc finger whereas the latter has an RanBP-type zinc finger very similar to the one found in bacterial Fpg. Some shared conserved structural features suggest that the zincless fingers evolved independently of the zinc finger motifs. Recent evidence suggests that the Neil2 and Neil3 proteins evolved from a common ancestor while Neil1 evolved separately (Barrantes-Reynolds, unpublished data).

We speculate that horizontal gene transfer, a common occurrence in bacteria, seems to be a likely event in the initial evolution of EcoNei proteins from a common ancestor, which contained at least one Fpg/Nei homolog and exhibited features similar to EcoFpg.[53] Vertical evolution may have been responsible for the transfer of an early Fpg/Nei gene to early eukaryotes in which these Fpg/Nei homologs led to the diversification of the Fpg/Nei proteins in higher eukaryotes.[34,52,53]

III. Fpg/Nei Structures

A. Introduction

Over the past decade, there has been a significant increase in the number of crystal structures of Fpg/Nei glycosylases.[54–69] The advent of techniques such as reductive crosslinking using sodium borohydride has played an essential role in trapping stable protein–DNA complexes for the purposes of crystallization and for the elucidation of the mechanism and role of these intricate enzymes (Refs. 54–56,64–66,70,71; for reviews see Refs. 70,71). Other approaches successfully used to produce stable glycosylase–DNA complexes include the generation of site-directed mutants of active site residues to abolish catalysis and the use of noncleavable substrates such as tetrahydrofuran (THF), which mimics an abasic (AP) site,[60,72] and noncleavable cyclopentane FapyG (cFapyG).[68] A summary of all the currently available crystal structures of the Fpg/Nei family of glycosylases and their substrate preferences is listed in Table I. Crystal structures of Fpg proteins from various bacterial species such as *Thermus thermophilus* (Tth) Fpg (without DNA),[54] *Escherichia coli* (EcoFpg),[55] *Geobacillus stearothermophilus* Fpg (BstFpg),[57,59,64–66] and *Lactococcus lactis* Fpg (LlaFpg)[58,60,63,68,69] complexed with DNA substrates have been determined. The Fpg–DNA complexes include Schiff base intermediates, noncovalent complexes with AP site analogs, and recognition or end product complexes. Although the structure of EcoNei as a Schiff base intermediate in a complex with DNA was solved,[56] it wasn't until recently that the unliganded structure of EcoNei was determined which revealed a unique interdomain conformational change upon DNA binding.[62] Furthermore, the structures of unliganded human NEIL1,[61] unliganded Mimivirus Neil1 (MvNeil1), and MvNeil1 in a complex with THF were subsequently obtained.[67] The first crystal structures of an Nei bound to damaged bases were recently reported: MvNeil1 was captured in a complex with DNA containing either thymine glycol (Tg) or 5-hydroxyuracil (5-OHU).[104]

Overall, the structures of the Fpg and Nei proteins are similar, with a distinct 2-domain architecture connected by a flexible hinge region (Fig. 1A and B using EcoFpg and EcoNei as examples).[55,56,72] In general, the N-terminal region is predominantly β-sheet-rich and is composed of a β-sandwich flanked by α-helices. The C-terminal domain comprises α-helices, two of which form a conserved H2TH motif, as well as two antiparallel β-strands that fold into a zinc finger motif. These signature motifs are characteristic of both Fpg and Nei

TABLE I

Summary of Current Crystal Structures of the Fpg/Nei Glycosylases

Protein	Mutation	Complex	Substrate/opposite base	Resolution (Å)	PDB ID (Reference)	Substrate specificity
TthFpg	–		–	1.90	1EE8[54]	8-oxoG, FapyG, Me-FapyG, FapyA[54]
EcoFpg	–	SBI	–	2.10	1K82[55]	Sp = 8-oxoG, FapyG, Me-FapyG, FapyA, Gh[6,18,72–77]
BstFpg	–	SBI	–	1.70	1L1Z[57]	8-oxoG, FapyG, Me-FapyG, FapyA[57,59,64–66]
	–	SBI	–	2.40	1L2B	
	–	RC	rAb:C	1.80	1L1T	
	–	RC	rAb:T	2.20	1L2C	
	–	RC	rAb:G	2.00	1L2D	
	E3Q	LRC	8-oxoG:C	2.34	1R2Y[59]	
	E3Q	LRC	DHU:C	1.63	1R2Z[64]	
	E3Q-DXL	CC1	8-oxoG:C	2.35	2F5S	
	E3Q-DXL	CC2	8-oxoG:C	2.35	2F5Q	
	E3Q-DXL	IC1	A:T	2.00	2F5N	
	E3Q-DXL	IC2	A:T	2.00	2F5P	
	E3Q-DXL	IC3	G:C	2.05	2F5O	
	E3Q-DXL	LRC3	8-oxoG:C	1.85	3GPY[65]	
	Δ220-235-DXL	EC3	8-oxoG:C	1.89	3GO8	
	V222P-DXL	EC3	8-oxoG:C	2.05	3GP1	
	T224P-DXL	EC3	8-oxoG:C	2.15	3GPP	
	Δ220-235-DXL	EC4	8-oxoG:C	1.62	3GPU	
	Δ220-235-DXL	IC4	G:C	1.78	3GPX	
	E3Q-DXL	LRC5	8-oxoG:C	1.70	3GQ4	
	Δ220-235-DXL	EC5	8-oxoG:C	1.83	3GQ3	
	V222P-DXL	IC5	G:C	1.90	3GQ5	
	N174C-DXL	XGC	G:C	2.60	3JR4[66]	
	N174C-DXL	LRC	8-oxoG:C	1.70	3JR5	

(Continues)

TABLE I (*Continued*)

Protein	Mutation	Complex	Substrate/opposite base	Resolution (Å)	PDB ID (Reference)	Substrate specificity
LlaFpg	P1G	RC	Pr:C	2.55	1KFV[58]	8-oxoG, FapyG, Me-FapyG, FapyA, Hyd[58,60,63,68,69]
	ΔP1	LRC	cFapydG:C	1.80	1TDZ[60]	
	P1G	RC	Pr:C	1.90	1NNJ[63]	
	P1G	RC	THF:C	1.90	1PJJ	
	–	RC	Pr:C	1.90	1PJI	
	–	RC	THF:C	1.95	1PM5	
	–	RC	cFapydG:C	1.95	1XC8[68]	
	–	RC	N7-Benzyl-FapyG:C	1.90	3C58	
	–	LRC	cHyd:C	1.80	2XZF[69]	
	–	DPC	cHyd:C	1.80	2XZU	
EcoNei		SBI	–	1.42	1K3W[56]	Tg, 5-OHU, 5-OHC, DHT, DHU, Sp, Gh[34,72,78,79]
		SBI	–	1.25	1K3X	
	–	–	–	2.80	1Q39[62]	
	E2A	–	–	2.30	1Q3C	
	R252A	–	–	2.05	1Q3B	
NEIL1	ΔC-terminal56	–	–	2.10	1TDH[61]	Sp = Gh > Tg > DHU > 5-OHU > 5-OHC > DHT > FapyG = FapyA ≫ 8-oxoG[30,36,39,44,61,80,81]
MvNei1	–	RC	THF:C	2.20	3A46[67]	Sp = Gh > Tg > 5-OHU > 5-OHC > DHT = DHU ≫ 8-oxoG[39,67,82]
	–		–	2.60	3A42	
	–		–	2.30	3A45	
MvNeil1	E3Q	RC	Tg:C	2.1Å	3VK7[104]	
	E3Q	RC	5-OHU:G	2.0Å	3VK8[104]	

SBI, Schiff-base intermediate; DXL, disulfide cross-linking; RC, recognition complex; LRC, lesion recognition complex; DPC, DNA–protein covalent; IC, interrogation complex; CC, control complex; EC, encounter complex; XGC, extrahelical G complex; Hyd or cHyd, where c refers to a carbanucleoside, 5-hydroxy-5-methylhydantoin.

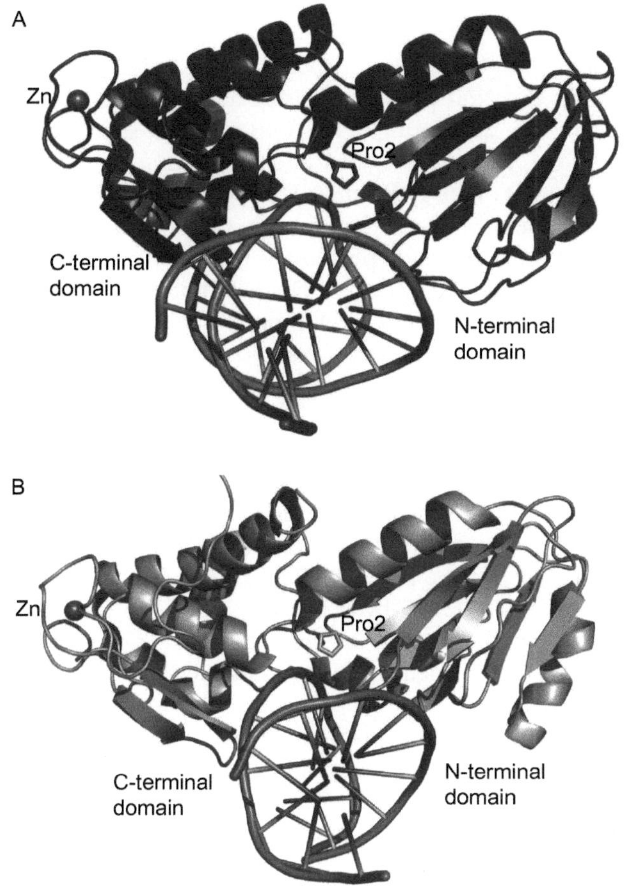

FIG. 1. Overall structural comparison between EcoFpg and EcoNei. (A). EcoFpg bound to DNA. Both N- and C-terminal domains are colored in purple (PDB ID 1K82).[55] (B) EcoNei bound to DNA. N- and C-terminal domains are shown in pale pink (IK3W from the PDB).[56] Zinc atoms are shown as gray spheres in both cases and the DNA is displayed as a ribbon. PyMol was used to generate the figures (DeLano Scientific, The PyMOL Molecular Graphics System, Version 1.2r3pre, Schrödinger, LLC). (See Color Insert.)

subfamilies. The zinc finger and H2TH motifs have been shown to be absolutely necessary for Fpg to bind to DNA.[72,83,84] In addition to structural similarity, the members of this superfamily exhibit a similar multistep catalytic mechanism that generally involves a nucleophilic attack at the C1′ position of the target nucleotide by an N-terminal proline residue (in the case of Fpg, Nei, and NEIL1).[78,85] A comparison of these structures is discussed further below.

B. Substrate Preference

Bacterial Fpg proteins characteristically excise oxidized purines, whereas bacterial Nei and the Neil proteins excise oxidized pyrimidines (for reviews see Refs.[28,29,34,71,72]). Interestingly, the Fpg proteins share similar substrate specificity with Ogg, whereas the substrate specificity of the Nei family members overlaps with that of the Nth family. However, all Fpg/Nei family members share sequence homology. Glycosylases vary in their discrimination of the bases opposite the lesion as well as their preference for the nature of the DNA, that is, single-stranded DNA, double-stranded DNA, or bubble-containing substrates. Fpg primarily exhibits a substrate preference for purines such as 8-oxoG and FapyG, although oxidized pyrimidines are also removed,[86,87] whereas bacterial Nei and the rest of the eukaryotic family members such as NEIL1 and NEIL2 recognize a wider array of substrates.

Recently, it was determined that EcoFpg is more efficient at removing spiroiminodihydantoin (Sp), a further oxidation product of 8-oxoG, from double-stranded DNA substrates than 8-oxoG itself.[73,88] EcoNei, like endonuclease III, recognizes Tg, dihydrothymine (DHT), β-ureidoisobutyric acid, and urea residues (for reviews see Refs. 28,29,34,72,78). EcoNei can also recognize 5-hydroxycytosine (5-OHC), 5-hydroxyuracil (5-OHU), and uracil glycol.[79] MvNeil and NEIL1 share substrate preferences for oxidized pyrimidines in duplex DNA and also recognize and process lesions from single-stranded DNA.[80,82] Although 8-oxoG is not a preferred substrate for NEIL1, its further oxidation products guanidinohydantoin (Gh) and Sp are both excellent substrates for these enzymes when paired opposite C rather than A.[39,40,80] The NEIL1 protein also excises Tg, 5,6-dihydrouracil (DHU), FapyA, and FapyG, as well as 5-OHU, 5-OHC, and oxanine.[30,31,81,89,90] Bacterial Fpg[91,92] and all the eukaryotic members of the Fpg/Nei family recognize lesions in single-stranded DNA.[31,36,80] NEIL2 and MmuNeil3 prefer to excise lesions present in single-stranded, bubble, or forked DNA structures over those in duplex DNA.[24,36,39,42–44,48]

C. Comparison of Structures of the Fpg/Nei Family

As mentioned earlier, currently there are crystal structures of Fpg proteins from four bacterial species, namely TthFpg, EcoFpg, BstFpg, and LlaFpg.[54,55,57–60,63–66,68,69] All four proteins share the same domain structure and considerable sequence homology. Structures of intermediates covalently linked to duplex DNA indicate that the DNA binds to the enzyme in a positively charged groove that runs roughly orthogonal to the DNA axis.[55,59] Bacterial Fpg binds DNA in the minor groove and the damaged base is extruded through the major groove. The DNA appears to be severely kinked at the lesion point (~66° roll angle in the case of EcoFpg[55]) upon enzyme binding, thereby allowing the extruded base to be positioned in the active site for catalysis. The minor groove is widened considerably at the lesion site; however, the rest of the DNA duplex surrounding the lesion retains canonical B-form.[55,59] Upon nucleotide eversion, three highly conserved residues in the bacterial Fpg proteins, namely Met74, Arg109, and Phe111 (in EcoFpg), fill the void that is created and stabilize the opposite base (Fig. 2A).[55] Met74 is part of the β4/5 loop and occupies the position of the extruded base by entering through the minor groove while Arg109 and Phe111 are part of a loop

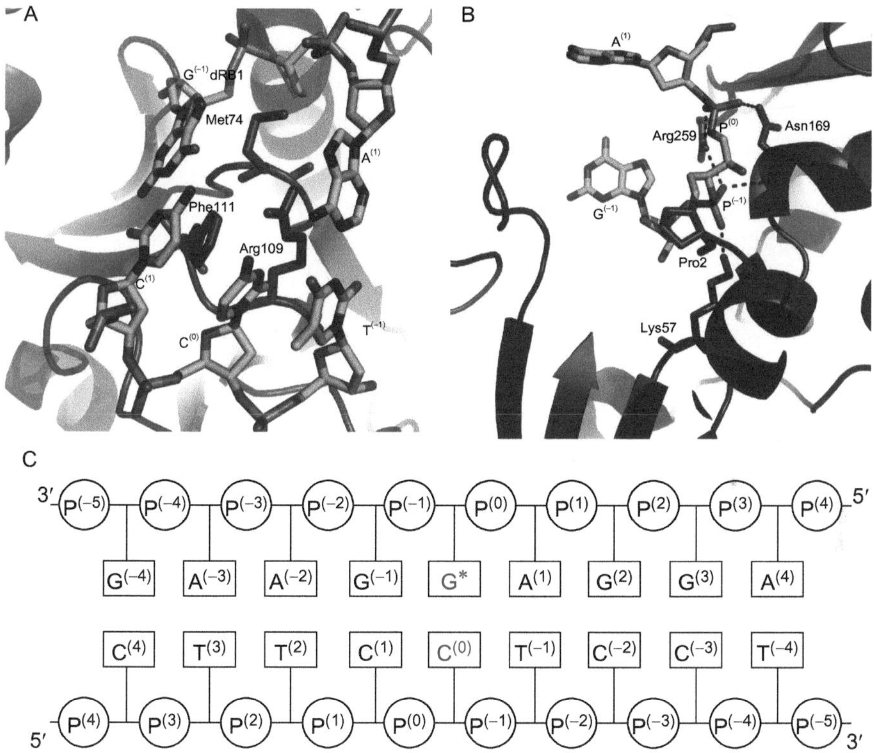

FIG. 2. Specific interactions between EcoFpg and DNA. (A) Triad of void-filling residues Met74, Phe111, and Arg109 that intercalate into the DNA, causing severe kinking at the site of the damage. (B) Interaction of conserved residues Lys57, Asn169, and Arg259 with DNA phosphates surrounding the ring-opened deoxyribitol moiety (dRb1) (PDB ID code 1K82[55]). (C) DNA sequence context present in the crystal structure of EcoFpg bound to DNA, indicating the typical nomenclature used to describe the phosphates and the bases surrounding the lesion. The lesion is indicated by G* while $C^{(0)}$ is the opposite base, both of which are indicated in red lettering. (See Color Insert.)

connecting strands β7 and β8. Phe111 is wedged between the base opposite the lesion (a cytosine) and the neighboring base, and causes unstacking of these bases, leading to the severe kinking of the DNA. Additionally, Arg109 forms H-bonds with the opposite base, leading to discrimination against A as the opposite base.[55,74]

The analogous residues in MvNeil (Leu84, Arg114, and Phe116),[61] human NEIL1 (Met81, Arg118, and Phe120),[67] and *Arabidopsis thaliana* Fpg (AthFpg) (Met78, Arg126, and Phe128) (Stéphanie Duclos, Pierre Aller, Pawel Jaruga, Miral Didzaroglu, Susan S. Wallace and Sylvie Doublié, Manuscript submitted to DNA Repair) are similar to those seen in the bacterial Fpg proteins, which discriminate against A as a base opposite the lesion.[74] In contrast, EcoNei inserts three consecutive residues Gln69, Leu70, and Tyr71 into the void created upon base extrusion.[56] The three residues are located on a loop connecting β4 and β5. Tyr71 is wedged between the orphaned base and its 3′ neighbor and stabilizes the severely kinked DNA. Two of the void-filling residues are lacking in MmuNeil3, which appears to be related to the preference of this enzyme for single-stranded DNA (Minmin Liu, Kayo Imamura, Sylbie Doublie and Susan S. Wallace, manuscript in preparation).

In addition to the void-filling residues, the Fpg/Nei proteins contain an absolutely conserved Lys residue (Lys57 in EcoFpg) and a conserved Asn (Asn169 in EcoFpg), which is part of the H2TH motif.[55] Lys57 forms salt-bridges with P^{-1} and P^{-2} while Asn169 forms bonds through backbone and side-chain amides to P^{-1} and P^{0} of the DNA (Fig. 2B and C). Another highly conserved residue in the Fpg/Nei family of proteins is Arg259 (in EcoFpg), which is part of the zinc finger motif and is involved in the formation of salt bridges with the phosphodiester backbone[55] (Fig. 2B). In the bacterial Fpg and eukaryotic Fpg/Nei proteins, there is no gross conformational change in the overall domain structure upon DNA binding.[55,57,59,64,72] On the other hand, the side-chains of the void-filling residues and conserved residues in the hallmark motifs show small changes. The development of disulfide-crosslinking techniques used to study the structure of BstFpg bound to lesion-containing DNA and undamaged DNA indicates that the enzyme possesses intrahelical recognition of the damage and can detect the subtle differences between the damaged base and its undamaged counterpart even at an initial encounter (Refs. 64,65 and see below). Comparing 8-oxoG with guanine in DNA suggests that the enzyme induces a local conformational change in the DNA backbone in which the sugar pucker (C2′-endo) adopts a different conformation (C4′-exo) to prevent a steric clash between the 8-oxo group of 8-oxoG and the C2′ of the sugar.[64]

A loop region (called the αF-β9/10 loop) in the α-helical C-terminal domain of the Fpg proteins has been presumed to be involved in lesion recognition. In the unliganded structure of TthFpg, this lesion recognition loop is ordered,[54] but in structures of BstFpg bound to DNA containing an AP site, the density for this loop disappears, suggesting that this region is disordered.[57] In the presence of lesion-containing DNA in a complex with catalytically inactive enzymes, the density for this loop resurfaces, indicating conformational mobility upon catalysis.[59,64,65] This loop plays a key role in the recognition of 8-oxoG: a projection from the loop wraps around the damaged base forming an extensive network of hydrogen bonds.[59] This same loop was shown to wrap around FapyG in the LlaFpg structure.[60] The major difference between the binding of Fpg to 8-oxoG and FapyG lies in the fact that binding in the extrahelical base-binding pocket of Fpg for the former lesion occurs in the *syn* conformation, whereas FapyG is in the *anti* conformation. However, despite the difference in base conformations, a similar type of interaction exists between the main-chain carbonyl carbons of conserved residues S218 (in LlaFpg) and S221 (in BstFpg), which bind to the protonated N1 and N7 of FapyG and 8-oxoG, respectively. Similarly, the conserved I220 (in LlaFpg) and the analogous V223 (in BstFpg) use their main-chain amide group for hydrogen bonding with the carbonyl moiety at position 6 of both lesions.[60]

The αF-β9/10 loops of bacterial Fpgs are functionally similar and are of comparable length (~27 residues) and conformation. In contrast, in the eukaryotic members of the family, which do not recognize 8-oxoG, this putative lesion-recognition loop is generally shorter or even missing as in the case of NEIL1,[61] AthFpg (Stéphanie Duclos, Pierre Aller, Pawel Jaruga, Miral Didzaroglu, Susan S. Wallace and Sylvie Doublié, Manuscript submitted to DNA Repair). A superposition of BstFpg bound to DNA (containing 8-oxoG:C[59]) with the unliganded human NEIL1[61] and the MvNei1 enzymes illustrates that this loop wraps around the lesion only in the case of BstFpg (Fig. 3). In contrast, in the case of NEIL1, the loop is replaced by an α-helix and the loop is shorter in MvNei1; in both cases, this segment is unable to wrap around the lesion (Fig. 3). These data are consistent with the fact that 8-oxoG is not a good substrate for NEIL1 or any of the eukaryotic and mimivirus enzymes that are missing this loop. In fact, deletion of the αF-β9/10 loop in EcoFpg yielded a variant that retains catalytic ability on oxidized pyrimidines and FapyG, but not 8-oxoG, implying that this loop is important for stabilizing 8-oxoG and not the other lesions (Stéphanie Duclos, Pierre Aller, Pawel Jaruga, Miral Didzaroglu, Susan S. Wallace and Sylvie Doublié, Manuscript submitted to DNA Repair).

The bacterial Nei proteins are composed of members that share several characteristics with Fpg but also provide some unique variations. The crystal structure of EcoNei reveals a similar structural fold and conservation of motifs present in the Fpg proteins.[56,62] A striking difference between the unliganded and DNA-bound EcoNei structures is a large conformational change of about 50° between the N- and C-terminal domains. This is the only DNA glycosylase to date that has been reported to display a DNA-induced global conformational change, in which the glycosylase transforms from an elongated "open" form to a "closed" DNA-bound form.[56,62] This conformational change was not observed for the MvNei1 protein upon DNA binding.[67]

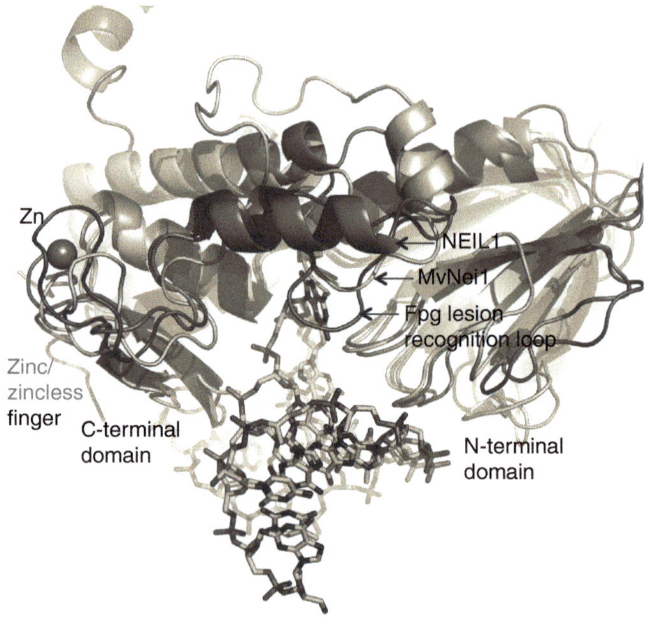

FIG. 3. Superposition of BstFpg (E3Q mutant, green) bound to DNA containing 8-oxoG with human NEIL1 (orange) and MvNei1 (beige). Overall, the three proteins superimpose well with an RMSD of 1.1–1.3 Å upon aligning C-α of analogous residues from the NEIL1 and MvNei1 with BstFpg (performed using COOT[93] and Superimpose; M. Rould, personal communication). Key differences among the three enzymes are the zinc/zincless finger in Fpg versus NEIL1 and MvNei1, and the presence of the lesion-recognition loop in Fpg. The 8-oxoG containing DNA is displayed as a ball and stick model. (PDB ID codes for the BstFpg protein, NEIL1, and MvNei1 are 1R2Y, 1TDH, and 3A42, respectively). (See Color Insert.)

The structure of human NEIL1 reveals the presence of a structural motif composed of two antiparallel β-strands that mimic the zinc finger fold. This motif superimposes well with the zinc finger of EcoNei and the bacterial Fpg proteins[59,61] (Fig. 3). However, the canonical Cys residues and the loops connecting the β-strands of the zinc finger are missing in NEIL1, which prevents the coordination of a zinc atom (Fig. 3). This motif, termed "zincless finger," contains a highly conserved Arg277 residue, which, when mutated, significantly diminishes glycosylase activity.[56,61] This zincless motif is also harbored by MvNei1,[61,67] and the plant and fungal Fpg glycosylases (Stéphanie Duclos, Pierre Aller, Pawel Jaruga, Miral Didzaroglu, Susan S. Wallace and Sylvie Doublié, Manuscript submitted to DNA Repair).

In summary, members of the Fpg/Nei family are structurally similar, but display significant variations in conserved domains/motifs involved in DNA interactions. One of the main differences between EcoNei and the bacterial

Fpg proteins is the composition of the void-filling, intercalation triad. In EcoNei, all three residues are located on the same β4/β5 loop and are consecutive, that is, Gln69, Leu70, and Tyr71. In bacterial Fpg proteins, NEIL1, MvNei1, and AthFpg, as mentioned earlier, the residues that constitute the triad reside in two different loops. Another difference between EcoFpg and the eukaryotic family members lies in the lesion-recognition loop located in the C-terminal domain of both proteins. In bacterial Fpg, the damaged base is everted from the DNA helix and is enveloped in a deep pocket, which is capped by the αF/β9 loop at one end (Fig. 3). This loop is missing in EcoNei and the eukaryotic members for which a structure exists, including NEIL1, AthFpg, and Neil3, and appears to be required for excising 8-oxoG. The vast repertoire of substrates of the Fpg/Nei family members and their different preferences for opposite bases and DNA substrates (single-stranded, double-stranded, or bubble DNA substrates) warrants further structural and biochemical scrutiny.

IV. Glycosylases Search for Lesions

It has long been a question in the field as to how DNA glycosylases locate the lesions they recognize in a sea of undamaged bases. This issue is complicated by the fact that a glycosylase flips out the damage from the DNA helix into its active-site pocket in order to perform its enzymatic function. Furthermore, glycosylases do not use biochemical energy and rely on thermal energy so that lesions are found through random collisions between the glycosylase and the DNA molecule. Because of this, three-dimensional diffusion is considered to be too inefficient to account for the number of lesions the glycosylase must excise. Glycosylases are thought to bind to a nonspecific site on the DNA molecule and slide along the DNA by one-dimensional diffusion until the enzyme finds the lesion or disassociates from DNA. There have been a number of hypotheses proposed for the lesion search itself. One model suggests that the glycosylase binds to an extruded DNA lesion and then moves along the DNA, testing every single base.[94] This appears to be unlikely as both kinetics[95] and single-molecule studies[96,97] have shown that glycosylases scan DNA close to diffusion limits, making it thermodynamically impossible for them to sequentially extrude and examine every base. In the second model, the DNA glycosylase traps a randomly extruded damaged base. This extrusion is more likely with lesions as hydrogen bonding and stacking interactions would be altered compared to the normal bases. This appears to be the mechanism used by uracil DNA glycosylase.[98] In the third model, glycosylases slide along the DNA molecule and are able to recognize their particular substrate by specific interactions between the glycosylase and the DNA molecule. This model has been suggested by structural studies[64,65] and by a recent single-molecule study.[99]

Two groups have attempted to address the question of how glycosylases search for a lesion using single-molecule approaches.[96,97,99] In the first study,[96] human OGG1 (oxoguanine DNA glycosylase, a member of the HhH superfamily) labeled with Cy3 was observed to undergo one-dimensional sliding along DNA that was stretched by shear flow. A similar diffusive motion was observed with BstFpg. These same authors showed that the one-dimensional diffusion constants measured were consistent with the glycosylases diffusing along the DNA helix in a rotational manner.[97] In a recent study from our laboratory,[99] quantum dot-labeled *E. coli* Fpg, Nei, and Nth were imaged in the absence of flow. In this study, the glycosylases were shown to diffuse along the DNA with a broad distribution of rates that ranged over two orders of magnitude. This broad distribution was common to all three glycosylases, suggesting that both the Fpg/Nei family and HhH superfamily scan using a similar mechanism. When the diffusive behavior was analyzed further, the three glycosylases were shown to exhibit a continuum of motion that was in keeping with rotational diffusion along the DNA molecule and that ranged from a slow, subdiffusive to a faster, unrestricted diffusive behavior.

As described earlier, members of the Fpg/Nei family of DNA glycosylases have three void-filling residues that are inserted into the DNA helix and aid in flipping out the damaged base in addition to stabilizing the DNA helix.[55,56,59,64–67] The HhH superfamily uses a similar mechanism.[100,101] Interestingly, a crystal structure of BstFpg crosslinked to undamaged DNA revealed that one of these void-filling residues, a phenylalanine, was found to be wedged into the helix, occupying a position analogous to its position in the Fpg complex bound to damage-containing DNA[64] (Fig. 2A). These data, together with kinetics data,[102,103] suggest that the phenylalanine may be acting as a wedge that scans for deformability of the base pair such as in the sugar pucker. Interestingly, when the corresponding *E. coli* wedge residue Phe111 was mutated to an alanine, there was a significant increase in the mean diffusion constant compared to the wild-type protein.[99] Moreover, the diffusive properties characteristic of wild type were altered, that is, the slow, subdiffusive population of glycosylases was selectively lost. Similar results were observed when the analogous residues in Nei and Nth were mutated (Dunn *et al.*, unpublished observations), suggesting that the slow subdiffusive glycosylases are those interrogating the DNA for damages. Taken together, the data support the idea that the Fpg/Nei family of DNA glycosylases diffuse one-dimensionally along the DNA molecule with diffusion constants that are consistent with rotation around the DNA molecule, presumably in the minor groove where they bind. It also appears that at least part of the glycosylase search mechanism may be accomplished by insertion of a particular wedge residue that senses the topography of the minor groove and pauses either to check for damage at random locations or in response to subtle deformations of the DNA helix.

V. Concluding Remarks

Advances in the structural biology and biochemistry of glycosylases have led to a better understanding of how these complex enzymes recognize and excise damaged bases. Based on current *in vitro* studies, we can speculate on the mechanisms of specific lesion recognition. However, despite the vast knowledge gained, several unanswered questions still remain. For instance, we know that the Fpg/Nei family members recognize a broad range of substrates but it is not clear how these enzymes discriminate among each of these lesions and how they distinguish these from undamaged bases. Moreover, as some glycosylases are active at different times in the cell cycle and interact with a number of protein partners, how are these enzymes involved in processes such as DNA replication or transcription? Additionally, it remains difficult to classify certain members of the Fpg/Nei family under a specific subfamily. For example, even though members of the Fpg/Nei family of proteins are structurally similar, some elements such as the intercalation triad present in NEIL1 and MvNei1 suggest that these members could be classified under the Fpg subfamily rather than the Nei subfamily after which they were originally named. Many aspects of phylogenetic characterization, lesion recognition, substrate specificities, and the biological functions of this glycosylase family still remain to be elucidated.

Acknowledgments

We would like to thank Minmin Liu, Drs. Scott Kathe, Stéphanie Duclos, and Ramiro Barrantes-Reynolds for their helpful discussions. Dr. Andrew Dunn performed the single molecule studies in our laboratory in collaboration with Dr. David Warshaw. This work was supported by NIH/NCI P01 CA098993 (S. S. W., PI).

References

1. Chetsanga CJ, Lindahl T. Release of 7-methylguanine residues whose imidazole rings have been opened from damaged DNA by a DNA glycosylase from *Escherichia coli*. *Nucleic Acids Res* 1979;**6**:3673–84.
2. Boiteux S, O'Connor TR, Laval J. Formamidopyrimidine-DNA glycosylase of *Escherichia coli*: cloning and sequencing of the fpg structural gene and overproduction of the protein. *EMBO J* 1987;**6**:3177–83.

3. Chetsanga CJ, Lozon M, Makaroff C, Savage L. Purification and characterization of *Escherichia coli* formamidopyrimidine-DNA glycosylase that excises damaged 7-methylguanine from deoxyribonucleic acid. *Biochemistry* 1981;**20**:5201–7.
4. Breimer LH. Enzymatic excision from gamma-irradiated polydeoxyribonucleotides of adenine residues whose imidazole rings have been ruptured. *Nucleic Acids Res* 1984;**12**:6359–67.
5. O'Connor TR, Laval J. Physical association of the 2,6-diamino-4-hydroxy-5N-formamidopyrimidine-DNA glycosylase of *Escherichia coli* and an activity nicking DNA at apurinic/apyrimidinic sites. *Proc Natl Acad Sci USA* 1989;**86**:5222–6.
6. Boiteux S, O'Connor TR, Lederer F, Gouyette A, Laval J. Homogeneous *Escherichia coli* FPG protein. A DNA glycosylase which excises imidazole ring-opened purines and nicks DNA at apurinic/apyrimidinic sites. *J Biol Chem* 1990;**265**:3916–22.
7. Cabrera M, Nghiem Y, Miller JH. mutM, a second mutator locus in *Escherichia coli* that generates G.C—T.A transversions. *J Bacteriol* 1988;**170**:5405–7.
8. Michaels ML, Pham L, Cruz C, Miller JH. MutM, a protein that prevents G.C—T.A transversions, is formamidopyrimidine-DNA glycosylase. *Nucleic Acids Res* 1991;**19**:3629–32.
9. Tchou J, Kasai H, Shibutani S, Chung MH, Laval J, Grollman AP, et al. 8-oxoguanine (8-hydroxyguanine) DNA glycosylase and its substrate specificity. *Proc Natl Acad Sci USA* 1991;**88**:4690–4.
10. Tchou J, Bodepudi V, Shibutani S, Antoshechkin I, Miller J, Grollman AP, et al. Substrate specificity of Fpg protein. Recognition and cleavage of oxidatively damaged DNA. *J Biol Chem* 1994;**269**:15318–24.
11. Michaels ML, Miller JH. The GO system protects organisms from the mutagenic effect of the spontaneous lesion 8-hydroxyguanine (7,8-dihydro-8-oxoguanine). *J Bacteriol* 1992;**174**:6321–5.
12. Brieba LG, Eichman BF, Kokoska RJ, Doublié S, Kunkel TA, Ellenberger T. Structural basis for the dual coding potential of 8-oxoguanosine by a high-fidelity DNA polymerase. *EMBO J* 2004;**23**:3452–61.
13. Shibutani S, Takeshita M, Grollman AP. Insertion of specific bases during DNA synthesis past the oxidation-damaged base 8-oxodG. *Nature* 1991;**349**:431–4.
14. Moriya M, Ou C, Bodepudi V, Johnson F, Takeshita M, Grollman AP. Site-specific mutagenesis using a gapped duplex vector: a study of translesion synthesis past 8-oxodeoxyguanosine in E. coli. *Mutat Res* 1991;**254**:281–8.
15. Wood ML, Dizdaroglu M, Gajewski E, Essigmann JM. Mechanistic studies of ionizing radiation and oxidative mutagenesis: genetic effects of a single 8-hydroxyguanine (7-hydro-8-oxoguanine) residue inserted at a unique site in a viral genome. *Biochemistry* 1990;**29**:7024–32.
16. Michaels ML, Cruz C, Grollman AP, Miller JH. Evidence that MutY and MutM combine to prevent mutations by an oxidatively damaged form of guanine in DNA. *Proc Natl Acad Sci USA* 1992;**89**:7022–5.
17. Maki H, Sekiguchi M. MutT protein specifically hydrolyses a potent mutagenic substrate for DNA synthesis. *Nature* 1992;**355**:273–5.
18. Boiteux S, Gajewski E, Laval J, Dizdaroglu M. Substrate specificity of the *Escherichia coli* Fpg protein (formamidopyrimidine-DNA glycosylase): excision of purine lesions in DNA produced by ionizing radiation or photosensitization. *Biochemistry* 1992;**31**:106–10.
19. Krishnamurthy N, Haraguchi K, Greenberg MM, David SS. Efficient removal of formamidopyrimidines by 8-oxoguanine glycosylases. *Biochemistry* 2008;**47**:1043–50.
20. Patro JN, Wiederholt CJ, Jiang YL, Delaney JC, Essigmann JM, Greenberg MM. Studies on the replication of the ring opened formamidopyrimidine, Fapy.dG in *Escherichia coli*. *Biochemistry* 2007;**46**:10202–12.
21. Wiederholt CJ, Greenberg MM. Fapy.dG instructs Klenow exo(-) to misincorporate deoxyadenosine. *J Am Chem Soc* 2002;**124**:7278–9.

22. Wiederholt CJ, Delaney MO, Pope MA, David SS, Greenberg MM. Repair of DNA containing Fapy.dG and its beta-C-nucleoside analogue by formamidopyrimidine DNA glycosylase and MutY. *Biochemistry* 2003;**42**:9755–60.
23. Dizdaroglu M, Jaruga P, Birincioglu M, Rodriguez H. Free radical-induced damage to DNA: mechanisms and measurement. *Free Radic Biol Med* 2002;**32**:1102–15.
24. Liu M, Bandaru V, Bond JP, Jaruga P, Zhao X, Christov PP, et al. The mouse ortholog of NEIL3 is a functional DNA glycosylase in vitro and in vivo. *Proc Natl Acad Sci USA* 2010;**107**: 4925–30.
25. Melamede RJ, Hatahet Z, Kow YW, Ide H, Wallace SS. Isolation and characterization of endonuclease VIII from *Escherichia coli*. *Biochemistry* 1994;**33**:1255–64.
26. Jiang D, Hatahet Z, Melamede RJ, Kow YW, Wallace SS. Characterization of *Escherichia coli* endonuclease VIII. *J Biol Chem* 1997;**272**:32230–9.
27. Jiang D, Hatahet Z, Blaisdell JO, Melamede RJ, Wallace SS. *Escherichia coli* endonuclease VIII: cloning, sequencing, and overexpression of the nei structural gene and characterization of nei and nei nth mutants. *J Bacteriol* 1997;**179**:3773–82.
28. Wallace SS, et al. Oxidative stress and the molecular biology of antioxidant defenses. In: Cold Spring Harbor: Cold Spring Harbor Press; 1997. pp. 49–90.
29. Wallace SS. Enzymatic processing of radiation-induced free radical damage in DNA. *Radiat Res* 1998;**150**:S60–79.
30. Bandaru V, Sunkara S, Wallace SS, Bond JP. A novel human DNA glycosylase that removes oxidative DNA damage and is homologous to *Escherichia coli* endonuclease VIII. *DNA Repair (Amst)* 2002;**1**:517–29.
31. Hazra TK, Izumi T, Boldogh I, Imhoff B, Kow YW, Jaruga P, et al. Identification and characterization of a human DNA glycosylase for repair of modified bases in oxidatively damaged DNA. *Proc Natl Acad Sci USA* 2002;**99**:3523–8.
32. Hazra TK, Kow YW, Hatahet Z, Imhoff B, Boldogh I, Mokkapati SK, et al. Identification and characterization of a novel human DNA glycosylase for repair of cytosine-derived lesions. *J Biol Chem* 2002;**277**:30417–20.
33. Morland I, Rolseth V, Luna L, Rognes T, Bjoras M, Seeberg E. Human DNA glycosylases of the bacterial Fpg/MutM superfamily: an alternative pathway for the repair of 8-oxoguanine and other oxidation products in DNA. *Nucleic Acids Res* 2002;**30**:4926–36.
34. Wallace SS, Bandaru V, Kathe SD, Bond JP. The enigma of endonuclease VIII. *DNA Repair (Amst)* 2003;**2**:441–53.
35. Takao M, Kanno S, Kobayashi K, Zhang QM, Yonei S, van der Horst GT, et al. A back-up glycosylase in Nth1 knock-out mice is a functional Nei (endonuclease VIII) homologue. *J Biol Chem* 2002;**277**:42205–13.
36. Dou H, Mitra S, Hazra TK. Repair of oxidized bases in DNA bubble structures by human DNA glycosylases NEIL1 and NEIL2. *J Biol Chem* 2003;**278**:49679–84.
37. Katafuchi A, Nakano T, Masaoka A, Terato H, Iwai S, Hanaoka F, et al. Differential specificity of human and *Escherichia coli* endonuclease III and VIII homologues for oxidative base lesions. *J Biol Chem* 2004;**279**:14464–71.
38. Jaruga P, Birincioglu M, Rosenquist TA, Dizdaroglu M. Mouse NEIL1 protein is specific for excision of 2,6-diamino-4-hydroxy-5-formamidopyrimidine and 4,6-diamino-5-formamidopyrimidine from oxidatively damaged DNA. *Biochemistry* 2004;**43**:15909–14.
39. Hailer MK, Slade PG, Martin BD, Rosenquist TA, Sugden KD. Recognition of the oxidized lesions spiroiminodihydantoin and guanidinohydantoin in DNA by the mammalian base excision repair glycosylases NEIL1 and NEIL2. *DNA Repair (Amst)* 2005;**4**:41–50.
40. Krishnamurthy N, Zhao X, Burrows CJ, David SS. Superior removal of hydantoin lesions relative to other oxidized bases by the human DNA glycosylase hNEIL1. *Biochemistry* 2008;**47**:7137–46.

41. Wiederhold L, Leppard JB, Kedar P, Karimi-Busheri F, Rasouli-Nia A, Weinfeld M, et al. AP endonuclease-independent DNA base excision repair in human cells. *Mol Cell* 2004;**15**:209–20.
42. Dou H, Theriot CA, Das A, Hegde ML, Matsumoto Y, Boldogh I, et al. Interaction of the human DNA glycosylase NEIL1 with proliferating cell nuclear antigen. The potential for replication-associated repair of oxidized bases in mammalian genomes. *J Biol Chem* 2008;**283**:3130–40.
43. Hegde ML, Theriot CA, Das A, Hegde PM, Guo Z, Gary RK, et al. Physical and functional interaction between human oxidized base-specific DNA glycosylase NEIL1 and flap endonuclease 1. *J Biol Chem* 2008;**283**:27028–37.
44. Theriot CA, Hegde ML, Hazra TK, Mitra S. RPA physically interacts with the human DNA glycosylase NEIL1 to regulate excision of oxidative DNA base damage in primer-template structures. *DNA Repair (Amst)* 2010;**9**:643–52.
45. Banerjee D, Mandal SM, Das A, Hegde ML, Das S, Bhakat KK, et al. Preferential repair of oxidized base damage in the transcribed genes of mammalian cells. *J Biol Chem* 2011;**286**: 6006–16.
46. Takao M, Kanno S, Shiromoto T, Hasegawa R, Ide H, Ikeda S, et al. Novel nuclear and mitochondrial glycosylases revealed by disruption of the mouse Nth1 gene encoding an endonuclease III homolog for repair of thymine glycols. *EMBO J* 2002;**21**:3486–93.
47. Krokeide SZ, Bolstad N, Laerdahl JK, Bjoras M, Luna L. Expression and purification of NEIL3, a human DNA glycosylase homolog. *Protein Expr Purif* 2009;**65**:160–4.
48. Takao M, Oohata Y, Kitadokoro K, Kobayashi K, Iwai S, Yasui A, et al. Human Nei-like protein NEIL3 has AP lyase activity specific for single-stranded DNA and confers oxidative stress resistance in *Escherichia coli* mutant. *Genes Cells* 2009;**14**:261–70.
49. Hildrestrand GA, Neurauter CG, Diep DB, Castellanos CG, Krauss S, Bjoras M, et al. Expression patterns of Neil3 during embryonic brain development and neoplasia. *BMC Neurosci* 2009;**10**:45.
50. Rolseth V, Runden-Pran E, Luna L, McMurray C, Bjoras M, Ottersen OP. Widespread distribution of DNA glycosylases removing oxidative DNA lesions in human and rodent brains. *DNA Repair (Amst)* 2008;**7**:1578–88.
51. Torisu K, Tsuchimoto D, Ohnishi Y, Nakabeppu Y. Hematopoietic tissue-specific expression of mouse Neil3 for endonuclease VIII-like protein. *J Biochem* 2005;**138**:763–72.
52. Kathe SD, Barrantes-Reynolds R, Jaruga P, Newton MR, Burrows CJ, Bandaru V, et al. Plant and fungal Fpg homologs are formamidopyrimidine DNA glycosylases but not 8-oxoguanine DNA glycosylases. *DNA Repair (Amst)* 2009;**8**:643–53.
53. Pumo DE, Barrantes-Reynolds R, Kathe S, Wallace SS, Bond JP. Evolution of the Fpg/Nei family of DNA glycosylases. In: Kovalchuk I, Kovalchuk O, editors. *Genome instability and transgenerational effects*. New York: Nova Science Publishers, Inc; 2010. pp. 71–88.
54. Sugahara M, Mikawa T, Kumasaka T, Yamamoto M, Kato R, Fukuyama K, et al. Crystal structure of a repair enzyme of oxidatively damaged DNA, MutM (Fpg), from an extreme thermophile, *Thermus thermophilus* HB8. *EMBO J* 2000;**19**:3857–69.
55. Gilboa R, Zharkov DO, Golan G, Fernandes AS, Gerchman SE, Matz E, et al. Structure of formamidopyrimidine-DNA glycosylase covalently complexed to DNA. *J Biol Chem* 2002;**277**:19811–6.
56. Zharkov DO, Golan G, Gilboa R, Fernandes AS, Gerchman SE, Kycia JH, et al. Structural analysis of an *Escherichia coli* endonuclease VIII covalent reaction intermediate. *EMBO J* 2002;**21**:789–800.
57. Fromme JC, Verdine GL. Structural insights into lesion recognition and repair by the bacterial 8-oxoguanine DNA glycosylase MutM. *Nat Struct Biol* 2002;**9**:544–52.

58. Serre L, Pereira de Jesus K, Boiteux S, Zelwer C, Castaing B. Crystal structure of the Lactococcus lactis formamidopyrimidine-DNA glycosylase bound to an abasic site analogue-containing DNA. *EMBO J* 2002;**21**:2854–65.
59. Fromme JC, Verdine GL. DNA lesion recognition by the bacterial repair enzyme MutM. *J Biol Chem* 2003;**278**:51543–8.
60. Coste F, Ober M, Carell T, Boiteux S, Zelwer C, Castaing B. Structural basis for the recognition of the FapydG lesion (2,6-diamino-4-hydroxy-5-formamidopyrimidine) by formamidopyrimidine-DNA glycosylase. *J Biol Chem* 2004;**279**:44074–83.
61. Doublié S, Bandaru V, Bond JP, Wallace SS. The crystal structure of human endonuclease VIII-like 1 (NEIL1) reveals a zincless finger motif required for glycosylase activity. *Proc Natl Acad Sci USA* 2004;**101**:10284–9.
62. Golan G, Zharkov DO, Feinberg H, Fernandes AS, Zaika EI, Kycia JH, et al. Structure of the uncomplexed DNA repair enzyme endonuclease VIII indicates significant interdomain flexibility. *Nucleic Acids Res* 2005;**33**:5006–16.
63. Pereira de Jesus K, Serre L, Zelwer C, Castaing B. Structural insights into abasic site for Fpg specific binding and catalysis: comparative high-resolution crystallographic studies of Fpg bound to various models of abasic site analogues-containing DNA. *Nucleic Acids Res* 2005;**33**:5936–44.
64. Banerjee A, Santos WL, Verdine GL. Structure of a DNA glycosylase searching for lesions. *Science* 2006;**311**:1153–7.
65. Qi Y, Spong MC, Nam K, Banerjee A, Jiralerspong S, Karplus M, et al. Encounter and extrusion of an intrahelical lesion by a DNA repair enzyme. *Nature* 2009;**462**:762–6.
66. Qi Y, Spong MC, Nam K, Karplus M, Verdine GL. Entrapment and structure of an extrahelical guanine attempting to enter the active site of a bacterial DNA glycosylase, MutM. *J Biol Chem* 2010;**285**:1468–78.
67. Imamura K, Wallace SS, Doublié S. Structural characterization of a viral NEIL1 ortholog unliganded and bound to abasic site-containing DNA. *J Biol Chem* 2009;**284**:26174–83.
68. Coste F, Ober M, Le Bihan YV, Izquierdo MA, Hervouet N, Mueller H, et al. Bacterial base excision repair enzyme Fpg recognizes bulky N7-substituted-FapydG lesion via unproductive binding mode. *Chem Biol* 2008;**15**:706–17.
69. Le Bihan YV, Angeles Izquierdo M, Coste F, Aller P, Culard F, Gehrke TH, et al. 5-Hydroxy-5-methylhydantoin DNA lesion, a molecular trap for DNA glycosylases. *Nucleic Acids Res* 2011;**39**:6277–90.
70. Verdine GL, Norman DP. Covalent trapping of protein-DNA complexes. *Annu Rev Biochem* 2003;**72**:337–66.
71. Fromme JC, Banerjee A, Verdine GL. DNA glycosylase recognition and catalysis. *Curr Opin Struct Biol* 2004;**14**:43–9.
72. Zharkov DO, Shoham G, Grollman AP. Structural characterization of the Fpg family of DNA glycosylases. *DNA Repair (Amst)* 2003;**2**:839–62.
73. Guo Y, Bandaru V, Jaruga P, Zhao X, Burrows CJ, Iwai S, et al. The oxidative DNA glycosylases of *Mycobacterium tuberculosis* exhibit different substrate preferences from their *Escherichia coli* counterparts. *DNA Repair (Amst)* 2010;**9**:177–90.
74. Zaika EI, Perlow RA, Matz E, Broyde S, Gilboa R, Grollman AP, et al. Substrate discrimination by formamidopyrimidine-DNA glycosylase: a mutational analysis. *J Biol Chem* 2004;**279**: 4849–61.
75. Michaels ML, Tchou J, Grollman AP, Miller JH. A repair system for 8-oxo-7,8-dihydrodeoxyguanine. *Biochemistry* 1992;**31**:10964–8.
76. Zharkov DO, Ishchenko AA, Douglas KT, Nevinsky GA. Recognition of damaged DNA by *Escherichia coli* Fpg protein: insights from structural and kinetic data. *Mutat Res* 2003;**531**: 141–56.

77. Hamm ML, Gill TJ, Nicolson SC, Summers MR. Substrate specificity of Fpg (MutM) and hOGG1, two repair glycosylases. *J Am Chem Soc* 2007;**129**:7724–5.
78. Kropachev KY, Zharkov DO, Grollman AP. Catalytic mechanism of *Escherichia coli* endonuclease VIII: roles of the intercalation loop and the zinc finger. *Biochemistry* 2006;**45**: 12039–49.
79. Purmal AA, Lampman GW, Bond JP, Hatahet Z, Wallace SS. Enzymatic processing of uracil glycol, a major oxidative product of DNA cytosine. *J Biol Chem* 1998;**273**:10026–35.
80. Zhao X, Krishnamurthy N, Burrows CJ, David SS. Mutation versus repair: NEIL1 removal of hydantoin lesions in single-stranded, bulge, bubble, and duplex DNA contexts. *Biochemistry* 2010;**49**:1658–66.
81. Yeo J, Goodman RA, Schirle NT, David SS, Beal PA. RNA editing changes the lesion specificity for the DNA repair enzyme NEIL1. *Proc Natl Acad Sci USA* 2010;**107**:20715–9.
82. Bandaru V, Zhao X, Newton MR, Burrows CJ, Wallace SS. Human endonuclease VIII-like (NEIL) proteins in the giant DNA Mimivirus. *DNA Repair (Amst)* 2007;**6**:1629–41.
83. O'Connor TR, Graves RJ, de Murcia G, Castaing B, Laval J. Fpg protein of *Escherichia coli* is a zinc finger protein whose cysteine residues have a structural and/or functional role. *J Biol Chem* 1993;**268**:9063–70.
84. Tchou J, Michaels ML, Miller JH, Grollman AP. Function of the zinc finger in *Escherichia coli* Fpg protein. *J Biol Chem* 1993;**268**:26738–44.
85. Tchou J, Grollman AP. The catalytic mechanism of Fpg protein. Evidence for a Schiff base intermediate and amino terminus localization of the catalytic site. *J Biol Chem* 1995;**270**: 11671–7.
86. Hatahet Z, Kow YW, Purmal AA, Cunningham RP, Wallace SS. New substrates for old enzymes. 5-Hydroxy-2′-deoxycytidine and 5-hydroxy-2′-deoxyuridine are substrates for *Escherichia coli* endonuclease III and formamidopyrimidine DNA N-glycosylase, while 5-hydroxy-2′-deoxyuridine is a substrate for uracil DNA N-glycosylase. *J Biol Chem* 1994;**269**:18814–20.
87. D'Ham C, Romieu A, Jaquinod M, Gasparutto D, Cadet J. Excision of 5,6-dihydroxy-5,6-dihydrothymine, 5,6-dihydrothymine, and 5-hydroxycytosine from defined sequence oligonucleotides by *Escherichia coli* endonuclease III and Fpg proteins: kinetic and mechanistic aspects. *Biochemistry* 1999;**38**:3335–44.
88. Luo W, Muller JG, Rachlin EM, Burrows CJ. Characterization of spiroiminodihydantoin as a product of one-electron oxidation of 8-Oxo-7,8-dihydroguanosine. *Org Lett* 2000;**2**:613–6.
89. Dong L, Meira LB, Hazra TK, Samson LD, Cao W. Oxanine DNA glycosylase activities in mammalian systems. *DNA Repair (Amst)* 2008;**7**:128–34.
90. Nakano T, Katafuchi A, Shimizu R, Terato H, Suzuki T, Tauchi H, et al. Repair activity of base and nucleotide excision repair enzymes for guanine lesions induced by nitrosative stress. *Nucleic Acids Res* 2005;**33**:2181–91.
91. Ishchenko AA, Koval VV, Fedorova OS, Douglas KT, Nevinsky GA. Structural requirements of double and single stranded DNA substrates and inhibitors, including a photoaffinity label, of Fpg protein from *Escherichia coli*. *J Biomol Struct Dyn* 1999;**17**:301–10.
92. Ishchenko AA, Bulychev NV, Maksakova GA, Johnson F, Nevinsky GA. Single-stranded oligodeoxyribonucleotides are substrates of Fpg protein from *Escherichia coli*. *IUBMB Life* 1999;**48**:613–8.
93. Emsley P, Cowtan K. Coot: model-building tools for molecular graphics. *Acta Crystallogr D Biol Crystallogr* 2004;**60**:2126–32.
94. Verdine GL, Bruner SD. How do DNA repair proteins locate damaged bases in the genome? *Chem Biol* 1997;**4**:329–34.
95. Porecha RH, Stivers JT. Uracil DNA glycosylase uses DNA hopping and short-range sliding to trap extrahelical uracils. *Proc Natl Acad Sci USA* 2008;**105**:10791–6.

96. Blainey PC, van Oijen AM, Banerjee A, Verdine GL, Xie XS. A base-excision DNA-repair protein finds intrahelical lesion bases by fast sliding in contact with DNA. *Proc Natl Acad Sci USA* 2006;**103**:5752–7.
97. Blainey PC, Luo G, Kou SC, Mangel WF, Verdine GL, Bagchi B, et al. Nonspecifically bound proteins spin while diffusing along DNA. *Nat Struct Mol Biol* 2009;**16**:1224–9.
98. Cao C, Jiang YL, Stivers JT, Song F. Dynamic opening of DNA during the enzymatic search for a damaged base. *Nat Struct Mol Biol* 2004;**11**:1230–6.
99. Dunn AR, Kad NM, Nelson SR, Warshaw DM, Wallace SS. Single Qdot-labeled glycosylase molecules use a wedge amino acid to probe for lesions while scanning along DNA. *Nucleic Acids Res* 2011;**39**:7487–98.
100. Bruner SD, Norman DP, Verdine GL. Structural basis for recognition and repair of the endogenous mutagen 8-oxoguanine in DNA. *Nature* 2000;**403**:859–66.
101. Fromme JC, Verdine GL. Structure of a trapped endonuclease III-DNA covalent intermediate. *EMBO J* 2003;**22**:3461–71.
102. Kuznetsov NA, Zharkov DO, Koval VV, Buckle M, Fedorova OS. Reversible chemical step and rate-limiting enzyme regeneration in the reaction catalyzed by formamidopyrimidine-DNA glycosylase. *Biochemistry* 2009;**48**:11335–43.
103. Koval VV, Kuznetsov NA, Ishchenko AA, Saparbaev MK, Fedorova OS. Real-time studies of conformational dynamics of the repair enzyme *E. coli* formamidopyrimidine-DNA glycosylase and its DNA complexes during catalytic cycle. *Mutat Res* 2010;**685**:3–10.
104. Imamura K, Averill A, Wallace SS, Doublié S. Structural Characterization of a Viral Ortholog of the Human DNA Glycosylase NEIL1 Bound to Thymine Glycol of 5-Hydroxyuracil-Containing DNA. *J Biol Chem* 2012;**287**:4288–98.

Regulation of Base Excision Repair in Eukaryotes by Dynamic Localization Strategies

DANIEL B. SWARTZLANDER,[*,†] NICHOLAS C. BAUER,[*,‡] ANITA H. CORBETT,[*,§] AND PAUL W. DOETSCH[*,§,¶,||]

[*]*Department of Biochemistry, Emory University School of Medicine, Atlanta, Georgia, USA*

[†]*Graduate Program in Genetics and Molecular Biology, Emory University School of Medicine, Atlanta, Georgia, USA*

[‡]*Graduate Program in Biochemistry, Cell, and Developmental Biology, Emory University School of Medicine, Atlanta, Georgia, USA*

[§]*Winship Cancer Institute, Emory University School of Medicine, Atlanta, Georgia, USA*

[¶]*Department of Radiation Oncology, Emory University School of Medicine, Atlanta, Georgia, USA*

[||]*Hematology and Medical Oncology, Emory University School of Medicine, Atlanta, Georgia, USA*

This chapter discusses base excision repair (BER) and the known mechanisms defined thus far regulating BER in eukaryotes. Unlike the situation with nucleotide excision repair and double-strand break repair, little is known about

Progress in Molecular Biology
and Translational Science, Vol. 110
DOI: 10.1016/B978-0-12-387665-2.00005-5

how BER is regulated to allow for efficient and accurate repair of many types of DNA base damage in both nuclear and mitochondrial genomes. Regulation of BER has been proposed to occur at multiple, different levels including transcription, posttranslational modification, protein–protein interactions, and protein localization; however, none of these regulatory mechanisms characterized thus far affect a large spectrum of BER proteins. This chapter discusses a recently discovered mode of BER regulation defined in budding yeast cells that involves mobilization of DNA repair proteins to DNA-containing organelles in response to genotoxic stress.

I. Base Excision Repair

A. Requirements and Limitations of Base Excision Repair

There are more than 100,000 oxidative lesions and 200,000 apurinic/apyrimidinic (AP) sites generated per human cell per day.[1,2] It is the primary responsibility of the base excision repair (BER) pathway to police cellular DNA for such endogenous DNA damage and to efficiently repair these lesions or face the accumulation of mutations and eventual loss of cell viability or oncogenic transformation. Furthermore, a cell must also be prepared to respond to any acute stresses such as those caused by oxidative bursts during infections or chronic stresses caused by the accumulation of mutations, which occurs during aging. Both infection and aging raise the level of DNA damage and require mechanisms of regulation to appropriately increase DNA repair capacity.[3,4]

Organisms have evolved partially redundant sets of DNA repair pathways to manage a wide range of damaged DNA, the BER pathway being one of them. Much of our understanding about the inner workings of eukaryotic BER pathways comes from knowledge obtained through work performed in the budding yeast, *Saccharomyces cerevisiae*. Numerous studies reveal that investigation of this evolutionarily conserved pathway in yeast provides valuable insight into the process in humans.[5,6] Through studies done in yeast in part, we appreciate that recognition of base damage by BER proteins is partially redundant with certain *N*-glycosylases having complementary base specificities, that BER proteins are present at relatively low copy numbers per cell, and finally, that BER proteins localize to nuclei, mitochondria, or both.[7] The relative expression level of each protein in the BER pathway is particularly important when we consider that the excision of damaged bases by *N*-glycosylases often generates mutagenic intermediates. If such intermediates are not processed and repaired, mutations may occur, resulting in cell death or other deleterious consequences.[8–10] As such, to ensure the maintenance of genome stability and overall organism health, the expression of BER proteins must be tightly controlled and regulated, with many proteins of the BER pathway showing temporarily increased protein expression in response to cellular DNA damage or the cell cycle.[11]

There are many examples in primary cancers where the activity, expression, or localization of specific BER proteins is altered or lost, providing evidence not just for the regulation of BER but also for the biological consequences associated with its deregulation. For example, in a subset of human lung cancer tumors, nuclei show increased thymine glycol DNA glycosylase (NTHL1) and 8-oxoguanine DNA glycosylase (OGG1) repair activity, while the same is reduced in their mitochondria.[12,13] In gastric cancers, the enzyme NTHL1 is often lost in the nucleus, resulting either from the global downregulation of steady-state protein levels or, as observed in a significant number of primary gastric cancers, from NTHL1 becoming abnormally localized to the cytoplasm rather than the nucleus.[14,15] There are further examples of BER dysfunction in neurodegenerative disorders.[16,17] For example, the activities of both the human uracil DNA glycosylase (UNG) and OGG1 are decreased in Alzheimer's patients.[18]

B. Regulation of BER: Current Concepts and Observations

As aberrant BER activity occurs in both primary cancers and neurodegenerative disorders, perhaps contributing to the development of disease, regulation of BER protein function must be tightly regulated. This regulation of BER proteins has been proposed to occur through a variety of different mechanisms including transcriptional control, posttranslational modification, protein localization, and protein–protein interactions (Table I).[163] Each of these mechanisms can contribute to regulation and therefore their integration can result in a diverse range of effects on the proteins involved.[19–59,61–99,102–162]

Changes in the steady-state transcript level of several BER proteins occur in a cell cycle-dependent manner while other BER proteins show such a response under increased oxidative stress (Table I).[11] For example, changes in transcript level have been documented for *N*-methylpurine DNA glycosylase (MPG), NTHL1, UNG, and human AP site-specific endonuclease (APEX1), in a cell cycle-dependent manner, increasing 2.5- to 3.5-fold during G_1 and decreasing to basal levels following mitosis.[26] On the other hand, the mRNA levels of NTHL1, OGG1, and APEX1 are increased by exposure to hydrogen peroxide and other reactive oxygen species (ROS) by as much as threefold.[68,108,164]

Protein–protein interactions involving BER proteins are numerous and have varied effects including increasing the efficacy of repair in mammals or linking BER to different pathways such as nucleotide excision repair (NER) or

TABLE I

EUKARYOTIC BASE EXCISION REPAIR PROTEINS AND MECHANISMS OF THEIR REGULATION

			Level of regulation			
Protein	Organism	Gene	PTM	Protein interactions	Transcript level	Localization
Uracil DNA glycosylase	*S. cerevisiae*	*UNG1*	–	–	↑ in G_1-phase[19]	Nuc and Mito[20]
	Human	*UNG*	Phos (Thr6, Thr126)[21]	APEX1 ↑ activity[22] Associated with RPA, XRCC1, and PCNA[23–25]	↑ in G_1- and S-phase[26,27] Mitochondrial form ↑ in response to ROS[28]	Nuc (UNG2) and Mito (UNG1) Forms[28,29]
		SMUG1	–	APEX1 ↑ activity[30]	-	Nuc[31]
3-methyl adenine DNA glycosylase	*S. cerevisiae*	*MAG1*	–	–	↑ in response to alkylating agents and UV[32,33]	Nuc[34]
	Human	*MPG*	Acetyl[35]	Associated with hHR23, XRCC1, and ERα[35–37]	↑ in G_1-phase[26]	Nuc[38]
Endonuclease III/ thymine glycol DNA glycosylase	*S. cerevisiae*	*NTG1*	SUMO[39]	–	↑ in response to menadione[40,41]	Nuc and Mito[39,42]
		NTG2	SUMO[39]	Associated with MLH1[43,44]	–	Nuc[39,42]
	Human	*NTHL1*	–	APEX1 ↑ activity[45] Associated with XPG, XRCC1, and YB-1[36,46–48]	↑ in G_1- and S-phase[26,49]	Nuc and Mito[49–51]
Endonuclease VIII-like	Human	*NEIL1*	–	Associated with FEN1 and XRCC1[52,53]	↑ in S-phase[54]	Nuc and Mito[55]
		NEIL2	Acetyl (Lys49, Lys153)[56]	Associated with p300 and XRCC1[36,56]	No cell cycle dependence[57]	Nuc[57]
		NEIL3	–	Associated with RPA[58]	–	Nuc[59]
8-oxoguanine DNA glycosylase	*S. cerevisiae*	*OGG1*	–	–	–	Nuc and Mito[60]
	Human	*hOGG1*	Phos (Ser326)[61,62] Acetyl (Lys41, Lys338)[63] Nitro[64]	APEX1 ↑ activity[65,66] Associated with XRCC1[67]	No cell cycle dependence[26] ↑ in response to ROS[68]	Nuc and Mito[51,69]

A-G-mismatch DNA glycosylase	Human	*MUTYH*	Phos (Ser524)[70,71]	Associated with APEX1, PCNA, RPA, RAD9-RAD1-HUS1 complex, and MSH6[72–74]	↑ in S-phase[75]	Nuc and Mito[51,69,75,76]
G-T-mismatch DNA glycosylase	Human	*MBD4*	–	Associated with MLH1, SIN3A, and HDAC1[77–79]	–	Nuc[80]
		TDG	Acetyl[81] SUMO[82]	APEX1 ↑ Activity[83,84] Associated with APEX1, CPB, p300, DMNT3A, DMNT3B, RXR, RAR, SCR1, p53, p73α, p63γ, ERα, RAD9, and XPC[78,81,85–91]	No cell cycle dependence[26]	Nuc[92]
AP endonuclease	*S. cerevisiae*	*APN1*	–	Associated with PIR1[93]	–	Nuc and Mito[93,94]
		APN2	–	Associated with POL30[95]	↑ in response to alkylating agents[96]	Nuc[34]
	Human	*APEX1*	Acetyl by p300[97] Phos[98] Ub (K24, K25, K27)[99] Nitro[100,101]	↑ Glycosylase activity of UNG1, SMUG, NTHL1, OGG1, and TDG[22,30,45,65,66] Associated with MUTYH,TDG, XRCC1, HSP70, p53, PCNA, LIG1, FEN1, and WRN[74,81,102–107]	↑ in G_1-phase[26] ↑ in response to ROS[108]	Nuc and Mito[109–111]
		APEX2	–	Associated with PCNA[112,113]	–	Nuc and Mito[112–115]
X-ray repair cross-complementing group 1	Human	*XRCC1*	Phos[116,117]	Associated with OGG1, NTHL1, NEIL1, NEIL2, MPG, POL*β*, LIG3, PARP1, PARP2, UNG2, PNK, CK2, TDP1, ATPX, PCNA, and APEX1[25,36,53,66,67,107,116–130]	↑ in S-phase[127]	Nuc[103,119,120,131]

(*Continues*)

TABLE I (*Continued*)

			Level of regulation			
Protein	Organism	Gene	PTM	Protein interactions	Transcript level	Localization
Poly (ADP-ribose) polymerase	Human	*PARP1*	Phos[132] SUMO (K486)[133,134] Acetyl (K498, K505, K508, K521, K524)[135]	Activity ↑ in response to ROS[136] Associated with LIG3, XRCC1, POL*β*, and numerous non-BER proteins[126,129,132,133,136]	–	Nuc[130]
Flap endonuclease	*S. cerevisiae*	*RAD27*	–	Associated with POL4 and DNL4/LIF1[137]	↑ in G_1-phase ↑ in response to alkylating agents[138]	Nuc and Mito[139]
	Human	*FEN1*	Phos (Ser187)[140] Acetyl[141] Methyl (R192)[142]	Associated with PCNA, WRN, CDK1, CDK2 NEIL1, LIG1, p300, and APEX1[25,52,103,113–115,131,140,141,143]	↑ in mitotic cycling cells ↓in stationary cells[144]	Nuc and Mito[145–147]
Proliferating cell nuclear antigen	*S. cerevisiae*	*POL30*	Ub (K164)[148] SUMO (K127, K164)[148]	Associated with APN2[95]	↑ in G_1- and S-phase[149]	Nuc[150]
	Human	*PCNA*	Ub (K164)[151] Acetyl[152] SUMO (K127, K164)[148,153]	Associated with MUTYH, p21, POLβ, XRCC1, FEN1, LIG1, APEX1, and numerous non-BER proteins[25,74,103,113–115,121,154–160]	↑ in S phase[161,162]	Nuc[112]

PTM, posttranslational modification; Phos, phosphorylation; SUMO, sumoylation; Ub, ubiquitination; acetyl, acetylation; methyl, methylation; nitro, nitrosylation; ↑, increases; ↓, decreases; Nuc, nuclear; Mito, mitochondrial.

mismatch repair (Table I).[98] For example, several human DNA glycosylases have increased base excision activity in the presence of APEX1 (Table I).[98] Physical interactions also occur. For example, specific glycosylases associate with X-ray repair cross-complementing group 1 (XRCC1) to enhance base excision activity (Table I), while other BER proteins are known to interact with proliferating cell nuclear antigen (PCNA) or replication protein A (RPA), suggesting that DNA glycosylases are involved in postreplication BER responses (Table I).[23,24,58,67,74]

Known posttranslational modifications of BER proteins include phosphorylation, acetylation, nitrosylation, ubiquitination, and sumoylation (Table I). Phosphorylation of OGG1, A/G-specific adenine DNA glycosylase (MUTYH), and UNG affects enzyme excision activity in different ways (Table I).[21,61,71] Acetylation by the histone acetyltransferase, p300, regulates OGG1, thymine DNA glycosylase (TDG), endonuclease VIII-like 2 (NEIL2), polymerase beta (polβ), and APEX1, each of them in a unique manner (Table I).[11] For instance, acetylation of OGG1 causes a decrease in affinity for AP sites while increasing both enzyme turnover and stimulation by APEX1. In contrast, acetylation of NEIL2 significantly reduces both glycosylase and AP lyase activities.[56,63] Nitrosylation of BER proteins occurs for APEX1 and OGG1, in both cases inhibiting repair activity, but doing so in different manners: triggering nuclear export or inhibiting excision activity, respectively.[100,101,165] Ubiquitination of a protein, which typically leads to degradation, is used as a method of controlling steady-state protein levels, such as reported for Polβ.[166] Ubiquitination, however, can also have other roles such as with APEX1 ubiquitination which does not affect protein degradation, but whose consequence is thus far unknown.[167] Lastly, sumoylation of a few BER proteins has been documented including TDG, and the *S. cerevisiae* dual functional endonuclease III-like thymine glycol DNA glycosylase/AP lyases (Ntg1) and (Ntg2).[39,168] The effect of TDG sumoylation is best understood where the small ubiquitin-related modifier (SUMO), competes with the regulatory domain of TDG for DNA binding, allowing TDG to overcome product inhibition and ultimately increasing enzyme turnover.[168,169]

Protein localization is perhaps the most crucial form of regulation in eukaryotic BER as BER proteins must be given access to the genomic DNA in order to carry out their functions. As such, almost all BER proteins have evolved a means for localizing to the nucleus, most via the classical nuclear import pathway (Table I).[170] Classical nuclear import operates through the recognition of a classical nuclear localization signal (NLS) by the NLS receptor, importin α, which binds the NLS-containing cargo protein in the cytoplasm and imports it into the nucleus through nuclear pores in complex with importin β.[171] Functional mitochondria are also important for cell survival and fitness, making the maintenance of mitochondrial DNA by BER proteins critical.[172]

BER proteins with known mechanisms of mitochondrial localization contain mitochondrial matrix-targeting sequences (MTSs). These sequences are usually located at the N terminus of a protein and consist of 10–80 amino acids that form amphipathic α-helices, which are recognized by the mitochondrial outer membrane translocase.[173,174] Proteins with an MTS enter into the mitochondrial matrix after being passed from the outer membrane translocase to the inner membrane translocase.[175–177] Protein localization may also be controlled through protein–protein interactions and posttranslational modifications in response to some stimuli, representing a higher level of protein localization regulation.[178,179]

C. Dynamic Localization

While none of the regulatory mechanisms mentioned above are responsible for the regulation of each and every protein in the BER pathway, there are some common threads that all BER proteins must share in order to effectively maintain the genome. Foremost is the ability to localize to compartments containing damaged DNA coupled with the ability to sense and play their part in repairing damaged DNA in an efficient and accurate manner. As a result, we believe the regulation of BER proteins is best carried out through a combination of regulatory mechanisms with the end result being dynamic localization of BER proteins.

Dynamic localization simply means that a protein can be mobilized to a specific cellular compartment in response to environmental stimuli. There are many examples of dynamic localization, including some proteins involved in DNA repair (Table II).[39,180–188] The xeroderma pigmentosum, complementation group G (XPGC) protein relocalizes from the perinucleolar sites to the nucleoplasm following UV irradiation, a process that requires a region of amino acids containing one of the nuclear localization signals within XPGC.[187,189] The E3 ubiquitin ligase (RAD18) protein localizes to double-strand breaks caused by ionizing radiation independently of the cell cycle, but has differential cell cycle-dependent localization, with RAD18 exhibiting nucleolar localization during late G_2 before the G_2/M transition.[182] Lastly, sumoylation is required for localization of topoisomerase II to centromeres prior to anaphase, and in the absence of this modification, nondisjunction may occur.[181]

With respect to BER, dynamic localization is defined as the ability to sense base damage and to recruit the appropriate enzymes and other proteins to neutralize the threat to genome integrity of a compartmentalized genome. Situations where dynamic localization of BER proteins is disrupted or inaccurately activated could lead to mutagenesis and subsequent deleterious biological consequences.

TABLE II

Dynamic Localization of Proteins Involved in DNA Repair

Protein(s)	Organism	Gene(s)	Reported mode of dynamic localization	Stimulus
Cockayne syndrome complementation group B	Human	*CSB*	Localization shifts to the mitochondria in response to mitochondrial oxidative stress where CSB affects incision activity of mitochondrial extracts for oxidative base damage[180]	Mitochondrial oxidative stress
DNA topoisomerase 2	Human	*TOP2*	Sumoylation of TOP2I causes it to localize to centromeres during mitotic chromosome assembly where it promotes a final step in decatenation[181]	Cell cycle
E3 ubiquitin–protein ligase RAD18	Human	*RAD18*	Cell cycle-dependent localization with one or two foci forming during G_1-phase; in S-phase, more foci form while decreasing in size, and in G_2-phase, foci disappear and Rad18 is relocated to the nucleoli[182]	Cell cycle
Endonuclease III/thymine glycol DNA glycosylase	*S. cerevisiae*	*NTG1*	Localization shifts to nuclei or mitochondria in response to organelle-specific oxidative DNA damage[39,183]	Oxidative DNA damage
Flap structure-specific endonuclease 1	Human	*FEN1*	Phosphorylation at Ser187 in response to UV irradiation drives translocation of FEN1 from the nucleoli to the general nucleus[184]	UV irradiation
Ribonucleotide reductase heterodimer subunit	*S. cerevisiae*	*Rnr2-Rnr4*	Increased cytoplasmic localization upon DNA damage through decreased nuclear retention and decreased nuclear import lowers the level of dNTPs in the nucleus[185]	DNA damage
Werner syndrome, RecQ helicase-like	Human	*WRN*	In response to UV exposure, WRN translocates from the nucleolus to nucleoplasmic foci; the rate and extent of WRN redistribution are correlated with the UV dose and linked to protein acetylation[186]	UV irradiation
Xeroderma pigmentosum, complementation group G	Human	*XPGC*	Relocalizes from the perinucleolar sites to the general nucleus following UV irradiation in a dose-dependent manner[187]	UV irradiation
X-ray repair complementing defective repair in Chinese hamster cells 1 and DNA ligase III alpha	Human	*XRCC1* and *LIGIIIα*	During metaphase in response to hydrogen peroxide, both XRCC1 and LIGIIIα translocate from the centrosomes to the mitotic chromosomes in order to repair SSBs[188]	Oxidative DNA damage

SSBs, single-stranded breaks; dNTPs, deoxyribonucleotide triphosphates; UV, ultraviolet.

II. Dynamic Localization of BER Proteins

A. General Pathway

In general, dynamic localization of BER proteins functions through the generation of an organellar-specific signal which recruits the required BER proteins to the organelle where DNA damage has occurred (Fig. 1). As BER proteins are responsible for maintenance of both nuclear and mitochondrial genomes, there is competition between the nucleus and the mitochondria for a rather small pool of BER proteins. The advantage of such a mechanism is an extremely rapid response to increased cellular stress which does not require new transcription and translation of BER genes and instead utilizes the overall low level of BER proteins normally present.

B. Requirements to Dynamically Localize

In order for dynamic localization to occur, a few requirements must be met. The first requirement is the generation of a signaling event to initiate dynamic localization. Such an event originates in response to different types of DNA damage including, but not limited to, oxidative DNA damage and alkylation damage.[39] The specific type of DNA damage may dictate which proteins are mobilized by dynamic localization. However, for proteins in the BER pathway to initially respond and be regulated by dynamic localization, they must first

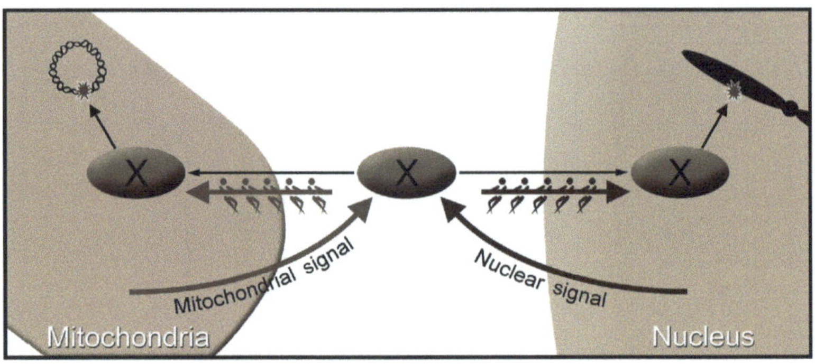

FIG. 1. General model of dynamic localization for base excision repair (BER) proteins. Protein X is any BER protein able to localize to both nuclei and mitochondria (black arrows). Under DNA-damaging conditions, organelle-specific signals are generated, originating from either the mitochondria (red arrows) or nuclei (blue arrows) depending upon the relative threat to organelle genome stability. These signals result in competition between mitochondria (red arrows) and nuclei (blue arrows) for protein X driving economical localization BER protein for efficient DNA repair. (See Color Insert.)

have the capacity to localize to more than one cellular compartment containing damaged DNA. In eukaryotic cells, BER proteins must thus be capable of dual localization to both the nucleus and the mitochondria. Localization to the nucleus for many BER proteins relies on the classical nuclear import pathway.[183] This pathway may be critical for dynamically targeting BER proteins to the nucleus as in one instance, when the NLS of a protein is abolished, the ability to respond to nuclear DNA damage is lost.[170] Mitochondrial localization follows the same pattern with both basal and dynamic mitochondrial localization relying upon the MTS.[183] The details of translating the initial signal into a change in localization are as yet unknown; however, it would involve a mechanism utilizing standard targeting signals (Fig. 2).

C. Examples of Dynamic Localization in Response to Genotoxic Stress

To date, the best characterized example of dynamic localization in response to DNA damage is the *S. cerevisiae* dual functional glycosylase/AP lyase, Ntg1.[39,183] Ntg1 possesses all the basic components required for dynamic localization including a bipartite classical nuclear localization signal and a mitochondrial MTS.[183] Dynamic localization of Ntg1 changes the basal localization of Ntg1 from both nuclei and mitochondria to a primarily nuclear localization in response to nuclear oxidative and alkylation stress, and to a more mitochondrial localization in response to mitochondrial oxidative stress.[39] Like budding yeast Ntg1, the human ortholog of Ntg1, NTHL1, has all the necessary components to respond to organelle-specific DNA damage signals as it contains both a classical nuclear import signal and a mitochondrial MTS. This aspect of NTHL1, however, has yet to be examined.

Another repair protein which dynamically localizes is the human Cockayne syndrome complementation group B (CSB) protein.[180] CSB has a role in transcription coupled repair (TCR), which is initiated when RNA polymerase II progression is blocked by DNA damage in the transcribed strand, triggering recruitment of NER proteins.[190] CSB, however, also operates outside of TCR and may have roles in the BER pathway in nuclei and mitochondria because CSB interacts with poly (ADP-ribose) polymerase (PARP1), APEX1, and endonuclease VIII-like 1 (NEIL1), and because CSB-deficient cells accumulate oxidative DNA damage.[191–194] Dynamic localization of CSB increases mitochondrial localization of CSB in response to mitochondrial oxidative stress where CSB affects incision activity of mitochondrial extracts for oxidative base damage.[180] This work reveals that dynamic localization is an important mechanism for the regulation of BER in humans and underlines the importance of understanding this process further.

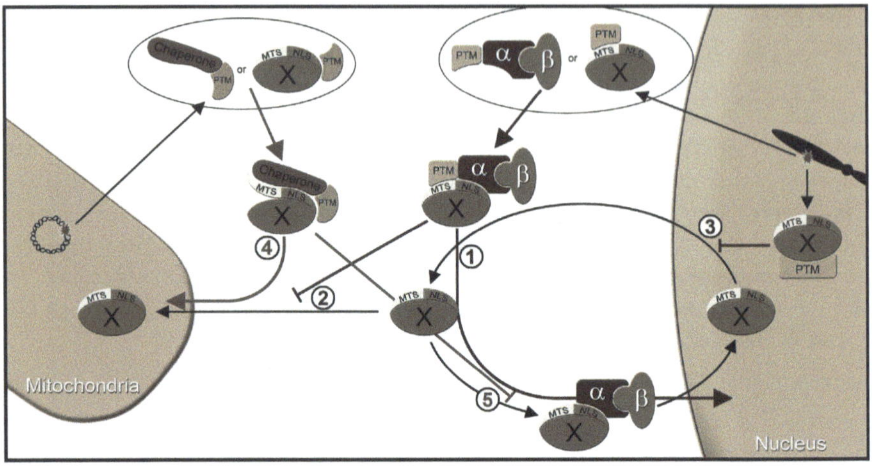

FIG. 2. Dynamic localization of base excision repair (BER) proteins requires localization signals. Protein X is any BER protein able to localize to both nuclei and mitochondria. Under nonstress conditions, protein X localizes to both nuclei and mitochondria (black arrows). Under nuclear or mitochondrial stress conditions, nuclear oxidative DNA damage signals (NODDS) or mitochondrial oxidative DNA damage signals (MODDS) are generated respectively, resulting in posttranslational modification (PTM) to either BER proteins or to transport machinery. NODDS lead to an increased nuclear concentration of BER proteins through one of three mechanisms: (1) increasing nuclear import through PTM of either protein X or the nuclear import machinery, enhancing the association of importin α/β with the classical nuclear localization signal (NLS) of protein X; (2) decreasing mitochondrial import by PTMs, which directly or indirectly disrupt the mitochondria matrix-targeting sequence (MTS); or (3) decreasing nuclear export by PTMs that block a nuclear export signal or create a nuclear retention signal (blue lines). MODDS lead to increased mitochondrial BER protein concentration through one of two mechanisms: (1) PTMs causing chaperones to associate with protein X and facilitating mitochondrial import by maintaining protein X in an unfolded state and enabling mitochondrial import state or (2) decreasing nuclear import through PTMs, which directly disrupts importin α/β binding to the NLS of protein X (red lines). (See Color Insert.)

One of the intriguing features of the *S. cerevisiae* BER system is that most of the early pathway proteins (glycosylases, AP lyases, AP endonucleases) are shared by the nucleus and mitochondria, making budding yeast an ideal system to interrogate the mechanisms regulating BER protein levels, activity, and localization.[195] The proteins shared by nuclei and mitochondria include Ntg1, the uracil DNA glycosylase (Ung1), the 8-oxoguanine DNA glycosylase/AP lyase (Ogg1), and the AP endonuclease (Apn1). In contrast, Ntg2, the 3-methyladenine DNA glycosylase (Mag1) and the AP endonuclease (Apn2) are strictly nuclear.[34,39,42] Ntg1/2 and Apn1/2 are the result of a genome duplication in the recent evolutionary past of *S. cerevisiae*, and as such, each pair has largely similar substrate specificities, which means that even if all of the mitochondrially shared proteins (Ntg1, Apn1) relocated to the mitochondria, a basal level of Ntg and Apn activity would remain in the nucleus.[196–198] Each of these early BER proteins has at least one predicted or confirmed NLS, and each of the shared proteins contains a predicted or confirmed MTS (Fig. 3). The nuclear protein Ntg2 contains an iron–sulfur center in the C terminus that may play a role in DNA damage detection and/or allow sensing of the cellular redox state.[210,211] Various posttranslational modification sites are strongly predicted for each of these proteins, including phosphorylation, methylation,

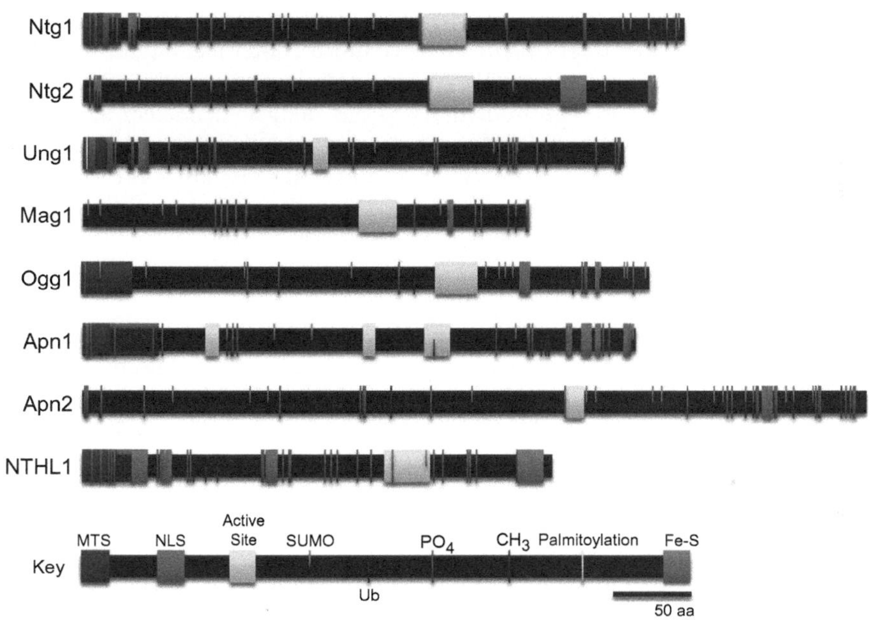

FIG. 3. Predicted motifs and posttranslational modification sites in *Saccharomyces cerevisiae* BER proteins and the human Ntg1 homolog NTHL1. The diagram shows the primary sequence locations of predicted motifs and posttranslational modification sites. Nuclear localization sequences were predicted using NUCDISC/PSORTII[199] (NLS, orange box); mitochondrial matrix-targeting sequences used iPSORT[200] and MitoProt[201] (MTS, blue box). The enzyme active site (yellow box) is as reported in PROSITE. Modification sites were predicted using SUMOsp 2.0,[202] PCI-SUMO,[203] SUMOplot for SUMOylation (SUMO, light green mark); UbPred[204] for ubiquitination (Ub, red mark); MASA[205] for methylation (CH_3, purple mark); NetPhosYeast 1.0[206] and Phosida[207] for phosphorylation (PO_4, light blue mark); CSS-Palm 2.0[208] for palmitoylation (yellow mark); and Metal Detector[209] for metal-binding sites (Fe-S, green box). Arbitrary score thresholds were used for each predictor so that only higher-scoring predictions are shown. Ntg1 localization sequences have been experimentally confirmed, and the NLS of Apn1 has been partially confirmed.[93,94,183] (See Color Insert.)

ubiquitination, sumoylation, and even palmitoylation (though the last mentioned is probably an anomaly). Some of these modifications are predicted to occur within or adjacent to the localization sequences and may play a role in altering protein localization. Most of these modification sites have not yet been experimentally verified but have the potential to play any number of roles in regulating the activity of these proteins, including, but not limited to, the alteration of protein localization by interfering with localization sequences or creating new protein interfaces; enzyme specificity for AP sites; or substrate turnover rates. Sumoylation of Ntg1 has been confirmed, as will be discussed later, and phosphorylation of Mag1 and Apn1 by various kinases, including those involved in DNA damage signaling, has been observed in a number of high-throughput screens.[212–214] The human homolog NTHL1, like Ntg1, contains both an NLS and an MTS, and an iron–sulfur center like Ntg2.[51] NTHL1 is relatively unique among human BER proteins in having a single isoform with both localization signals. The human UNG has dedicated nuclear and mitochondrial isoforms. Still others, like nudix (nucleoside diphosphate linked moiety X)-type motif 1 (NUDT1), OGG1, and MUTYH, are more complex; some of their multiple isoforms have only an MTS, some only have an NLS, and some have both.[69]

D. Insight into Dynamic Localization

Through the study of Ntg1, insight has been gained as to how dynamic localization may result through the targeting of Ntg1 to the nucleus in response to DNA damage. Ntg1 requires a classical nuclear localization signal to localize to the nucleus in response to DNA damage; the importance of nuclear localization also extends beyond the NLS to include the requirement that the classical nuclear import pathway remain intact.[183] These requirements suggest that classical nuclear import is the means by which proteins are targeted for nuclear dynamic localization and not some other alternative import pathway, signifying that the classical nuclear import pathway is a key regulator of genomic stability. A possible missing link between the DNA damage signal and a change in protein localization could be the posttranslational modification of Ntg1 by the SUMO.[39] Sumoylated Ntg1 is likely restricted to the nuclear compartment with no sumoylated Ntg1 detected in the mitochondria; in fact, the nuclear level of sumoylated Ntg1 increases fivefold upon treatment of cells with hydrogen peroxide.[39] This change in Ntg1 status suggests a potential mechanism for how Ntg1 may be targeted to the nucleus with Ntg1 sumoylation targeting Ntg1 to or sequestering Ntg1 within the nucleus in response to nuclear oxidative DNA damage.

What event drives Ntg1 to the mitochondria in response to mitochondrial DNA damage remains unclear. However, if Ntg1 is not able to dynamically localize to the mitochondria or to the nucleus, spontaneous mutation rates increase and cells become sensitive to both hydrogen peroxide and methyl methanesulfonate.[183] This finding provides evidence that the regulation of BER proteins through dynamic localization is crucial in the maintenance of genomic stability and suggests a possible mechanism for how BER proteins become mislocalized in certain cancers.[12–15]

III. Hypotheses on the Orchestration of Dynamic Localization

There are many possible mechanisms one might imagine initiating dynamic localization that fit with the requirements described.[39,183] Here, we discuss our two-part hypothesis detailing the orchestration of dynamic localization (Fig. 4).

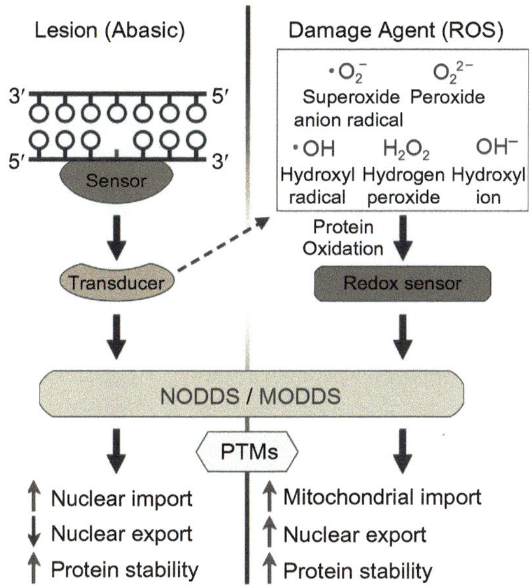

Fig. 4. Proposed model for the transduction of nuclear oxidative DNA damage signals (NODDS) and mitochondrial oxidative DNA damage signals (MODDS). The diagram shows two potential pathways that could be responsible for the generation of NODDS and MODDS. The left half shows how individual lesions (abasic sites) may be detected by sensor proteins and transduced into NODDS and MODDS. The right half shows how DNA damage agents that induce the formation of reactive oxygen species (ROS) may lead to protein oxidation of redox sensor proteins and subsequent generation of NODDS and MODDS. The dashed arrow between the left and right panels represents how signal transduction may occur through increasing ROS, potentially connecting the two pathways. NODDS and MODDS lead to posttranslational modifications (PTMs) of base excision repair (BER) proteins, ultimately causing an increase in either nuclear or mitochondrial BER protein concentration through mechanisms described in the text. (See Color Insert.)

First, we hypothesize that dynamic localization is initiated through sensing either base damage or DNA-damaging agents, which results in the generation of nuclear oxidative DNA damage signals (NODDS) or mitochondrial oxidative DNA damage signals (MODDS). Second, NODDS and MODDS lead to an increase in nuclear or mitochondrial BER protein levels, respectively.

There are a finite number of ways in which a cell can detect DNA base damage: the lesions themselves repair intermediate lesions, or released damage products. Detection of strand breaks, the most dangerous lesions to a cell, operates by sensing the lesions themselves: double-strand breaks are detected by the MRN complex (comprised of Mre11, Rad50, and Nbs1), which leads to ATM kinase pathway signaling, while single-stranded DNA (an extended single-strand break) is recognized by RPA, leading to ATR kinase pathway signaling.[215,216] Direct sensing of DNA lesions is a common theme in DNA damage signaling, so if a signaling mechanism for BER-repairable damage exists, as is suggested by the dynamic localization of Ntg1, the base lesions themselves are candidates for the proximate DNA base damage signal. We propose that of the base lesions repaired by BER, abasic sites (intact, 3′-, or 5′-cleaved) are responsible for creating NODDS and MODDS and initiating dynamic localization of BER proteins. Abasic sites are uniquely suited to a general signaling role for BER as all base lesions repaired by BER have an abasic site as an intermediate.[6] Furthermore, all cellular compartments contain a basal level of their set of glycosylases, allowing an abasic site to be quickly generated from the initial base lesion.[7] Each abasic site sensed by the cell would lead to the generation of an oxidative DNA damage signal, which would then recruit BER proteins to the appropriate organelle.

Dynamic localization may also be driven through organellar detection of the threats responsible for causing DNA base lesions themselves. Agents that cause DNA base damage such as ROS and alkylating agents also react with other cellular products like proteins. Posttranslational modification of key proteins by these DNA-damaging agents may be the cause of BER protein dynamic localization. ROS are uniquely suited to fill this role because of the existence of complex mechanisms for controlling cellular ROS levels and because of their roles in signaling to redox sensor proteins.[217,218] An alternative hypothesis to the one presented is that DNA-damaging agents themselves functioning through redox-sensing proteins are responsible for creating NODDS and MODDS, initiating dynamic localization of BER proteins. There are many examples of posttranslational modifications of protein by ROS, driving changes in activity, abundance, localization, and interaction with other proteins.[218] Furthermore, nonoxidative DNA-damaging agents induce an increase in ROS, allowing for many types of base damage to trigger redox signaling in this manner.[219] Redox sensor proteins would initiate dynamic localization of BER proteins by sensing ROS levels in either the nuclei or mitochondria acting as NODDS and MODDS. This response would be rapid, in some cases occurring before or as DNA base lesions are forming, and would be applicable to the recruitment of all of BER proteins.

A number of mechanisms could translate NODDS and MODDS into mobilization of a protein to nuclei or mitochondria. There are two primary ways to increase the nuclear concentrations of a fixed quantity of BER proteins. The first is by increasing the rate of nuclear import, which can be achieved through posttranslation modification of a number of targets. Posttranslation modification of the BER protein itself could increase nuclear protein localization by creating a new protein–protein interaction that allows for coimport to the nucleus; by interfering with the mitochondrial MTS; or by enhancing the association for receptors of the classical nuclear import pathway. Posttranslation modifications could also occur on other proteins enhancing BER protein nuclear import. The second method of increasing nuclear concentrations of BER protein is decreasing nuclear protein loss, which may be achieved by reducing nuclear protein export and/or by increasing nuclear protein stability through posttranslational modifications on the BER proteins themselves and/or proteins involved in nuclear export/protein degradation. Increasing mitochondrial concentrations would occur in a similar manner but is likely limited to altering protein stability and mitochondrial targeting as no known pathway exists to export proteins from the mitochondria. A third possibility arises if we consider new protein synthesis as a means to increase either nuclear or mitochondrial protein concentrations. Transcriptional regulation may also occur in response to base damage as sensed through abasic sites and might lead to increased nuclear and mitochondrial BER protein levels. This mechanism would, however, be a slower response and would not fall under the purview of dynamic localization.

Dynamic localization represents a novel mechanism for the regulation of BER proteins, integrating many different levels of regulation into a consummate solution flexible enough to maintain both the nuclear and mitochondrial genomes in ever-changing and complex environments. Understanding the underlying pathway driving dynamic localization and the extent to which this regulatory mechanism coordinates BER repair in humans is critical given hints that the process may have roles in cancer and other neurodegenerative disorders.

References

1. Fraga CG, Shigenaga MK, Park JW, Degan P, Ames BN. Oxidative damage to DNA during aging: 8-hydroxy-2′-deoxyguanosine in rat organ DNA and urine. *Proc Natl Acad Sci USA* 1990;**87**:4533–7.
2. Nakamura J, Swenberg JA. Endogenous apurinic/apyrimidinic sites in genomic DNA of mammalian tissues. *Cancer Res* 1999;**59**:2522–6.
3. Ferguson LR. Chronic inflammation and mutagenesis. *Mutat Res* 2010;**690**:3–11.

4. Van Remmen H, Hamilton ML, Richardson A. Oxidative damage to DNA and aging. *Exerc Sport Sci Rev* 2003;**31**:149–53.
5. Eisen JA, Hanawalt PC. A phylogenomic study of DNA repair genes, proteins, and processes. *Mutat Res* 1999;**435**:171–213.
6. Memisoglu A, Samson L. Base excision repair in yeast and mammals. *Mutat Res* 2000;**451**:39–51.
7. Ghaemmaghami S, Huh WK, Bower K, Howson RW, Belle A, Dephoure N, et al. Global analysis of protein expression in yeast. *Nature* 2003;**425**:737–41.
8. Coquerelle T, Dosch J, Kaina B. Overexpression of N-methylpurine-DNA glycosylase in Chinese hamster ovary cells renders them more sensitive to the production of chromosomal aberrations by methylating agents—a case of imbalanced DNA repair. *Mutat Res* 1995;**336**:9–17.
9. Calleja F, Jansen JG, Vrieling H, Laval F, van Zeeland AA. Modulation of the toxic and mutagenic effects induced by methyl methanesulfonate in Chinese hamster ovary cells by overexpression of the rat N-alkylpurine-DNA glycosylase. *Mutat Res* 1999;**425**:185–94.
10. Canitrot Y, Cazaux C, Frechet M, Bouayadi K, Lesca C, Salles B, et al. Overexpression of DNA polymerase beta in cell results in a mutator phenotype and a decreased sensitivity to anticancer drugs. *Proc Natl Acad Sci USA* 1998;**95**:12586–90.
11. Tudek B. Base excision repair modulation as a risk factor for human cancers. *Mol Aspects Med* 2007;**28**:258–75.
12. Karahalil B, Bohr VA, De Souza-Pinto NC. Base excision repair activities differ in human lung cancer cells and corresponding normal controls. *Anticancer Res* 2010;**30**:4963–71.
13. Radak Z, Goto S, Nakamoto H, Udud K, Papai Z, Horvath I. Lung cancer in smoking patients inversely alters the activity of hOGG1 and hNTH1. *Cancer Lett* 2005;**219**:191–5.
14. Goto M, Shinmura K, Igarashi H, Kobayashi M, Konno H, Yamada H, et al. Altered expression of the human base excision repair gene NTH1 in gastric cancer. *Carcinogenesis* 2009;**30**:1345–52.
15. Koketsu S, Watanabe T, Nagawa H. Expression of DNA repair protein: MYH, NTH1, and MTH1 in colorectal cancer. *Hepatogastroenterology* 2004;**51**:638–42.
16. Coppede F, Migliore L. DNA damage and repair in Alzheimer's disease. *Curr Alzheimer Res* 2009;**6**:36–47.
17. Jeppesen DK, Bohr VA, Stevnsner T. DNA repair deficiency in neurodegeneration. *Prog Neurobiol* 2011;**94**:166–200.
18. Weissman L, Jo DG, Sorensen MM, de Souza-Pinto NC, Markesbery WR, Mattson MP, et al. Defective DNA base excision repair in brain from individuals with Alzheimer's disease and amnestic mild cognitive impairment. *Nucleic Acids Res* 2007;**35**:5545–55.
19. Johnston LH, Johnson AL. The DNA repair genes RAD54 and UNG1 are cell cycle regulated in budding yeast but MCB promoter elements have no essential role in the DNA damage response. *Nucleic Acids Res* 1995;**23**:2147–52.
20. Chatterjee A, Singh KK. Uracil-DNA glycosylase-deficient yeast exhibit a mitochondrial mutator phenotype. *Nucleic Acids Res* 2001;**29**:4935–40.
21. Lu X, Bocangel D, Nannenga B, Yamaguchi H, Appella E, Donehower LA. The p53-induced oncogenic phosphatase PPM1D interacts with uracil DNA glycosylase and suppresses base excision repair. *Mol Cell* 2004;**15**:621–34.
22. Parikh SS, Mol CD, Slupphaug G, Bharati S, Krokan HE, Tainer JA. Base excision repair initiation revealed by crystal structures and binding kinetics of human uracil-DNA glycosylase with DNA. *EMBO J* 1998;**17**:5214–26.
23. Nagelhus TA, Haug T, Singh KK, Keshav KF, Skorpen F, Otterlei M, et al. A sequence in the N-terminal region of human uracil-DNA glycosylase with homology to XPA interacts with the C-terminal part of the 34-kDa subunit of replication protein A. *J Biol Chem* 1997;**272**:6561–6.

24. Otterlei M, Warbrick E, Nagelhus TA, Haug T, Slupphaug G, Akbari M, et al. Post-replicative base excision repair in replication foci. *EMBO J* 1999;**18**:3834–44.
25. Wu X, Li J, Li X, Hsieh CL, Burgers PM, Lieber MR. Processing of branched DNA intermediates by a complex of human FEN-1 and PCNA. *Nucleic Acids Res* 1996;**24**:2036–43.
26. Bouziane M, Miao F, Bates SE, Somsouk L, Sang BC, Denissenko M, et al. Promoter structure and cell cycle dependent expression of the human methylpurine-DNA glycosylase gene. *Mutat Res* 2000;**461**:15–29.
27. Slupphaug G, Olsen LC, Helland D, Aasland R, Krokan HE. Cell cycle regulation and in vitro hybrid arrest analysis of the major human uracil-DNA glycosylase. *Nucleic Acids Res* 1991;**19**:5131–7.
28. Akbari M, Otterlei M, Pena-Diaz J, Krokan HE. Different organization of base excision repair of uracil in DNA in nuclei and mitochondria and selective upregulation of mitochondrial uracil-DNA glycosylase after oxidative stress. *Neuroscience* 2007;**145**:1201–12.
29. Nilsen H, Otterlei M, Haug T, Solum K, Nagelhus TA, Skorpen F, et al. Nuclear and mitochondrial uracil-DNA glycosylases are generated by alternative splicing and transcription from different positions in the UNG gene. *Nucleic Acids Res* 1997;**25**:750–5.
30. Kavli B, Sundheim O, Akbari M, Otterlei M, Nilsen H, Skorpen F, et al. hUNG2 is the major repair enzyme for removal of uracil from U:A matches, U:G mismatches, and U in single-stranded DNA, with hSMUG1 as a broad specificity backup. *J Biol Chem* 2002;**277**:39926–36.
31. Haushalter KA, Todd Stukenberg MW, Kirschner MW, Verdine GL. Identification of a new uracil-DNA glycosylase family by expression cloning using synthetic inhibitors. *Curr Biol* 1999;**9**:174–85.
32. Chen J, Derfler B, Samson L. Saccharomyces cerevisiae 3-methyladenine DNA glycosylase has homology to the AlkA glycosylase of E. coli and is induced in response to DNA alkylation damage. *EMBO J* 1990;**9**:4569–75.
33. Chen J, Samson L. Induction of S. cerevisiae MAG 3-methyladenine DNA glycosylase transcript levels in response to DNA damage. *Nucleic Acids Res* 1991;**19**:6427–32.
34. Huh WK, Falvo JV, Gerke LC, Carroll AS, Howson RW, Weissman JS, et al. Global analysis of protein localization in budding yeast. *Nature* 2003;**425**:686–91.
35. Likhite VS, Cass EI, Anderson SD, Yates JR, Nardulli AM. Interaction of estrogen receptor alpha with 3-methyladenine DNA glycosylase modulates transcription and DNA repair. *J Biol Chem* 2004;**279**:16875–82.
36. Campalans A, Marsin S, Nakabeppu Y, O'Connor TR, Boiteux S, Radicella JP. XRCC1 interactions with multiple DNA glycosylases: a model for its recruitment to base excision repair. *DNA Repair (Amst)* 2005;**4**:826–35.
37. Miao F, Bouziane M, Dammann R, Masutani C, Hanaoka F, Pfeifer G, et al. 3-Methyladenine-DNA glycosylase (MPG protein) interacts with human RAD23 proteins. *J Biol Chem* 2000;**275**:28433–8.
38. Engelward BP, Boosalis MS, Chen BJ, Deng Z, Siciliano MJ, Samson LD. Cloning and characterization of a mouse 3-methyladenine/7-methyl-guanine/3-methylguanine DNA glycosylase cDNA whose gene maps to chromosome 11. *Carcinogenesis* 1993;**14**:175–81.
39. Griffiths LM, Swartzlander D, Meadows KL, Wilkinson KD, Corbett AH, Doetsch PW. Dynamic compartmentalization of base excision repair proteins in response to nuclear and mitochondrial oxidative stress. *Mol Cell Biol* 2009;**29**:794–807.
40. Eide L, Bjoras M, Pirovano M, Alseth I, Berdal KG, Seeberg E. Base excision of oxidative purine and pyrimidine DNA damage in Saccharomyces cerevisiae by a DNA glycosylase with sequence similarity to endonuclease III from Escherichia coli. *Proc Natl Acad Sci USA* 1996;**93**:10735–40.
41. You HJ, Swanson RL, Doetsch PW. Saccharomyces cerevisiae possesses two functional homologues of Escherichia coli endonuclease III. *Biochemistry* 1998;**37**:6033–40.

42. You HJ, Swanson RL, Harrington C, Corbett AH, Jinks-Robertson S, Senturker S, et al. Saccharomyces cerevisiae Ntg1p and Ntg2p: broad specificity N-glycosylases for the repair of oxidative DNA damage in the nucleus and mitochondria. *Biochemistry* 1999;**38**:11298–306.
43. Dherin C, Gueneau E, Francin M, Nunez M, Miron S, Liberti SE, et al. Characterization of a highly conserved binding site of Mlh1 required for exonuclease I-dependent mismatch repair. *Mol Cell Biol* 2009;**29**:907–18.
44. Gellon L, Werner M, Boiteux S. Ntg2p, a Saccharomyces cerevisiae DNA N-glycosylase/apurinic or apyrimidinic lyase involved in base excision repair of oxidative DNA damage, interacts with the DNA mismatch repair protein Mlh1p. Identification of a Mlh1p binding motif. *J Biol Chem* 2002;**277**:29963–72.
45. Marenstein DR, Chan MK, Altamirano A, Basu AK, Boorstein RJ, Cunningham RP, et al. Substrate specificity of human endonuclease III (hNTH1). Effect of human APE1 on hNTH1 activity. *J Biol Chem* 2003;**278**:9005–12.
46. Bessho T. Nucleotide excision repair 3′ endonuclease XPG stimulates the activity of base excision repairenzyme thymine glycol DNA glycosylase. *Nucleic Acids Res* 1999;**27**:979–83.
47. Klungland A, Hoss M, Gunz D, Constantinou A, Clarkson SG, Doetsch PW, et al. Base excision repair of oxidative DNA damage activated by XPG protein. *Mol Cell* 1999;**3**:33–42.
48. Marenstein DR, Ocampo MT, Chan MK, Altamirano A, Basu AK, Boorstein RJ, et al. Stimulation of human endonuclease III by Y box-binding protein 1 (DNA-binding protein B). Interaction between a base excision repair enzyme and a transcription factor. *J Biol Chem* 2001;**276**:21242–9.
49. Luna L, Bjoras M, Hoff E, Rognes T, Seeberg E. Cell-cycle regulation, intracellular sorting and induced overexpression of the human NTH1 DNA glycosylase involved in removal of formamidopyrimidine residues from DNA. *Mutat Res* 2000;**460**:95–104.
50. Ikeda S, Kohmoto T, Tabata R, Seki Y. Differential intracellular localization of the human and mouse endonuclease III homologs and analysis of the sorting signals. *DNA Repair (Amst)* 2002;**1**:847–54.
51. Takao M, Aburatani H, Kobayashi K, Yasui A. Mitochondrial targeting of human DNA glycosylases for repair of oxidative DNA damage. *Nucleic Acids Res* 1998;**26**:2917–22.
52. Hegde ML, Theriot CA, Das A, Hegde PM, Guo Z, Gary RK, et al. Physical and functional interaction between human oxidized base-specific DNA glycosylase NEIL1 and flap endonuclease 1. *J Biol Chem* 2008;**283**:27028–37.
53. Wiederhold L, Leppard JB, Kedar P, Karimi-Busheri F, Rasouli-Nia A, Weinfeld M, et al. AP endonuclease-independent DNA base excision repair in human cells. *Mol Cell* 2004;**15**:209–20.
54. Hazra TK, Izumi T, Boldogh I, Imhoff B, Kow YW, Jaruga P, et al. Identification and characterization of a human DNA glycosylase for repair of modified bases in oxidatively damaged DNA. *Proc Natl Acad Sci USA* 2002;**99**:3523–8.
55. Hu J, de Souza-Pinto NC, Haraguchi K, Hogue BA, Jaruga P, Greenberg MM, et al. Repair of formamidopyrimidines in DNA involves different glycosylases: role of the OGG1, NTH1, and NEIL1 enzymes. *J Biol Chem* 2005;**280**:40544–51.
56. Bhakat KK, Hazra TK, Mitra S. Acetylation of the human DNA glycosylase NEIL2 and inhibition of its activity. *Nucleic Acids Res* 2004;**32**:3033–9.
57. Hazra TK, Kow YW, Hatahet Z, Imhoff B, Boldogh I, Mokkapati SK, et al. Identification and characterization of a novel human DNA glycosylase for repair of cytosine-derived lesions. *J Biol Chem* 2002;**277**:30417–20.
58. Morland I, Rolseth V, Luna L, Rognes T, Bjoras M, Seeberg E. Human DNA glycosylases of the bacterial Fpg/MutM superfamily: an alternative pathway for the repair of 8-oxoguanine and other oxidation products in DNA. *Nucleic Acids Res* 2002;**30**:4926–36.

59. Torisu K, Tsuchimoto D, Ohnishi Y, Nakabeppu Y. Hematopoietic tissue-specific expression of mouse Neil3 for endonuclease VIII-like protein. *J Biochem* 2005;**138**:763–72.
60. Singh KK, Sigala B, Sikder HA, Schwimmer C. Inactivation of Saccharomyces cerevisiae OGG1 DNA repair gene leads to an increased frequency of mitochondrial mutants. *Nucleic Acids Res* 2001;**29**:1381–8.
61. Hu J, Imam SZ, Hashiguchi K, de Souza-Pinto NC, Bohr VA. Phosphorylation of human oxoguanine DNA glycosylase (alpha-OGG1) modulates its function. *Nucleic Acids Res* 2005;**33**:3271–82.
62. Luna L, Rolseth V, Hildrestrand GA, Otterlei M, Dantzer F, Bjoras M, et al. Dynamic relocalization of hOGG1 during the cell cycle is disrupted in cells harbouring the hOGG1-Cys326 polymorphic variant. *Nucleic Acids Res* 2005;**33**:1813–24.
63. Bhakat KK, Mokkapati SK, Boldogh I, Hazra TK, Mitra S. Acetylation of human 8-oxoguanine-DNA glycosylase by p300 and its role in 8-oxoguanine repair in vivo. *Mol Cell Biol* 2006;**26**:1654–65.
64. Jaiswal M, LaRusso NF, Nishioka N, Nakabeppu Y, Gores GJ. Human Ogg1, a protein involved in the repair of 8-oxoguanine, is inhibited by nitric oxide. *Cancer Res* 2001;**61**:6388–93.
65. Hill JW, Hazra TK, Izumi T, Mitra S. Stimulation of human 8-oxoguanine-DNA glycosylase by AP-endonuclease: potential coordination of the initial steps in base excision repair. *Nucleic Acids Res* 2001;**29**:430–8.
66. Whitehouse CJ, Taylor RM, Thistlethwaite A, Zhang H, Karimi-Busheri F, Lasko DD, et al. XRCC1 stimulates human polynucleotide kinase activity at damaged DNA termini and accelerates DNA single-strand break repair. *Cell* 2001;**104**:107–17.
67. Marsin S, Vidal AE, Sossou M, Menissier-de Murcia J, Le Page F, Boiteux S, et al. Role of XRCC1 in the coordination and stimulation of oxidative DNA damage repair initiated by the DNA glycosylase hOGG1. *J Biol Chem* 2003;**278**:44068–74.
68. Kim HN, Morimoto Y, Tsuda T, Ootsuyama Y, Hirohashi M, Hirano T, et al. Changes in DNA 8-hydroxyguanine levels, 8-hydroxyguanine repair activity, and hOGG1 and hMTH1 mRNA expression in human lung alveolar epithelial cells induced by crocidolite asbestos. *Carcinogenesis* 2001;**22**:265–9.
69. Nakabeppu Y. Regulation of intracellular localization of human MTH1, OGG1, and MYH proteins for repair of oxidative DNA damage. *Prog Nucleic Acid Res Mol Biol* 2001;**68**:75–94.
70. Kundu S, Brinkmeyer MK, Eigenheer RA, David SS. Ser 524 is a phosphorylation site in MUTYH and Ser 524 mutations alter 8-oxoguanine (OG): a mismatch recognition. *DNA Repair (Amst)* 2010;**9**:1026–37.
71. Parker AR, O'Meally RN, Sahin F, Su GH, Racke FK, Nelson WG, et al. Defective human MutY phosphorylation exists in colorectal cancer cell lines with wild-type MutY alleles. *J Biol Chem* 2003;**278**:47937–45.
72. Gu Y, Parker A, Wilson TM, Bai H, Chang DY, Lu AL. Human MutY homolog, a DNA glycosylase involved in base excision repair, physically and functionally interacts with mismatch repair proteins human MutS homolog 2/human MutS homolog 6. *J Biol Chem* 2002;**277**:11135–42.
73. Luncsford PJ, Chang DY, Shi G, Bernstein J, Madabushi A, Patterson DN, et al. A structural hinge in eukaryotic MutY homologues mediates catalytic activity and Rad9-Rad1-Hus1 checkpoint complex interactions. *J Mol Biol* 2010;**403**:351–70.
74. Parker A, Gu Y, Mahoney W, Lee SH, Singh KK, Lu AL. Human homolog of the MutY repair protein (hMYH) physically interacts with proteins involved in long patch DNA base excision repair. *J Biol Chem* 2001;**276**:5547–55.
75. Boldogh I, Milligan D, Lee MS, Bassett H, Lloyd RS, McCullough AK. hMYH cell cycle-dependent expression, subcellular localization and association with replication foci: evidence

suggesting replication-coupled repair of adenine:8-oxoguanine mispairs. *Nucleic Acids Res* 2001;**29**:2802–9.

76. Ohtsubo T, Nishioka K, Imaiso Y, Iwai S, Shimokawa H, Oda H, et al. Identification of human MutY homolog (hMYH) as a repair enzyme for 2-hydroxyadenine in DNA and detection of multiple forms of hMYH located in nuclei and mitochondria. *Nucleic Acids Res* 2000;**28**:1355–64.
77. Bellacosa A, Cicchillitti L, Schepis F, Riccio A, Yeung AT, Matsumoto Y, et al. MED1, a novel human methyl-CpG-binding endonuclease, interacts with DNA mismatch repair protein MLH1. *Proc Natl Acad Sci USA* 1999;**96**:3969–74.
78. Boland MJ, Christman JK. Characterization of Dnmt3b: thymine-DNA glycosylase interaction and stimulation of thymine glycosylase-mediated repair by DNA methyltransferase(s) and RNA. *J Mol Biol* 2008;**379**:492–504.
79. Kondo E, Gu Z, Horii A, Fukushige S. The thymine DNA glycosylase MBD4 represses transcription and is associated with methylated p16(INK4a) and hMLH1 genes. *Mol Cell Biol* 2005;**25**:4388–96.
80. Hendrich B, Bird A. Identification and characterization of a family of mammalian methyl-CpG binding proteins. *Mol Cell Biol* 1998;**18**:6538–47.
81. Tini M, Benecke A, Um SJ, Torchia J, Evans RM, Chambon P. Association of CBP/p300 acetylase and thymine DNA glycosylase links DNA repair and transcription. *Mol Cell* 2002;**9**:265–77.
82. Steinacher R, Schar P. Functionality of human thymine DNA glycosylase requires SUMO-regulated changes in protein conformation. *Curr Biol* 2005;**15**:616–23.
83. Privezentzev CV, Saparbaev M, Laval J. The HAP1 protein stimulates the turnover of human mismatch-specific thymine-DNA-glycosylase to process 3, N(4)-ethenocytosine residues. *Mutat Res* 2001;**480–481**:277–84.
84. Waters TR, Gallinari P, Jiricny J, Swann PF. Human thymine DNA glycosylase binds to apurinic sites in DNA but is displaced by human apurinic endonuclease 1. *J Biol Chem* 1999;**274**:67–74.
85. Chen D, Lucey MJ, Phoenix F, Lopez-Garcia J, Hart SM, Losson R, et al. T:G mismatch-specific thymine-DNA glycosylase potentiates transcription of estrogen-regulated genes through direct interaction with estrogen receptor alpha. *J Biol Chem* 2003;**278**:38586–92.
86. Guan X, Madabushi A, Chang DY, Fitzgerald ME, Shi G, Drohat AC, et al. The human checkpoint sensor Rad9-Rad1-Hus1 interacts with and stimulates DNA repair enzyme TDG glycosylase. *Nucleic Acids Res* 2007;**35**:6207–18.
87. Kim EJ, Um SJ. Thymine-DNA glycosylase interacts with and functions as a coactivator of p53 family proteins. *Biochem Biophys Res Commun* 2008;**377**:838–42.
88. Li YQ, Zhou PZ, Zheng XD, Walsh CP, Xu GL. Association of Dnmt3a and thymine DNA glycosylase links DNA methylation with base-excision repair. *Nucleic Acids Res* 2007;**35**:390–400.
89. Lucey MJ, Chen D, Lopez-Garcia J, Hart SM, Phoenix F, Al-Jehani R, et al. T:G mismatch-specific thymine-DNA glycosylase (TDG) as a coregulator of transcription interacts with SRC1 family members through a novel tyrosine repeat motif. *Nucleic Acids Res* 2005;**33**:6393–404.
90. Shimizu Y, Iwai S, Hanaoka F, Sugasawa K. Xeroderma pigmentosum group C protein interacts physically and functionally with thymine DNA glycosylase. *EMBO J* 2003;**22**:164–73.
91. Um S, Harbers M, Benecke A, Pierrat B, Losson R, Chambon P. Retinoic acid receptors interact physically and functionally with the T:G mismatch-specific thymine-DNA glycosylase. *J Biol Chem* 1998;**273**:20728–36.

92. Mohan RD, Litchfield DW, Torchia J, Tini M. Opposing regulatory roles of phosphorylation and acetylation in DNA mispair processing by thymine DNA glycosylase. *Nucleic Acids Res* 2010;**38**:1135–48.
93. Vongsamphanh R, Fortier PK, Ramotar D. Pir1p mediates translocation of the yeast Apn1p endonuclease into the mitochondria to maintain genomic stability. *Mol Cell Biol* 2001;**21**:1647–55.
94. Ramotar D, Kim C, Lillis R, Demple B. Intracellular localization of the Apn1 DNA repair enzyme of Saccharomyces cerevisiae. Nuclear transport signals and biological role. *J Biol Chem* 1993;**268**:20533–9.
95. Unk I, Haracska L, Gomes XV, Burgers PM, Prakash L, Prakash S. Stimulation of 3′–>5′ exonuclease and 3′-phosphodiesterase activities of yeast apn2 by proliferating cell nuclear antigen. *Mol Cell Biol* 2002;**22**:6480–6.
96. Bennett RA. The Saccharomyces cerevisiae ETH1 gene, an inducible homolog of exonuclease III that provides resistance to DNA-damaging agents and limits spontaneous mutagenesis. *Mol Cell Biol* 1999;**19**:1800–9.
97. Bhakat KK, Izumi T, Yang SH, Hazra TK, Mitra S. Role of acetylated human AP-endonuclease (APE1/Ref-1) in regulation of the parathyroid hormone gene. *EMBO J* 2003;**22**:6299–309.
98. Fan J, Wilson 3rd. DM. Protein-protein interactions and posttranslational modifications in mammalian base excision repair. *Free Radic Biol Med* 2005;**38**:1121–38.
99. Busso CS, Iwakuma T, Izumi T. Ubiquitination of mammalian AP endonuclease (APE1) regulated by the p53-MDM2 signaling pathway. *Oncogene* 2009;**28**:1616–25.
100. Qu J, Liu GH, Huang B, Chen C. Nitric oxide controls nuclear export of APE1/Ref-1 through S-nitrosation of cysteines 93 and 310. *Nucleic Acids Res* 2007;**35**:2522–32.
101. Tang CH, Wei W, Liu L. Regulation of DNA repair by S-nitrosylation. *Biochim Biophys Acta* 2012;**1820**:730–5.
102. Ahn B, Harrigan JA, Indig FE, Wilson 3rd DM, Bohr VA. Regulation of WRN helicase activity in human base excision repair. *J Biol Chem* 2004;**279**:53465–74.
103. Dianova II, Bohr VA, Dianov GL. Interaction of human AP endonuclease 1 with flap endonuclease 1 and proliferating cell nuclear antigen involved in long-patch base excision repair. *Biochemistry* 2001;**40**:12639–44.
104. Gaiddon C, Moorthy NC, Prives C. Ref-1 regulates the transactivation and pro-apoptotic functions of p53 in vivo. *EMBO J* 1999;**18**:5609–21.
105. La Belle M, Linn S, Thompson LH. Apurinic/apyrimidinic endonuclease activities appear normal in the CHO-cell ethyl methanesulfonate-sensitive mutant, EM9. *Mutat Res* 1984;**141**:41–4.
106. Ranalli TA, Tom S, Bambara RA. AP endonuclease 1 coordinates flap endonuclease 1 and DNA ligase I activity in long patch base excision repair. *J Biol Chem* 2002;**277**:41715–24.
107. Vidal AE, Boiteux S, Hickson ID, Radicella JP. XRCC1 coordinates the initial and late stages of DNA abasic site repair through protein-protein interactions. *EMBO J* 2001;**20**:6530–9.
108. Pines A, Perrone L, Bivi N, Romanello M, Damante G, Gulisano M, et al. Activation of APE1/Ref-1 is dependent on reactive oxygen species generated after purinergic receptor stimulation by ATP. *Nucleic Acids Res* 2005;**33**:4379–94.
109. Fung H, Kow YW, Van Houten B, Taatjes DJ, Hatahet Z, Janssen YM, et al. Asbestos increases mammalian AP-endonuclease gene expression, protein levels, and enzyme activity in mesothelial cells. *Cancer Res* 1998;**58**:189–94.
110. Tell G, Crivellato E, Pines A, Paron I, Pucillo C, Manzini G, et al. Mitochondrial localization of APE/Ref-1 in thyroid cells. *Mutat Res* 2001;**485**:143–52.
111. Tomkinson AE, Bonk RT, Linn S. Mitochondrial endonuclease activities specific for apurinic/apyrimidinic sites in DNA from mouse cells. *J Biol Chem* 1988;**263**:12532–7.

112. Tsuchimoto D, Sakai Y, Sakumi K, Nishioka K, Sasaki M, Fujiwara T, et al. Human APE2 protein is mostly localized in the nuclei and to some extent in the mitochondria, while nuclear APE2 is partly associated with proliferating cell nuclear antigen. *Nucleic Acids Res* 2001;**29**:2349–60.
113. Warbrick E, Lane DP, Glover DM, Cox LS. Homologous regions of Fen1 and p21Cip1 compete for binding to the same site on PCNA: a potential mechanism to co-ordinate DNA replication and repair. *Oncogene* 1997;**14**:2313–21.
114. Gary R, Ludwig DL, Cornelius HL, MacInnes MA, Park MS. The DNA repair endonuclease XPG binds to proliferating cell nuclear antigen (PCNA) and shares sequence elements with the PCNA-binding regions of FEN-1 and cyclin-dependent kinase inhibitor p21. *J Biol Chem* 1997;**272**:24522–9.
115. Li X, Li J, Harrington J, Lieber MR, Burgers PM. Lagging strand DNA synthesis at the eukaryotic replication fork involves binding and stimulation of FEN-1 by proliferating cell nuclear antigen. *J Biol Chem* 1995;**270**:22109–12.
116. Loizou JI, El-Khamisy SF, Zlatanou A, Moore DJ, Chan DW, Qin J, et al. The protein kinase CK2 facilitates repair of chromosomal DNA single-strand breaks. *Cell* 2004;**117**:17–28.
117. Luo H, Chan DW, Yang T, Rodriguez M, Chen BP, Leng M, et al. A new XRCC1-containing complex and its role in cellular survival of methyl methanesulfonate treatment. *Mol Cell Biol* 2004;**24**:8356–65.
118. Caldecott KW. DNA single-strand break repair and spinocerebellar ataxia. *Cell* 2003;**112**:7–10.
119. Caldecott KW, McKeown CK, Tucker JD, Ljungquist S, Thompson LH. An interaction between the mammalian DNA repair protein XRCC1 and DNA ligase III. *Mol Cell Biol* 1994;**14**:68–76.
120. Caldecott KW, Tucker JD, Stanker LH, Thompson LH. Characterization of the XRCC1-DNA ligase III complex in vitro and its absence from mutant hamster cells. *Nucleic Acids Res* 1995;**23**:4836–43.
121. Fan J, Otterlei M, Wong HK, Tomkinson AE, Wilson 3rd. DM. XRCC1 co-localizes and physically interacts with PCNA. *Nucleic Acids Res* 2004;**32**:2193–201.
122. Gryk MR, Marintchev A, Maciejewski MW, Robertson A, Wilson SH, Mullen GP. Mapping of the interaction interface of DNA polymerase beta with XRCC1. *Structure* 2002;**10**:1709–20.
123. Marintchev A, Gryk MR, Mullen GP. Site-directed mutagenesis analysis of the structural interaction of the single-strand-break repair protein, X-ray cross-complementing group 1, with DNA polymerase beta. *Nucleic Acids Res* 2003;**31**:580–8.
124. Marintchev A, Mullen MA, Maciejewski MW, Pan B, Gryk MR, Mullen GP. Solution structure of the single-strand break repair protein XRCC1 N-terminal domain. *Nat Struct Biol* 1999;**6**:884–93.
125. Marintchev A, Robertson A, Dimitriadis EK, Prasad R, Wilson SH, Mullen GP. Domain specific interaction in the XRCC1-DNA polymerase beta complex. *Nucleic Acids Res* 2000;**28**:2049–59.
126. Masson M, Niedergang C, Schreiber V, Muller S, Menissier-de Murcia J, de Murcia G. XRCC1 is specifically associated with poly(ADP-ribose) polymerase and negatively regulates its activity following DNA damage. *Mol Cell Biol* 1998;**18**:3563–71.
127. Nash RA, Caldecott KW, Barnes DE, Lindahl T. XRCC1 protein interacts with one of two distinct forms of DNA ligase III. *Biochemistry* 1997;**36**:5207–11.
128. Plo I, Liao ZY, Barcelo JM, Kohlhagen G, Caldecott KW, Weinfeld M, et al. Association of XRCC1 and tyrosyl DNA phosphodiesterase (Tdp1) for the repair of topoisomerase I-mediated DNA lesions. *DNA Repair (Amst)* 2003;**2**:1087–100.
129. Schreiber V, Ame JC, Dolle P, Schultz I, Rinaldi B, Fraulob V, et al. Poly(ADP-ribose) polymerase-2 (PARP-2) is required for efficient base excision DNA repair in association with PARP-1 and XRCC1. *J Biol Chem* 2002;**277**:23028–36.

130. Wei YF, Robins P, Carter K, Caldecott K, Pappin DJ, Yu GL, et al. Molecular cloning and expression of human cDNAs encoding a novel DNA ligase IV and DNA ligase III, an enzyme active in DNA repair and recombination. *Mol Cell Biol* 1995;**15**:3206–16.
131. Brosh Jr. RM, von Kobbe C, Sommers JA, Karmakar P, Opresko PL, Piotrowski J, et al. Werner syndrome protein interacts with human flap endonuclease 1 and stimulates its cleavage activity. *EMBO J* 2001;**20**:5791–801.
132. Caldecott KW, Aoufouchi S, Johnson P, Shall S. XRCC1 polypeptide interacts with DNA polymerase beta and possibly poly (ADP-ribose) polymerase, and DNA ligase III is a novel molecular 'nick-sensor' in vitro. *Nucleic Acids Res* 1996;**24**:4387–94.
133. Leppard JB, Dong Z, Mackey ZB, Tomkinson AE. Physical and functional interaction between DNA ligase IIIalpha and poly(ADP-Ribose) polymerase 1 in DNA single-strand break repair. *Mol Cell Biol* 2003;**23**:5919–27.
134. Messner S, Schuermann D, Altmeyer M, Kassner I, Schmidt D, Schar P, et al. Sumoylation of poly(ADP-ribose) polymerase 1 inhibits its acetylation and restrains transcriptional coactivator function. *FASEB J* 2009;**23**:3978–89.
135. Hassa PO, Haenni SS, Buerki C, Meier NI, Lane WS, Owen H, et al. Acetylation of poly (ADP-ribose) polymerase-1 by p300/CREB-binding protein regulates coactivation of NF-kappaB-dependent transcription. *J Biol Chem* 2005;**280**:40450–64.
136. Hassa PO, Hottiger MO. The diverse biological roles of mammalian PARPS, a small but powerful family of poly-ADP-ribose polymerases. *Front biosci* 2008;**13**:3046–82.
137. Tseng HM, Tomkinson AE. Processing and joining of DNA ends coordinated by interactions among Dnl4/Lif1, Pol4, and FEN-1. *J Biol Chem* 2004;**279**:47580–8.
138. Vallen EA, Cross FR. Mutations in RAD27 define a potential link between G1 cyclins and DNA replication. *Mol Cell Biol* 1995;**15**:4291–302.
139. Kalifa L, Beutner G, Phadnis N, Sheu SS, Sia EA. Evidence for a role of FEN1 in maintaining mitochondrial DNA integrity. *DNA Repair (Amst)* 2009;**8**:1242–9.
140. Henneke G, Koundrioukoff S, Hubscher U. Phosphorylation of human Fen1 by cyclin-dependent kinase modulates its role in replication fork regulation. *Oncogene* 2003;**22**:4301–13.
141. Hasan S, Stucki M, Hassa PO, Imhof R, Gehrig P, Hunziker P, et al. Regulation of human flap endonuclease-1 activity by acetylation through the transcriptional coactivator p300. *Mol Cell* 2001;**7**:1221–31.
142. Guo Z, Zheng L, Xu H, Dai H, Zhou M, Pascua MR, et al. Methylation of FEN1 suppresses nearby phosphorylation and facilitates PCNA binding. *Nat Chem Biol* 2010;**6**:766–73.
143. Pascal JM, O'Brien PJ, Tomkinson AE, Ellenberger T. Human DNA ligase I completely encircles and partially unwinds nicked DNA. *Nature* 2004;**432**:473–8.
144. Kim IS. Down-regulation of human FEN-1 gene expression during differentiation of promyelocytic leukemia cells. *Exp Mol Med* 1998;**30**:252–6.
145. Liu P, Qian L, Sung JS, de Souza-Pinto NC, Zheng L, Bogenhagen DF, et al. Removal of oxidative DNA damage via FEN1-dependent long-patch base excision repair in human cell mitochondria. *Mol Cell Biol* 2008;**28**:4975–87.
146. Qiu J, Li X, Frank G, Shen B. Cell cycle-dependent and DNA damage-inducible nuclear localization of FEN-1 nuclease is consistent with its dual functions in DNA replication and repair. *J Biol Chem* 2001;**276**:4901–8.
147. Szczesny B, Tann AW, Longley MJ, Copeland WC, Mitra S. Long patch base excision repair in mammalian mitochondrial genomes. *J Biol Chem* 2008;**283**:26349–56.
148. Hoege C, Pfander B, Moldovan GL, Pyrowolakis G, Jentsch S. RAD6-dependent DNA repair is linked to modification of PCNA by ubiquitin and SUMO. *Nature* 2002;**419**:135–41.
149. Bauer GA, Burgers PM. Molecular cloning, structure and expression of the yeast proliferating cell nuclear antigen gene. *Nucleic Acids Res* 1990;**18**:261–5.

150. Miller A, Chen J, Takasuka TE, Jacobi JL, Kaufman PD, Irudayaraj JM, et al. Proliferating cell nuclear antigen (PCNA) is required for cell cycle-regulated silent chromatin on replicated and nonreplicated genes. *J Biol Chem* 2010;**285**:35142–54.
151. Lee KY, Myung K. PCNA modifications for regulation of post-replication repair pathways. *Mol Cells* 2008;**26**:5–11.
152. Naryzhny SN, Lee H. The post-translational modifications of proliferating cell nuclear antigen: acetylation, not phosphorylation, plays an important role in the regulation of its function. *J Biol Chem* 2004;**279**:20194–9.
153. Bergink S, Jentsch S. Principles of ubiquitin and SUMO modifications in DNA repair. *Nature* 2009;**458**:461–7.
154. Jonsson ZO, Hindges R, Hubscher U. Regulation of DNA replication and repair proteins through interaction with the front side of proliferating cell nuclear antigen. *EMBO J* 1998;**17**:2412–25.
155. Kedar PS, Kim SJ, Robertson A, Hou E, Prasad R, Horton JK, et al. Direct interaction between mammalian DNA polymerase beta and proliferating cell nuclear antigen. *J Biol Chem* 2002;**277**:31115–23.
156. Levin DS, Bai W, Yao N, O'Donnell M, Tomkinson AE. An interaction between DNA ligase I and proliferating cell nuclear antigen: implications for Okazaki fragment synthesis and joining. *Proc Natl Acad Sci USA* 1997;**94**:12863–8.
157. Levin DS, McKenna AE, Motycka TA, Matsumoto Y, Tomkinson AE. Interaction between PCNA and DNA ligase I is critical for joining of Okazaki fragments and long-patch base-excision repair. *Curr Biol* 2000;**10**:919–22.
158. Montecucco A, Rossi R, Levin DS, Gary R, Park MS, Motycka TA, et al. DNA ligase I is recruited to sites of DNA replication by an interaction with proliferating cell nuclear antigen: identification of a common targeting mechanism for the assembly of replication factories. *EMBO J* 1998;**17**:3786–95.
159. Naryzhny SN. Proliferating cell nuclear antigen: a proteomics view. *Cell Mol Life Sci* 2008;**65**:3789–808.
160. Waga S, Stillman B. Cyclin-dependent kinase inhibitor p21 modulates the DNA primer-template recognition complex. *Mol Cell Biol* 1998;**18**:4177–87.
161. Bravo R, Celis JE. A search for differential polypeptide synthesis throughout the cell cycle of HeLa cells. *J Cell Biol* 1980;**84**:795–802.
162. Bravo R, Fey SJ, Bellatin J, Larsen PM, Arevalo J, Celis JE. Identification of a nuclear and of a cytoplasmic polypeptide whose relative proportions are sensitive to changes in the rate of cell proliferation. *Exp Cell Res* 1981;**136**:311–9.
163. Saha T, Rih JK, Roy R, Ballal R, Rosen EM. Transcriptional regulation of the base excision repair pathway by BRCA1. *J Biol Chem* 2010;**285**:19092–105.
164. Meyer F, Fiala E, Westendorf J. Induction of 8-oxo-dGTPase activity in human lymphoid cells and normal fibroblasts by oxidative stress. *Toxicology* 2000;**146**:83–92.
165. Hussain SP, Hofseth LJ, Harris CC. Radical causes of cancer. *Nat Rev Cancer* 2003;**3**:276–85.
166. Dianov GL, Meisenberg C, Parsons JL. Regulation of DNA repair by ubiquitylation. *Biochemistry (Mosc)* 2011;**76**:69–79.
167. Busso CS, Lake MW, Izumi T. Posttranslational modification of mammalian AP endonuclease (APE1). *Cell Mol Life Sci* 2010;**67**:3609–20.
168. Hardeland U, Steinacher R, Jiricny J, Schar P. Modification of the human thymine-DNA glycosylase by ubiquitin-like proteins facilitates enzymatic turnover. *EMBO J* 2002;**21**:1456–64.
169. Smet-Nocca C, Wieruszeski JM, Leger H, Eilebrecht S, Benecke A. SUMO-1 regulates the conformational dynamics of thymine-DNA Glycosylase regulatory domain and competes with its DNA binding activity. *BMC Biochem* 2011;**12**:4.

170. Knudsen NO, Andersen SD, Lutzen A, Nielsen FC, Rasmussen LJ. Nuclear translocation contributes to regulation of DNA excision repair activities. *DNA Repair (Amst)* 2009;**8**:682–9.
171. Gorlich D, Kostka S, Kraft R, Dingwall C, Laskey RA, Hartmann E, et al. Two different subunits of importin cooperate to recognize nuclear localization signals and bind them to the nuclear envelope. *Curr Biol* 1995;**5**:383–92.
172. Troen BR. The biology of aging. *Mt Sinai J Med* 2003;**70**:3–22.
173. Neupert W, Herrmann JM. Translocation of proteins into mitochondria. *Annu Rev Biochem* 2007;**76**:723–49.
174. Bruch MD, Hoyt DW. Conformational analysis of a mitochondrial presequence derived from the F1-ATPase beta-subunit by CD and NMR spectroscopy. *Biochim Biophys Acta* 1992;**1159**:81–93.
175. Wang Y, Weiner H. Evaluation of electrostatic and hydrophobic effects on the interaction of mitochondrial signal sequences with phospholipid bilayers. *Biochemistry* 1994;**33**:12860–7.
176. Dekker PJ, Martin F, Maarse AC, Bomer U, Muller H, Guiard B, et al. The Tim core complex defines the number of mitochondrial translocation contact sites and can hold arrested preproteins in the absence of matrix Hsp70-Tim44. *EMBO J* 1997;**16**:5408–19.
177. Dietmeier K, Honlinger A, Bomer U, Dekker PJ, Eckerskorn C, Lottspeich F, et al. Tom5 functionally links mitochondrial preprotein receptors to the general import pore. *Nature* 1997;**388**:195–200.
178. Liu W, Nichols AF, Graham JA, Dualan R, Abbas A, Linn S. Nuclear transport of human DDB protein induced by ultraviolet light. *J Biol Chem* 2000;**275**:21429–34.
179. Fitch ME, Cross IV, Ford JM. p53 responsive nucleotide excision repair gene products p48 and XPC, but not p53, localize to sites of UV-irradiation-induced DNA damage, in vivo. *Carcinogenesis* 2003;**24**:843–50.
180. Aamann MD, Sorensen MM, Hvitby C, Berquist BR, Muftuoglu M, Tian J, et al. Cockayne syndrome group B protein promotes mitochondrial DNA stability by supporting the DNA repair association with the mitochondrial membrane. *FASEB J* 2010;**24**:2334–46.
181. Lee MT, Bachant J. SUMO modification of DNA topoisomerase II: trying to get a CENse of it all. *DNA Repair (Amst)* 2009;**8**:557–68.
182. Inagaki A, van Cappellen WA, van der Laan R, Houtsmuller AB, Hoeijmakers JH, Grootegoed JA, et al. Dynamic localization of human RAD18 during the cell cycle and a functional connection with DNA double-strand break repair. *DNA Repair (Amst)* 2009;**8**:190–201.
183. Swartzlander DB, Griffiths LM, Lee J, Degtyareva NP, Doetsch PW, Corbett AH. Regulation of base excision repair: Ntg1 nuclear and mitochondrial dynamic localization in response to genotoxic stress. *Nucleic Acids Res* 2010;**38**:3963–74.
184. Guo Z, Qian L, Liu R, Dai H, Zhou M, Zheng L, et al. Nucleolar localization and dynamic roles of flap endonuclease 1 in ribosomal DNA replication and damage repair. *Mol Cell Biol* 2008;**28**:4310–9.
185. Lee YD, Wang J, Stubbe J, Elledge SJ. Dif1 is a DNA-damage-regulated facilitator of nuclear import for ribonucleotide reductase. *Mol Cell* 2008;**32**:70–80.
186. Blander G, Zalle N, Daniely Y, Taplick J, Gray MD, Oren M. DNA damage-induced translocation of the Werner helicase is regulated by acetylation. *J Biol Chem* 2002;**277**:50934–40.
187. Cloud KG, Shen B, Strniste GF, Park MS. XPG protein has a structure-specific endonuclease activity. *Mutat Res* 1995;**347**:55–60.
188. Okano S, Lan L, Tomkinson AE, Yasui A. Translocation of XRCC1 and DNA ligase IIIalpha from centrosomes to chromosomes in response to DNA damage in mitotic human cells. *Nucleic Acids Res* 2005;**33**:422–9.
189. Knauf JA, Pendergrass SH, Marrone BL, Strniste GF, MacInnes MA, Park MS. Multiple nuclear localization signals in XPG nuclease. *Mutat Res* 1996;**363**:67–75.

190. Licht CL, Stevnsner T, Bohr VA. Cockayne syndrome group B cellular and biochemical functions. *Am J Hum Genet* 2003;**73**:1217–39.
191. Tuo J, Muftuoglu M, Chen C, Jaruga P, Selzer RR, Brosh Jr. RM, et al. The Cockayne Syndrome group B gene product is involved in general genome base excision repair of 8-hydroxyguanine in DNA. *J Biol Chem* 2001;**276**:45772–9.
192. Muftuoglu M, de Souza-Pinto NC, Dogan A, Aamann M, Stevnsner T, Rybanska I, et al. Cockayne syndrome group B protein stimulates repair of formamidopyrimidines by NEIL1 DNA glycosylase. *J Biol Chem* 2009;**284**:9270–9.
193. Thorslund T, von Kobbe C, Harrigan JA, Indig FE, Christiansen M, Stevnsner T, et al. Cooperation of the Cockayne syndrome group B protein and poly(ADP-ribose) polymerase 1 in the response to oxidative stress. *Mol Cell Biol* 2005;**25**:7625–36.
194. Wong HK, Kim D, Hogue BA, McNeill DR, Wilson 3rd. DM. DNA damage levels and biochemical repair capacities associated with XRCC1 deficiency. *Biochemistry* 2005;**44**: 14335–43.
195. Larsen NB, Rasmussen M, Rasmussen LJ. Nuclear and mitochondrial DNA repair: similar pathways? *Mitochondrion* 2005;**5**:89–108.
196. Unk I, Haracska L, Johnson RE, Prakash S, Prakash L. Apurinic endonuclease activity of yeast Apn2 protein. *J Biol Chem* 2000;**275**:22427–34.
197. Senturker S, Auffret van der Kemp P, You HJ, Doetsch PW, Dizdaroglu M, Boiteux S. Substrate specificities of the ntg1 and ntg2 proteins of Saccharomyces cerevisiae for oxidized DNA bases are not identical. *Nucleic Acids Res* 1998;**26**:5270–6.
198. Kellis M, Birren BW, Lander ES. Proof and evolutionary analysis of ancient genome duplication in the yeast Saccharomyces cerevisiae. *Nature* 2004;**428**:617–24.
199. Nakai K, Horton P. PSORT: a program for detecting sorting signals in proteins and predicting their subcellular localization. *Trends Biochem Sci* 1999;**24**:34–5.
200. Bannai H, Tamada Y, Maruyama O, Nakai K, Miyano S. Extensive feature detection of N-terminal protein sorting signals. *Bioinformatics* 2002;**18**:298–305.
201. Claros MG, Vincens P. Computational method to predict mitochondrially imported proteins and their targeting sequences. *Eur J Biochem* 1996;**241**:779–86.
202. Ren J, Gao X, Jin C, Zhu M, Wang X, Shaw A, et al. Systematic study of protein sumoylation: development of a site-specific predictor of SUMOsp 2.0. *Proteomics* 2009;**9**:3409–12.
203. Green JR, Dmochowski GM, Golshani A. Prediction of protein sumoylation sites via parallel cascade identification, In: *Canadian medical and biological engineering conference, Vancouver, BC*; 2006.
204. Radivojac P, Vacic V, Haynes C, Cocklin RR, Mohan A, Heyen JW, et al. Identification, analysis, and prediction of protein ubiquitination sites. *Proteins* 2010;**78**:365–80.
205. Shien DM, Lee TY, Chang WC, Hsu JB, Horng JT, Hsu PC, et al. Incorporating structural characteristics for identification of protein methylation sites. *J Comput Chem* 2009;**30**: 1532–43.
206. Ingrell CR, Miller ML, Jensen ON, Blom N. NetPhosYeast: prediction of protein phosphorylation sites in yeast. *Bioinformatics* 2007;**23**:895–7.
207. Gnad F, Gunawardena J, Mann M. PHOSIDA 2011: the posttranslational modification database. *Nucleic Acids Res* 2011;**39**:D253–60.
208. Ren J, Wen L, Gao X, Jin C, Xue Y, Yao X. CSS-Palm 2.0: an updated software for palmitoylation sites prediction. *Protein Eng Des Sel* 2008;**21**:639–44.
209. Lippi M, Passerini A, Punta M, Rost B, Frasconi P. MetalDetector: a web server for predicting metal-binding sites and disulfide bridges in proteins from sequence. *Bioinformatics* 2008;**24**:2094–5.
210. Ren B, Duan X, Ding H. Redox control of the DNA damage-inducible protein DinG helicase activity via its iron-sulfur cluster. *J Biol Chem* 2009;**284**:4829–35.

211. Romano CA, Sontz PA, Barton JK. Mutants of the base excision repair glycosylase, endonuclease III: DNA charge transport as a first step in lesion detection. *Biochemistry* 2011;**50**:6133–45.
212. Fasolo J, Sboner A, Sun MG, Yu H, Chen R, Sharon D, et al. Diverse protein kinase interactions identified by protein microarrays reveal novel connections between cellular processes. *Genes Dev* 2011;**25**:767–78.
213. Ptacek J, Devgan G, Michaud G, Zhu H, Zhu X, Fasolo J, et al. Global analysis of protein phosphorylation in yeast. *Nature* 2005;**438**:679–84.
214. Krogan NJ, Cagney G, Yu H, Zhong G, Guo X, Ignatchenko A, et al. Global landscape of protein complexes in the yeast Saccharomyces cerevisiae. *Nature* 2006;**440**:637–43.
215. Stracker TH, Petrini JH. The MRE11 complex: starting from the ends. *Nat Rev Mol Cell Biol* 2011;**12**:90–103.
216. Zou L, Elledge SJ. Sensing DNA damage through ATRIP recognition of RPA-ssDNA complexes. *Science* 2003;**300**:1542–8.
217. Ma Q. Transcriptional responses to oxidative stress: pathological and toxicological implications. *Pharmacol Ther* 2010;**125**:376–93.
218. Wang Y, Yang J, Yi J. Redox sensing by proteins: oxidative modifications on cysteines and the consequent events. *Antioxid Redox Signal* 2012;**16**:649–57.
219. Rowe LA, Degtyareva N, Doetsch PW. DNA damage-induced reactive oxygen species (ROS) stress response in Saccharomyces cerevisiae. *Free Radic Biol Med* 2008;**45**:1167–77.

Oxidized Base Damage and Single-Strand Break Repair in Mammalian Genomes: Role of Disordered Regions and Posttranslational Modifications in Early Enzymes

Muralidhar L. Hegde,[*] Tadahide Izumi,[†] and Sankar Mitra[*]

[*]*Department of Biochemistry and Molecular Biology, University of Texas Medical Branch, Galveston, Texas, USA*
[†]*Graduate Center for Toxicology, University of Kentucky, Lexington, Kentucky, USA*

Oxidative genome damage induced by reactive oxygen species includes oxidized bases, abasic (AP) sites, and single-strand breaks, all of which are repaired via the evolutionarily conserved base excision repair/single-strand break repair (BER/SSBR) pathway. BER/SSBR in mammalian cells is complex,

Progress in Molecular Biology and Translational Science, Vol. 110
DOI: 10.1016/B978-0-12-387665-2.00006-7

with preferred and backup sub-pathways, and is linked to genome replication and transcription. The early BER/SSBR enzymes, namely, DNA glycosylases (DGs) and the end-processing proteins such as abasic endonuclease 1 (APE1), form complexes with downstream repair (and other noncanonical) proteins via pairwise interactions. Furthermore, a unique feature of mammalian early BER/SSBR enzymes is the presence of a disordered terminal extension that is absent in their *Escherichia coli* prototypes. These nonconserved segments usually contain organelle-targeting signals, common interaction interfaces, and sites of posttranslational modifications that may be involved in regulating their repair function including lesion scanning. Finally, the linkage of BER/SSBR deficiency to cancer, aging, and human neurodegenerative diseases, and therapeutic targeting of BER/SSBR are discussed.

I. Oxidative DNA Damage and Its Repair in Mammalian Cells

The sequence fidelity of the genome, essential for maintaining phenotypes of all organisms, is continuously challenged because of DNA's inherent instability, spontaneous chemical reactions, and replication errors. Moreover, all aerobic organisms continuously generate reactive oxygen species (ROS) as by-products of respiration. ROS are also produced by cellular oxidases in response to a variety of external insults including environmental chemicals found in cigarette smoke and chemotherapeutic drugs, ultraviolet (UV) light, or ionizing radiation, and during inflammatory response.[1–4] Intracellular ROS include superoxide anion (O^{2-}), the direct product of oxidases and respiration, while hydroxyl radical ($^{\bullet}OH$) and hydrogen peroxide (H_2O_2) are generated via Fenton reactions and enzymatic processes, respectively. The generated O^{2-} may be scavenged by NO and form peroxynitrite ($^{-}OONO$). Exogenous ROS including singlet O^{2-}, O_3, and $^{\bullet}OH$ radical are produced during radiolysis of H_2O by ionizing radiation. These reactive species produce multiple oxidative DNA damage in aerobic organisms, including oxidized DNA bases, oxidized sugar fragments, abasic (AP) sites, and single-strand breaks (SSBs). Furthermore, closely spaced SSBs, oxidized bases, or abasic (AP) sites (generated during repair) in the genome could form DNA double-strand breaks (DSBs). While unrepaired DSBs are lethal, oxidized base lesions could be mutagenic, cytotoxic, or both. Thus, ROS-induced genome damage represents the most pervasive insult to the genetic material and more than 10^5 such lesions are estimated to be generated by endogenous ROS in a mammalian cell per day.[3] Thus, it is not surprising that ROS-induced genome damage has been implicated in a multitude of diseases, including cardiovascular dysfunction, arthritis, and cancer, as well as in the aging process and age-related neurodegenerative disorders. However, DNA damage is induced sometimes on purpose, for example, for immunoglobulin gene diversification.[5,6]

The most abundant oxidative genome damage products are oxidation products of purines, namely, 8-oxoguanine (8-oxoG) and formamidopyrimidines (FapyG and FapyA), while the common oxidized pyrimidines are thymine glycol and 5-OHU; the latter is generated via oxidative deamination of C. Except for thymine glycol and the hydantoins, these abnormal bases do not block DNA replication. The less common oxidized base lesions such as 8-oxoA, 5-formylU, and 5-OHC could also be mutagenic/toxic[7] (Table I).

TABLE I

COMMON OXIDIZED BASES INDUCED BY REACTIVE OXYGEN SPECIES (ROS)

Normal DNA bases			
1. Adenine 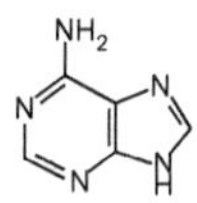	2. Guanine 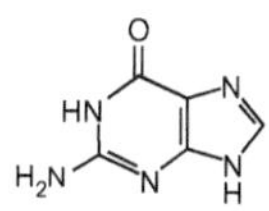	3. Cytosine 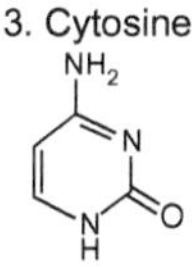	4. Thymine 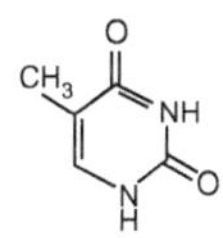
Common oxidized base lesions			
Formamidoadenine (FapyA) 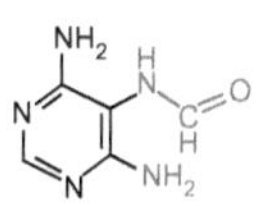	8-Oxoguanine 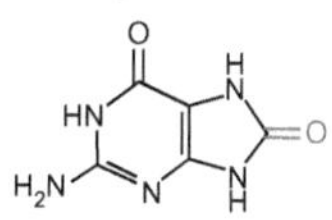	5-Hydroxycytosine (5-OHC) 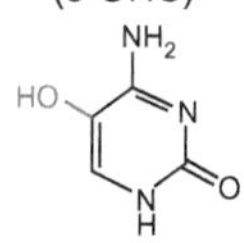	Thymine glycol 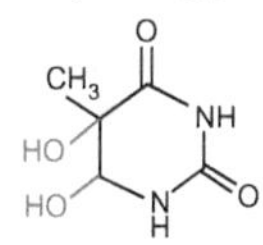
8-Oxoadenine (8-OxoA) 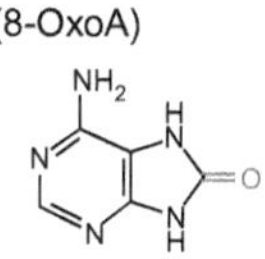	Formamidoguanine (FapyG) 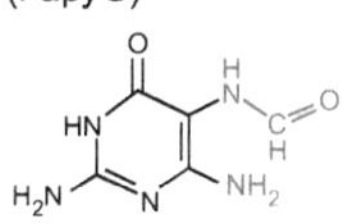	5-Hydroxyuracil (5-OHU) 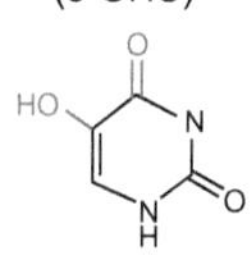	Dihydrothymine 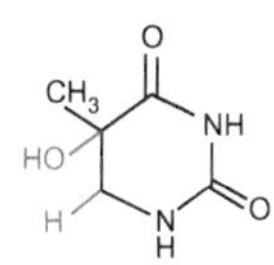
	Guanidinohydantoin 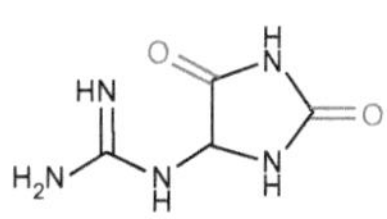	5-Formyluracil (5-FormylU) 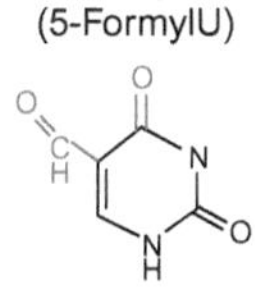	

Oxidative modifications are shown in *gray*.

A. BER of Oxidized Bases and AP sites in Mammalian Genomes

The major process for restoring genomic integrity by repairing oxidative as well as spontaneous alkylation base damage, and AP sites and abnormal bases such as U generated by spontaneous deamination of C is the base excision repair (BER) pathway. BER proteins are involved in the repair of SSBs as well, as discussed in Section I.C. While evolutionarily conserved organisms range from bacteria to mammals, BER is a versatile DNA repair process that handles many types of damage including base damage, AP sites and their oxidation products, and DNA SSBs. BER is initiated with lesion base excision by a mono- or bifunctional DNA glycosylase (DG), and consists of the following basic steps, as also outlined in Fig. 1.

(i) *Base lesion recognition, excision, and cleavage of AP site.* Monofunctional DGs including uracil-DNA glycosylase (UDG) and methylpurine-DNA glycosylase (MPG, also named alkyladenine-DNA glycosylase or AAG) recognize and excise uracil and alkylated bases, respectively, via *N*-glycosyl bond cleavage to generate AP sites. In contrast, all oxidized bases are excised by bifunctional DGs with intrinsic AP lyase activity.[8,9] These, in a concerted action, excise substrate bases and cleave the resulting AP sites via β or βδ elimination reaction.[8,10,11] Five oxidized base-specific DGs have so far been discovered in mammalian cells, which belong to two families named after the prototypes in *Escherichia coli*: OGG1 and NTH1 in the Nth family possess β lyase activity and generate SSBs with 3′ phospho αβ unsaturated aldehyde (3′PUA) and 5′P termini. In contrast, NEIL1 and NEIL2 in the Nei family possess βδ AP lyase activity generating an SSB with 3′P and 5′P termini.[11] NEIL3, another paralog in the Nei family, was identified at about the same time as NEIL1 and NEIL2, based on sequence homology, and it has recently been characterized as having predominantly β lyase (and weak βδ lyase) activity.[12,13] Another evolutionarily conserved DG, MYH (ortholog of *E. coli* MutY), excises the normal base A from an 8-oxoG•A pair in DNA.[14] AP sites generated directly by ROS or by monofunctional DG are hydrolyzed by AP endonuclease 1 (APE1) in mammalian cells to generate an SSB with 3′OH and 5′ deoxyribose phosphate (dRP) ends. Thus, the SSB products of DGs or APE1 have a 3′ or 5′ terminus that prevents subsequent gap-filling DNA synthesis or ligation. These blocked termini are removed in the second step of BER.

(ii) *End-processing of SSB termini to generate 3′OH/5′P.* SSB end-processing is an essential step in BER/SSBR, utilizing diverse enzymes to process various blocked termini. The two major 3′ end-processing enzymes involved in mammalian BER/SSBR are APE1 and polynucleotide kinase 3′ phosphatase (PNKP), which remove 3′dRP (β elimination product) and 3′P

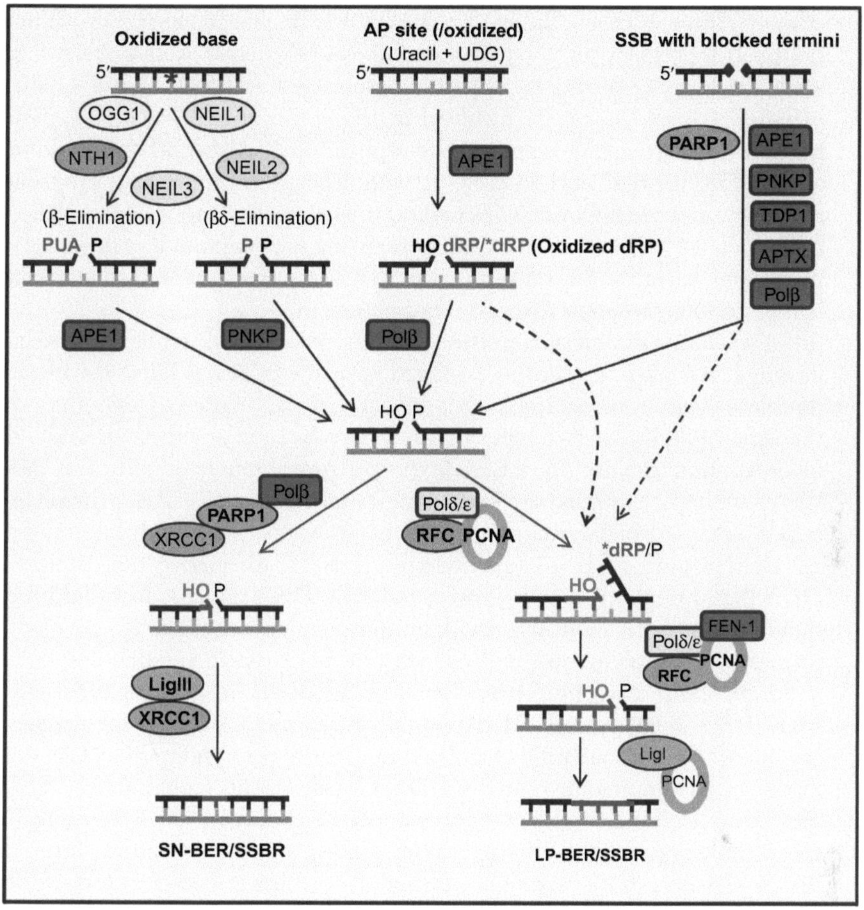

FIG. 1. Schematic representation of mammalian BER/SSBR pathways for repair of oxidized bases, AP sites, and SSBs. The BER and SSBR pathways converge at the end-processing step. The gap-filling step may involve synthesis of 1 nt (SN-BER/SSBR) by Polβ or 2–8 nt (LP-BER/SSBR) by Polδ/ε or Polβ in collaboration with FEN-1. Other details are in the text. (See Color Insert.)

(βδ elimination product), respectively, generating 3′OH.[11] Unlike its *E. coli* homolog Xth, mammalian APE1 has extremely weak 3' phosphatase activity.[11] Thus, while APE1 is required for OGG1/NTH1-initiated BER, PNKP participates in NEIL-initiated BER. Furthermore, 5′dRP residue (APE1's product of AP site cleavage) is removed by DNA polymerase β (Polβ) in mammalian cells, which possesses 5′dRP lyase activity.[15] However, Polβ is ineffective if the 5′dRP residue is oxidized, as often occurs during oxidative stress.[16] In such cases, the oxidized 5′dRP is displaced during gap-filling synthesis together with 2–8 additional nucleotides (nts) as an ssDNA flap, which is then removed by flap endonuclease I (FEN-1). FEN-1 is an essential enzyme normally involved in the removal of Okazaki fragment primers during lagging strand DNA replication.[17] A second flap endonuclease, DNA2, that acts in concert with FEN-1 in processing long flap structures has recently been discovered in both nuclei and mitochondria of human cells.[18,19] Several other end-processing enzymes are also involved in processing various blocked termini at SSBs directly generated in mammalian genomes; these are discussed in Section I.C.

(iii) *Gap-filling after lesion excision*. Lesion excision and termini processing usually leave a 1-nt gap at the damaged base site, which is filled in with the template strand-guided nucleotide by a DNA polymerase. Depending on the repair patch size, two types of BER have been characterized: single nt incorporation repair (SN-BER) involving replacement of only the base lesion with the parent base, and long-patch repair (LP-BER) involving repair patch size of 2–8 nt upstream of the lesion site. As already mentioned, LP-BER requires FEN-1 (and possibly DNA2) to remove the displaced DNA flap. Mammalian cells express multiple DNA polymerases that function in BER/SSBR pathways. Polβ, ubiquitous in mammalian tissues, is the primary BER polymerase and carries out SN-BER in nondividing cells, although Polβ may also participate in LP-BER in coordination with FEN-1.[20,21] However, LP-BER generally utilizes DNA replication machinery including replicative DNA pols, Polδ/ε, the sliding clamp PCNA, clamp loader replication factor-C (RF-C), DNA ligase I (LigI), and FEN-1.[16,22,23] The complex issue of choice between SN-BER and LP-BER is yet to be completely understood although initial studies suggested that the nature of the 5′-phosphoribose terminus (normal vs. oxidized) would be a deciding factor.[24] However, involvement of DNA replication proteins with LP-BER strongly suggests that LP-BER could be the preferred pathway during DNA replication, irrespective of the 5′ terminal group.

(iv) *Nick sealing by DNA ligases*. The sealing of the nick with 3′OH and 5′P termini by a DNA ligase to restore genomic integrity is the final step in BER/SSBR. DNA ligase IIIα (LigIIIα) and Lig I are the major DNA ligases (in addition to Ligase IV, which is involved in DSB repair) in human cells; the former is generally associated with SN-BER and the latter with LP-BER although the distinction may not be absolute.[25]

B. DG: The BER-initiating Enzyme

BER is unique among excision repair pathways in that the damaged bases are recognized by distinct damaged base-specific DGs. DGs are relatively small (~30–50 kDa) monomeric proteins that do not require cofactors for their activity. Although NTH1 contains an Fe–S cluster, it serves only as an architectural element to position a loop containing positively charged residues near the phosphodiester backbone of a target DNA molecule.[26,27] Nei family proteins in *E. coli* and mammals (except NEIL1) contain Zn finger motifs, which are involved in DNA binding.[28] Because DGs remove base lesions that usually cause minor, or no, distortion in the DNA duplex, damaged base sensing is a major challenge for the DG, particularly for lesions such as U with subtle modifications.[8] The difficulty is more severe in mammalian genome because the lesion is located in condensed chromatin. It should be mentioned that although DGs are involved in the recognition of base lesions owing to their affinity for a group of modified bases, other signaling factors may also be involved that have not been extensively addressed.

The mechanism of base lesion excision involves extrahelical flipping of the damaged nucleotide into the DG's recognition pocket.[29,30] All DGs studied so far bind to the minor groove of DNA, kinking it at the site of damage, and flip the lesion nucleotide out of the major groove of DNA.[29,30] Thus, only those lesions that could be accommodated in the binding pocket after nucleotide flipping to provide the necessary contacts and orientation for their excision are removed by the DGs. ROS produce more than 20 major oxidized base lesions that are repaired by only four (or five) DGs in human cells; thus, each DG acts on subsets of base lesions.[4,8,31] It is likely that the plasticity of the catalytic pockets of DGs allows an induced fit of diverse substrates. It appears that DGs invariably have low turnover to compensate for their promiscuity. The properties of oxidized base-specific DGs and their preferred substrates are listed in Table II.

C. SSBR: A DG-independent Variant of BER

Repair of ROS-induced SSBs shares the last three steps of the BER pathway, namely, end cleaning, gap-filling, and nick sealing, although SSBR could involve additional end-processing enzymes to remove various termini produced by ROS.

(i) *Diverse end-processing for SSBR.* The end-processing of SSBs has recently been shown to be more versatile than previously suggested. The most common block at ROS-induced SSB is 3′phosphoglycolate (or 3′phosphoglycolaldehyde), which is removed by APE1.[36,37] Tyrosylphosphodiesterase 1 (TDP1), another 3′ end-processing enzyme, cleaves Top1 (Tyr)-cross-linked to 3′P at the strand break generated by abortive topoisomerase 1 (Top1) reaction. TDP1 also processes 3′phosphoglycolate at DSBs.[38] The resulting 3′P is subsequently removed by PNKP,[39–42] which also phosphorylates the 5′OH generated at an SSB. A unique 5′ blocking group is formed as intermediates during abortive DNA ligation, namely, adenylate linked to the 5′P terminus at an SSB via a 5′•5′ pyrophosphate bond. Aprataxin releases 5′AMP to restore the 5′P terminus.[43,44] Some

TABLE II

OXIDIZED DNA BASE-SPECIFIC DNA GLYCOSYLASES IN MAMMALIAN CELLS AND THEIR PROPERTIES

Enzyme/property	OGG1	NTH1	NEIL1	NEIL2	NEIL3
Type	Nth type		Nei type		
AP lyase	β Elimination	β Elimination	βδ Elimination	βδ Elimination	β (weak βδ) elimination
Product	3′dRP	3′dRP	3′P	3′P	3′dRP/3′P
Downstream Enzyme	APE1	APE1	PNKP	PNKP	APE1/PNKP
DNA substrate	Only duplex	Only duplex	Duplex, ss, bubble	Duplex, ss, bubble	Duplex, ss, bubble
Conserved motif	HhH	HhH	H2TH	H2TH	H2TH
Presence of Zn finger	None	None	Zn less finger	Zn finger	Zn finger (2)
Fe–S cluster	None	Fe–S cluster	None	None	None
Catelytic residue	Lys249	Lys212	Pro1	Pro1	Val1
Cell cycle-dependent expression	None	None	S-phase specific	None	Unknown

Compiled from Refs. 8,9,11–13,28,32–35.

blocked 3′ termini could also be removed by ERCC1/XPF nuclease, a functional homolog of the yeast Rad1/Rad10 complex proposed for this role in the budding yeast.[45] The various end-processing enzymes involved in BER/SSBR and their substrates are listed in Table III.

Unlike repair of simple nicks, which may be generated by nucleases or caused by replication inhibition and usually contain 3′OH and 5′P, repair of most ROS-induced SSBs involves gap-filling repair synthesis. Polβ, which could carry out LP-BER in collaboration with FEN-1, likely provides the major DNA polymerase activity for SSBR, particularly in nonreplicating cells.[20,21]

(ii) *Role of XRCC1 and PARP-1 in BER/SSBR.* XRCC1 and PARPs are other key proteins with indirect roles in BER/SSBR. While XRCC1 acts as a scaffold to recruit BER proteins for excision or strand break repair, PARP acts as an SSB sensor protein.[46] XRCC1 physically interacts with NEILs, PNKP, APE1, and other BER proteins Polβ and LigIIIα, implicating it in the first step of BER.[11,32] Moreover, persistence of 5′-OH and 3′-P termini at SSBs in XRCC1-deficient cells underscores the latter's role in end-processing.[47] PARPs, expressed in mammalian cells but absent in *E. coli* or yeast, constitute a superfamily with regulatory functions in various cellular processes including development.[48,49] Only PARP-1, -2, and -3 are involved in DNA repair.[50] PARP-1 and -2 are activated by SSBs when they transfer ADP-ribose moiety from NAD to a variety of proteins including themselves. PARP-2 may serve as a backup for PARP-1 because inactivation of both *PARP 1* and *2* genes in mice is lethal, while these are individually dispensable for viability.[51,52] These proteins play structural and regulatory roles in SSB repair by acting as sensors and by recruiting other repair proteins to the strand break site.[53] However, their direct involvement in damage processing during BER has not been demonstrated so far. PARP-3 has recently been implicated in DNA DSB repair.[54]

TABLE III

DNA End-Processing Enzymes Involved in BER/SSBR

Enzyme	Substrate	Product
APE1	3′dRP (3′PUA)	3′OH
	3′phosphoglycolate	
	3′phosphoglycolaldehyde	
PNKP	3′P	3′OH
	5′OH	5′P
Polβ	5′dRP	5′P
TDP1	Top (Tyr) linked 3′P	3′P
Aprataxin	5′adenylate•5′P	5′P

Their substrates and products are indicated.

D. BER/SSBR in Mammalian Mitochondria

The mitochondria are the major site of ROS generation in aerobic organisms and thus their genomes are continuously exposed to oxidative damage. Furthermore, lack of protective histones makes the mitochondrial DNA more susceptible to oxidative damage than the nuclear genome. Repair of oxidized bases and SSBs via both SN-BER and LP-BER has been established in mammalian mitochondria.[55–57] The early mitochondrial BER proteins, all encoded by nuclear genes, are usually isoforms of the nuclear proteins, generated due to alternate RNA splicing or proteolytic cleavage. Among the DGs, OGG1, NTH1, and NEIL1 have been shown to localize in the mitochondria.[58–60] We have shown that an N-terminal truncation product of APE1 is present in mammalian mitochondria.[61] Recent studies have shown the presence of TDP1 and aprataxin in mitochondria.[62,63] In contrast to the sharing of early BER proteins between nucleus and mitochondria, DNA polymerase γ (Polγ), the only DNA polymerase in mammalian mitochondria, is required for both mt genome replication and repair.[64] A splice variant of nuclear LigIIIα functions similarly in both mt genome replication and repair.[25] Although mitochondrial BER/SSBR was initially thought to be a simple process involving only a few essential enzymes, recent studies have demonstrated the presence of several additional nuclear BER/SSBR components in mitochondria.[65]

II. Complexity and Sub-pathways of BER/SSBR

Complete nuclear BER/SSBR that requires only four or five enzymes could be demonstrated *in vitro*. However, recent studies have revealed that BER is far more complex, involving a network of distinct cell cycle dependent as well as genome region-specific repair sub-pathways and could also involve several non-BER proteins.[66,67]

A. Preferred and Backup Sub-pathways

The overlapping substrate specificity of DGs and the lack of strong phenotypes associated with their individual deficiency suggest that these enzymes have preferential and backup functions. Mouse mutants individually lacking OGG1, NTH1, NEIL1, or MYH and the cells derived thereof are viable, while combined deficiency of two DGs (e.g., NEIL1 and NTH1 or OGG1 and MYH) strongly increases cancer susceptibility.[68–73] We postulate that these enzymes with the unusual plasticity of their catalytic pockets can carry out excision of diverse damaged bases in DNA, which is consistent with their catalytic inefficiency. However, the choice of the DG-initiated BER sub-pathway for the same lesion may also depend on the cellular state.

Our initial discovery that OGG1 and NTH1 are active only with the duplex DNA while NEILs are more active with ss DNA substrates that mimic replicating or transcribing sequences led us to hypothesize that NEILs preferentially function in repair during DNA replication and/or transcription.[66] We subsequently characterized the role of NEIL2 in preferential repair of transcribed genes, where NEIL2 functionally interacts with RNA polymerase II.[74] Unlike NEIL2, NEIL1 is induced in S-phase cells and functionally interacts with DNA replication proteins including PCNA, replication protein A (RPA) FEN-1, and Werner's helicase (WRN), which suggests that NEIL1 is preferentially involved in replication-coordinated BER.[33,34,75–78]

B. Role of Noncanonical Proteins

Several noncanonical proteins have been shown to be involved in BER, adding another dimension to the BER complexity. However, their precise *in vivo* functions in BER/SSBR are yet to be unraveled. We showed that NEIL2 interacts with YB-1, a Y-box binding protein, and it was suggested that YB-1 may be required for the fine-tuning of repair.[79] NTH1 and APE1 were also shown to interact with, and be stimulated by, YB-1.[80,81] Recently, Banerjee *et al.* showed association of NEIL2 with hnRNP-U, an RNA-binding protein, and its role in transcription-coupled BER.[74] HMGB1 has been implicated in SSBR involving Polβ.[82] Tumor suppressor protein p53 interacts with APE1 and Polβ, and stimulates BER *in vitro*.[83] P53 was also shown to play a role in UV radiation-induced DNA damage repair.[84] Jaiswal *et al.* showed inhibition of both SN-BER and LP-BER or LP-BER alone by the adenomatous polyposis coli (APC) and by cyclin-dependent kinase inhibitor p21 genes, respectively.[85,86] The growing list of noncanonical proteins underscores the paradigm that BER/SSBR *in vivo* is far more complex than *in vitro* repair demonstrated with minimal components.

C. Repair Interactome: Preformed Complexes Versus Sequential Recruitment

The prevailing view of BER is that it comprises a sequence of steps with individual repair enzymes carrying out reactions independently of one another. This concept was initially proposed based on X-ray crystallographic studies, as a "hand-off" or "passing the baton" process, wherein the repair product of each enzyme in the BER pathway is handed over to the next enzyme, primarily based on differential bending of DNA in each intermediate step.[35,87,88] However, our recent studies as well as others' have shown that early BER/SSBR enzymes (e.g., NEIL1, APE1) stably interact with most downstream repair components including DNA ligase, via their common interacting domain.[11,32,77,78,89] Furthermore, NEIL1 and NEIL2 immunoprecipitates (IP) from human cells contain the SN-BER proteins PNKP, Polβ, LigIIIα, and XRCC1, as well as LP-BER-specific DNA replication proteins including PCNA and FEN-1,[11,32,76,77] with which the NEILs also interact in a pairwise fashion in the absence of DNA. Direct interaction of NEILs with LigIIIα, the last enzyme in SN-BER, indicates that the repair is regulated by the initiating DG, which appears to act as a hub protein. This has led us to propose a new BER paradigm in which collaboration of multiple proteins in a coordinated fashion involving dynamic protein–protein interactions enhances repair efficiency. Although the *in vivo* role of sequential engagement of individual BER proteins versus coordinated action of a preformed BER complex is not determined yet, we propose that the preformed BER complexes repair endogenous base lesions, while repair via hand-off mechanism by sequential recruitment could occur for induced DNA damage. Further characterization of the dynamics of such preformed "BERosomes" is required to unravel the precise repair processes.

III. Nonconserved Terminal Extensions in Mammalian Early BER Proteins

Mammalian DGs possess unique structural features absent in their homologs in lower organisms because of the presence of a nonconserved extension at the N or C terminus (reviewed in Ref. 90). These might have been acquired during evolution via terminal fusion of a non-BER gene (Fig. 2). The sequence alignment of human (h) NTH1 with its prototype Nth in *E. coli* and NTHs in other lower organisms clearly defines hNTH1's unique N-terminal extension. HNTH1 was also shown to have reduced activity compared to *E. coli* Nth prototype, which appears to be due to the inhibitory role of the N-terminal tail.[27,91] This extension reduces the rate of product release without affecting base excision or AP lyase activities.[92] Similarly, the crystal structure of the catalytically active deletion construct of hNEIL1 (lacking 56 C-terminal residues) indicated that hNEIL1 has C-terminal extension (~100 residues), which is absent in the *E. coli* prototype Nei.[90,93] Furthermore, PONDR, the predictor of naturally disordered regions in proteins, and other modeling studies showed that the terminal extensions in mammalian DGs are mostly disordered. A sequence comparison of hMYH and its prototype MutY in *E. coli* shows the presence of N-terminal disordered segment in the former.[90,94] Similar disordered terminal sequences may also be present in other human DGs, for example, UNG2 and TDG. In the case of APE, the nonconserved N-terminal segment (~65 residues) in hAPE1, absent in the *E. coli* prototype Xth, appears to be mostly disordered. Although the unfolded sequence generally exists at the N or C terminus, this could also exist internally as in hNEIL2, where it may serve as a linker of two domains.[90]

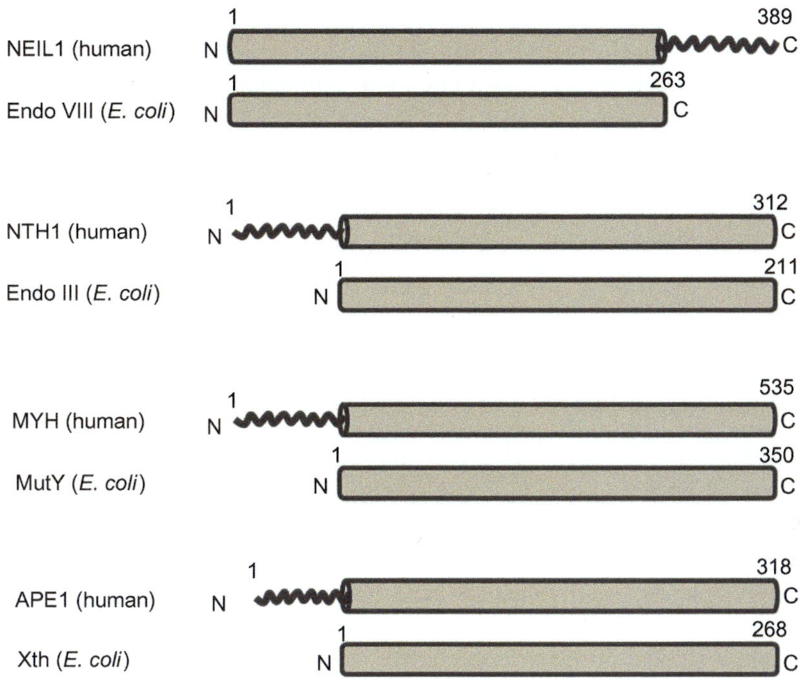

FIG. 2. Disordered terminal extensions in human (and other mammalian) early BER/SSBR proteins that are absent in their *E. coli* prototypes (*not drawn to scale*). In many cases, disordered segments were deleted for X-ray crystallographic structure analysis, which are consistent with PONDR prediction.[8,90] (For color version of this figure, the reader is referred to the online version of this chapter.)

The size range of disordered extensions in early BER proteins is 50–100 residues, with few exceptions, for example, hOGG1 and human Polβ, which have short ($\sim$10 residues) disordered segments at both termini.[90]

A. Functions of Disordered Terminal Extensions

The nonconserved, mostly disordered terminal peptide segments of early BER/SSBR proteins in mammals implicate these in important functions including damage sensing, protein–protein interactions, repair regulation via posttranslational modifications, and nuclear localization signal (NLS).[8,90] Furthermore, the disordered regions provide an opportunity for alternative splicing without perturbing the structured regions.[95] Disorder also provides size advantage in a polypeptide by providing a common interface for multiprotein binding and sites of covalent modifications. Thus, disorder may help higher organisms limit protein size and reduce intracellular crowding.[96]

(i) *Protein–protein interactions*. As mentioned earlier, the "hub" proteins such as NEIL1 or APE1 with multiple interacting partners (>10) utilize the disordered segment as a common interaction interface.[97] The "hub" protein complexes have been shown to be widely present in higher eukaryotes, invariably via interaction with disordered regions.[98] The crucial role of intrinsic disorder in hub proteins has been reviewed elsewhere.[99,100] Although it is intriguing how NEIL1 or APE1 could simultaneously bind multiple proteins with high affinity and specificity via a small common interaction interface, this phenomenon is quite common for mammalian "hub" proteins with disordered structures.[8,90] Further, disorder-mediated interactions may confer advantages over order-mediated interactions because rapid interconversion among diverse conformers allows for formation of dynamic complexes.[101] In addition, disorder-mediated interactions have steric advantages as they provide a large surface area for binding interface for wrapping around partners, resulting in stronger specificity.[96]

(ii) *Subcellular localization*. Organelle localization signals including NLS and mitochondrial transport signal (MTS) are typically contained in short disordered segments (generally <20 residues). The classical NLS consists of seven basic residues and the bipartite NLS has two strings of basic residues separated by a short intervening sequence.[102] Almost all NLS regions with mostly basic residues are disordered.[103] We mapped the NLS of hAPE1 to the disordered N-terminal 20 residues, the deletion of which markedly diminishes its translocation to the nucleus.[104] Our preliminary studies of GFP-fusion polypeptide of truncated NEIL1 suggest the presence of putative NLS at the disordered C-terminal region (Hegde ML and Mitra S, unpublished observation). Similarly, the disordered N-terminal tail in hNTH1, UNG2, and TDG contains putative NLS and MTS.[91,105,106] Taken together, these studies show that subcellular distribution of many human BER proteins is mediated through signals localized in their disordered regions.

(iii) *Target DNA scanning*. The ability of early BER/SSBR proteins to locate and then bind to substrate lesions in a large pool of DNA should strongly impact their repair efficiency. Such target DNA search could be achieved via facilitated diffusion comprising four mechanisms, namely, one-dimensional (1D) sliding, hopping, 3D search, and intersegmental transfer. An efficient search mechanism involves a combination of these different modes.[107] Recent studies have shown that the most efficient and rapid scanning of the DNA for the target site involves 80% hopping and intersegmental transfer and 20% sliding by the DNA-binding proteins, which invariably contain a disordered terminal extension or a disordered linker for multidomain or multi-subunit proteins.[108–110] Furthermore, nearly 70% of DNA-binding proteins have such disordered tails compared to about 25% for non-DNA-binding proteins.[109,111] Another unique characteristic of the disordered tails in DNA-binding proteins is clustering of positively charged residues in the distal region, which were shown to be important for the scanning. Mutating such residues in HOXD9 and NK-2 markedly decreases the scanning efficiency.[109,112] Similar results were obtained when the N-terminal segment in these proteins were deleted, suggesting that the initial scanning is mediated by a nonspecific, mostly electrostatic transient DNA binding via the basic disordered segment, which is followed by target DNA sequence binding by the active site.[109,113]

In light of these studies, we examined hNEIL1's C terminus, which possesses most of the characteristics required for DNA scanning, including the presence of clustered basic residues. Our recent biochemical studies using C-terminal deletion mutants of NEIL1 showed that the C terminus is important for NEIL1's substrate scanning and efficiency of damage recognition, via its nonspecific DNA binding (Hegde ML and Mitra S, unpublished data). Although limited information is available on the role of disordered regions in other early BER proteins, we predict a similar situation for these proteins as observed for NEIL1 or APE1.

IV. Posttranslational Modifications in Early BER Proteins

Posttranslational modifications of proteins, including acetylation, phosphorylation, ubiquitylation, ADP-ribosylation, sumoylation, and methylation, play a critical role in diverse cellular processes including DNA repair.[114] Such covalent modifications may have multiple physiological effects on these proteins, including stability, interaction with DNA or other proteins, organelle targeting, and enzymatic activity.[115] Furthermore, the modification sites are often localized in disordered regions, for example, in the N-terminal segment in hAPE1,[116,117] N and C termini of p53,[118] and the C-terminal region in hNEIL1 (Bhakat KK, Hegde ML and Mitra S, unpublished data).

A. Acetylation and Phosphorylation Modulate Repair Activity

The functions of DNA repair proteins are generally regulated via their acetylation or phosphorylation. These modifications could also regulate protein stability, interaction, and intracellular distribution (reviewed in Ref. 115). Our laboratories and others identified and characterized acetylation of hAPE1 at Lys6 or Lys7, Lys25, Lys27, and Lys31,[116,119] Lys 6 and 7 acetylation strongly affects APE1's transcriptional regulatory functions.[116] APE1 stably interacts with Y-box-binding protein 1 (YB-1) and acetylation further enhances its binding with YB-1 both *in vivo* and *in vitro*,[80] leading to the activation of the multidrug resistance (*MDR1*) gene. Thus, it appears likely that acetylation-mediated conformational change in the disordered N-terminal segment (~40 residues), which is dispensable for APE1's endonuclease activity, modulates protein–protein interactions. The physiological importance of APE1 acetylation became more evident with our findings that the levels of acetylated APE1 are increased in response to a variety of cellular stresses, or changes in intracellular Ca^{2+}, or bacterial infection to gastric epithelial cells.[120,121] We also showed that hNEIL2 is acetylated at Lys49 and Lys153 both *in vitro* and in cells.[122] Acetylation of Lys49 located in the disordered region inactivated NEIL2's base excision and AP lyase activity while acetylation of Lys150 had no effect. We thus proposed that acetylation of Lys49 could act as a regulatory switch in NEIL2.[122] TDG is acetylated at Lys70, 94, 95, and 98 in the N-terminal region, within the disordered segment (~100 residues).[123] Acetylation of TDG by CBP/p300 indirectly deregulates TDG-coupled repair by releasing CBP/p300 from the DNA-bound complex, leading to reduced interaction with APE1 and suppressing APE1-dependent repair, and could contribute to genomic instability,[123]

Acetylation of hOGG1 occurs at Lys338 and Lys341 within its short disordered C terminus, and the modification increases OGG1's turnover by reducing its affinity for the product AP site.[124] Moreover, oxidative stress increases the level of acetylated OGG1, most likely as a result of ROS-induced activation of p300. We have speculated that OGG1 acetylation provides a mechanism for rapid cellular response without requiring its de novo synthesis when enhanced repair is promptly needed for handling increased lesion load in cell genomes after exposure to oxidative stress. This is further supported by our observation that the repair of 8-oxoG is correlated with the level of acetylated OGG1.[124]

APE1 was also shown to be phosphorylated *in vitro* and in cells by protein kinase C.[117,125,126] APE1 can also be phosphorylated *in vitro* by casein kinase I and II (CKI and CKII) at various sites[126,127]; CKII-mediated phosphorylation abolished DNA repair activity *in vitro*, while phosphorylation by CKI or PKC had no such effect.[126] APE1 phosphorylation by CKII also enhances transactivation of the AP-1 transcription factor.[128] XRCC1 in human cells is phosphorylated by CKII and is required for its stability and efficient BER/SSBR.[129] More recently, Huang *et al.* found that APE1 was phosphorylated at its Thr233 residue by threonine kinase CDK5, a paralog of CDK2/4.[130] CDK5 is expressed in neurons and is speculated to be involved in cell death triggered by uncontrolled DNA replication. A high level of phosphorylated Thr233 was observed in brain tissues from Parkinson's and Alzheimer's patients,[130] and inactivation and degradation of APE1 by phosphorylation and the subsequent ubiquitylation may profoundly affect the severity of the diseases. Furthermore, phosphorylation of the DGs, UDG,[131,132] MYH,[133] and OGG1[134] has been reported to modulate their repair function. Thus acetylation and phosphorylation of early BER/SSBR proteins play a key role in regulating their repair functions and modulating their interactions with other repair proteins.

B. Ubiquitylation and BER Protein Turnover

Ubiquitylation has been demonstrated to play a role in regulating the steady-state level of BER/SSBR enzymes in mammalian cells by directing their degradation and turnover.[135] Ubiquitin is a highly conserved small protein (~8.5 kDa) whose covalent addition to a target protein as a polymer by ubiquitin ligases promotes the latter's degradation by the 26S proteasome.[136] However, ubiquitylated proteins may also be recycled by the removal of ubiquitin by deubiquitylases (DUBs) or ubiquitin-specific proteases.[137,138] Proteins may be ubiquitylated with a single ubiquitin moiety (monoubiquitylation) or with multiple ubiquitin molecules (polyubiquitylation). Polβ is ubiquitylated by the E3 ubiquitin ligase Mule at Lys 41, 61, and 81, and substitution of these Lys with Arg increased Polβ stability.[135] Furthermore, Mule-depleted cells contain higher levels of Polβ and increased DNA repair activity. It was recently demonstrated that APE1's ubiquitylation, which is enhanced by phosphorylation at Thr233,[139] could act as a signal for regulating the stability, subcellular localization, and gene regulatory functions of APE1.[117,137] The mammalian nuclear UDG, UNG2, is also ubiquitylated in cell. Blocking its nuclear export using a nuclear export inhibitor prevents its ubiquitylation and subsequent degradation.[140] This suggests that translocation of proteins from the nucleus to the cytoplasm is required for their turnover. In a distinct situation, PCNA is both mono- and polyubiquitylated, thus mediating translesion DNA synthesis and error-free lesion bypass, respectively.[141,142]

V. BER/SSBR Deficiency in Human Diseases

Maintaining genomic integrity through DNA repair is essential for the functioning and survival of an organism. Hence, defective or deficient repair leading to accumulation of unrepaired genome damage has been associated with a range of human disorders including cancer susceptibility, aging, and various neurodegenerative diseases. While compelling evidence exists linking deficiency in SSBR proteins to human diseases,[143] there is some controversy regarding a similar association with the early BER proteins, presumably due to the backup role of BER sub-pathways as discussed earlier.

A. Cancer

Varying degrees of cancer proneness have also been demonstrated with BER/SSBR defects. A connection between a repair defect and cancer risk is established for the hMYH.[144,145] The genetic locus of OGG1 (3p26.2) is frequently lost in various human cancers, including lung and renal cell cancer.[146,147] Polymorphic variants of OGG1 (S326C and R46Q) have also been identified within the population that demonstrate mild reductions in enzymatic activity.[148,149] However, there are conflicting epidemiological reports regarding the association of these variants with cancer susceptibility.[150] While an increase in the levels of 8-oxoG and mutation frequency was evident in OGG1-null mice, no significant enhancement of spontaneous tumor incidence was observed.[68,151,152] Interestingly, the NEIL1-knockout mouse was found to exhibit clinical features distinctly similar to the human metabolic syndrome, with extensive fat deposits in various tissues.[73] As already mentioned, because of the backup functions of mammalian DGs, mice lacking a single DG do not show a significant phenotype, and only when two or more DGs are simultaneously eliminated do symptoms appear. For example, when both MYH and OGG1 were deleted, the resulting double-knockout animal (unlike the single knockouts) exhibited an increased predisposition to spontaneous tumorigenesis, particularly lung and ovarian tumors, and lymphomas.[153] It is likely that single DG-null animals possess compensatory mechanisms for base damage tolerance that prevent disease manifestation,[154] which could explain why the association of BER defects with cancer proneness appears less compelling. Furthermore, specific environmental factor(s) may foster the observed disease susceptibility in DG-deficient animals/cells.[155]

In contrast to the weak association of DGs with cancer risk, deficiencies in downstream BER/SSBR proteins, for example, Polβ, APE1, or XRCC1, as well as FEN1 or a DNA ligase (I or III), lead to embryonic lethality in mice, suggesting an absolute requirement for the BER process during embryonic development.[156–159] Furthermore, inactivation of Polβ mutations has been strongly associated with cancer. Genetic analysis revealed that an astonishing ~30% of human tumors express Polβ variants,[160] many of which (e.g., K289M and I260M) exhibit reduced polymerase fidelity.[161] APE1 and XRCC1 variants have also been found in numerous epidemiology studies to associate with specific human cancers[162,163]; however, definitive conclusions could not be drawn because of the small sample sizes.[164,165]

B. Neurodegenerative Diseases

More than 200 neurological disorders with diverse etiologies and genetic characteristics have been reported so far in humans, many of which have been linked to inherited or acquired defects in one of the DNA repair pathways, leading to imbalance in damage and repair capacity.[166–169] This is consistent with accumulation of oxidative genome damage in the neurons for the majority of such diseases including Alzheimer's disease (AD), Parkinson's disease (PD), Huntington' disease, amyotrophic lateral sclerosis (ALS), and various ataxias. Mutations or altered expression of BER (e.g., OGG1, XRCC1[170–172]) and SSBR (e.g., TDP1, aprataxin, PNKP[39,43,173,174]) proteins has been observed in humans predisposed to various hereditary neurodegenerative diseases. Association between the XRCC1 Arg399Gln polymorphism and ALS risk was reported.[175] A common Ser326Cys polymorphism in OGG1 with reduced enzyme activity and increased risk of cancer was shown to be weakly associated with HD and ALS,[176] but not with AD or PD.[171,177,178] Asp148Glu polymorphism in APE1 with lower repair activity, was linked to increased ALS risk.[179] Further, defects in end-processing proteins aprataxin and TDP1 have been shown to be associated with ataxia with oculomotor apraxia1 and spinocerebellar ataxia with axonal neuropathy, respectively.[173,180,181] A recent study linked mutations in PNKP to autosomal-recessive disease characterized by severe neurological abnormalities including microcephaly, early-onset intractable seizures, and developmental delay (denoted MCSZ[174]).

Decreased repair capacity for oxidative genome damage was also observed in brain cell extracts in sporadic neurodegenerative diseases constituting more than 80% of most disease incidences, whose causes are not known. Weissman *et al.*[169] showed that the significant BER deficiencies in the brains of AD patients are due to limited DNA base damage processing by DGs and reduced repair synthesis by Polβ. We recently showed that transition metals iron (Fe) and copper (Cu) that invariably accumulate in neurons in these diseases significantly inhibit the activity of NEIL1 and NEIL2 at physiological levels in both neuronal cells and *in vitro*.[182,183] These metals affect both base excision and AP lyase activities of NEILs and inhibit NEIL1's interaction with downstream repair proteins including Polβ, and FEN-1, further inhibiting the overall repair. Inhibition of NEILs by Cu involved oxidation of cysteines as well as structural changes via direct binding. The lack of OGG1 inhibition under similar conditions suggests binding specificity of the NEILs and excludes metal ions' direct binding to DNA. These results showed for the first time that the Fe/Cu overload associated with neurodegenerative diseases could be a "double whammy" by increasing oxidative genome damage load and, at the same time, inhibiting its repair. Many other BER/SSBR proteins are also inhibited by transition metals. Whiteside *et al.* showed that Cd and Cu inhibit both phosphatase and kinase activities of PNKP with human cell extracts and recombinant protein.[184] Furthermore, elevated Fe levels cause reduction in FEN-1 and LigIII activities because of the interference of repair protein binding to their DNA substrates.[185]

Interestingly, curcumin, a natural spice component with both metal chelation and reducing activities, was able to reverse the metal-induced inhibition of NEILs both *in vitro* and in neuronal cells, suggesting its therapeutic potential.[182,183]

C. BER/SSBR as Cancer Therapeutic Targets: Are We at a Crossroad?

Cellular sensitivity to DNA-damaging agents, such as ionizing radiation and radiomimetic drugs, is generally believed to reflect their DNA repair competence. For example, high expression of APE1 in tumor tissues is linked to their resistance to radiation and chemotherapy.[186–188] Therefore, research targeting BER proteins, in particular APE1, for sensitizing cancer cells has been intensified in recent years.[86,189–191] Among BER/SSBR proteins, PARP-1 was shown to be an effective therapeutic target where PARP-1 inhibitors strongly enhanced susceptibility to radiation (IR)/chemotherapy of BRCA1-negative human breast cancer cells, which was proposed to be due to synthetic lethality.[192,193] Resistance of tumor cells to radiation and radiomimetic drugs results from efficient repair of induced DNA DSBs via homologous recombination (HR), which involves BRCA1. In the absence of HR, PARP-1 inhibitors were proposed to target SSBR, also induced by radiation and drugs.[194] However, we propose an alternative basis for drug sensitization by PARP inhibitors. Based on the underlying assumption that SSBs do not trigger cell death until they are converted into DSBs via DNA replication and that DSBs activate apoptotic signaling, we suggest that in the absence of HR, the DSBs that are produced as replication intermediates are repaired via the alternate end joining (Alt-EJ) pathway, which is distinct from classical nonhomologous end joining (NHEJ), by requiring SSBR proteins including PARP-1, LigIIIα, and XRCC1, but not Ku, DNA-PKcs, LigIV/XRCC4, etc.[195] Unraveling the detailed molecular mechanisms of BER/SSBR and their regulation has significant ramifications for future cancer therapy where not only PARP1 inhibitors but also other BER/SSBR proteins could be targeted to sensitize cancer cells, particularly when the surviving cells develop resistance to PARP inhibitors.

VI. Conclusions and Future Perspectives

Although BER/SSBR is universally conserved, the mammalian BER/SSBR has multiple sub-pathways, sometimes involving noncanonical proteins, commensurate with the structural complexity of the genome and its organization. Our recent studies have led us to propose a new paradigm for BER/SSBR whose key features are as follows: (a) The DG controls the BER sub-pathway by pairwise interaction with most downstream proteins via the disordered common interaction interface that is located in one of the polypeptide termini; (b) essentiality of common interface-mediated interaction for efficient in cell oxidized base repair is shown by its trans-dominant negative activity; (c) the DGs could be present in cell as megadalton complexes containing both repair and noncanonical proteins to direct repair via distinct sub-pathways; (d) the BER complex enhancement by oxidative stress could involve acetylation and other covalent modifications of DGs and other BER proteins. Contrary to the early perception of BER being the simplest among the DNA repair pathways, it is now clear that many profound questions are still unresolved including: (1) lesion scanning in chromatinized DNA, (2) repair of oxidized bases in heterochromatin versus euchromatin, (3) the role of the plethora of proteins involved in chromatin unfolding and remodeling and the impact of their posttranslational modifications. Future studies should also focus on a comprehensive characterization of the role of noncanonical proteins in the repair of endogenous versus induced oxidized bases, including chromatin-modifying enzymes involved in oxidized base repair. Understanding how DNA repair deficiency occurs and affects cellular functions could provide a rational therapeutic basis for ameliorating the genotoxic consequences in neurodegenerative diseases and for sensitizing cells for cancer therapy by blocking DNA repair.

Acknowledgments

The research in the authors' laboratory has been supported by USPHS grants R01 CA81063 (S. M.), R01 CA53791 (S. M.), P01CA92854 (S. M.), and R01 CA98664 (T. I.). This review with limited focus is not meant to provide a comprehensive coverage of the literature, and many appropriate references could not be included for which the authors apologize. We thank Drs. Tapas K. Hazra (Department of Internal Medicine) and Istvan Boldogh (Department of Microbiology and Immunology) at the University of Texas Medical Branch for careful reading of the manuscript.

References

1. Dawson TL, Gores GJ, Nieminen AL, Herman B, Lemasters JJ. Mitochondria as a source of reactive oxygen species during reductive stress in rat hepatocytes. *Am J Physiol* 1993;**264**: C961–7.
2. Griendling KK, Sorescu D, Ushio-Fukai M. NAD(P)H oxidase: role in cardiovascular biology and disease. *Circ Res* 2000;**86**:494–501.
3. Lindahl T. Instability and decay of the primary structure of DNA. *Nature* 1993;**362**:709–15.
4. Mitra S, Hegde ML, Theriot CA, Das A, Hegde PM, Hazra TK. Complexity in repair of oxidative genome damage and its regulation, In: *Proceedings of princess Takamatsu symposium, Tokyo, Japan*; 2009.

5. Di Noia JM, Neuberger MS. Molecular mechanisms of antibody somatic hypermutation. *Annu Rev Biochem* 2007;**76**:1–22.
6. Doseth B, Visnes T, Wallenius A, Ericsson I, Sarno A, Pettersen HS, et al. Uracil-DNA glycosylase in base excision repair and adaptive immunity: species differences between man and mouse. *J Biol Chem* 2011;**286**:16669–80.
7. Dizdaroglu M. Base-excision repair of oxidative DNA damage by DNA glycosylases. *Mutat Res* 2005;**591**:45–59.
8. Hegde ML, Hazra TK, Mitra S. Early steps in the DNA base excision/single-strand interruption repair pathway in mammalian cells. *Cell Res* 2008;**18**:27–47.
9. Mitra S, Izumi T, Boldogh I, Bhakat KK, Hill JW, Hazra TK. Choreography of oxidative damage repair in mammalian genomes. *Free Radic Biol Med* 2002;**33**:15–28.
10. Dodson ML, Lloyd RS. Mechanistic comparisons among base excision repair glycosylases. *Free Radic Biol Med* 2002;**32**:678–82.
11. Wiederhold L, Leppard JB, Kedar P, Karimi-Busheri F, Rasouli-Nia A, Weinfeld M, et al. AP endonuclease-independent DNA base excision repair in human cells. *Mol Cell* 2004;**15**:209–20.
12. Hildrestrand GA, Neurauter CG, Diep DB, Castellanos CG, Krauss S, Bjoras M, et al. Expression patterns of Neil3 during embryonic brain development and neoplasia. *BMC Neurosci* 2009;**10**:45.
13. Takao M, Oohata Y, Kitadokoro K, Kobayashi K, Iwai S, Yasui A, et al. Human Nei-like protein NEIL3 has AP lyase activity specific for single-stranded DNA and confers oxidative stress resistance in Escherichia coli mutant. *Genes cells* 2009;**14**:261–70.
14. Russo MT, De Luca G, Degan P, Bignami M. Different DNA repair strategies to combat the threat from 8-oxoguanine. *Mutat Res* 2007;**614**:69–76.
15. Sobol RW, Prasad R, Evenski A, Baker A, Yang XP, Horton JK, et al. The lyase activity of the DNA repair protein beta-polymerase protects from DNA-damage-induced cytotoxicity. *Nature* 2000;**405**:807–10.
16. Frosina G, Fortini P, Rossi O, Carrozzino F, Raspaglio G, Cox LS, et al. Two pathways for base excision repair in mammalian cells. *J Biol Chem* 1996;**271**:9573–8.
17. Liu Y, Kao HI, Bambara RA. Flap endonuclease 1: a central component of DNA metabolism. *Annu Rev Biochem* 2004;**73**:589–615.
18. Zheng L, Zhou M, Guo Z, Lu H, Qian L, Dai H, et al. Human DNA2 is a mitochondrial nuclease/helicase for efficient processing of DNA replication and repair intermediates. *Mol Cell* 2008;**32**:325–36.
19. Duxin JP, Dao B, Martinsson P, Rajala N, Guittat L, Campbell JL, et al. Human Dna2 is a nuclear and mitochondrial DNA maintenance protein. *Mol Cell Biol* 2009;**29**:4274–82.
20. Prasad R, Dianov GL, Bohr VA, Wilson SH. FEN1 stimulation of DNA polymerase beta mediates an excision step in mammalian long patch base excision repair. *J Biol Chem* 2000;**275**:4460–6.
21. Podlutsky AJ, Dianova II, Podust VN, Bohr VA, Dianov GL. Human DNA polymerase beta initiates DNA synthesis during long-patch repair of reduced AP sites in DNA. *EMBO J* 2001;**20**:1477–82.
22. Klungland A, Lindahl T. Second pathway for completion of human DNA base excision-repair: reconstitution with purified proteins and requirement for DNase IV (FEN1). *EMBO J* 1997;**16**:3341–8.
23. Matsumoto Y, Kim K, Hurwitz J, Gary R, Levin DS, Tomkinson AE, et al. Reconstitution of proliferating cell nuclear antigen-dependent repair of apurinic/apyrimidinic sites with purified human proteins. *J Biol Chem* 1999;**274**:33703–8.
24. Matsumoto Y, Kim K. Excision of deoxyribose phosphate residues by DNA polymerase beta during DNA repair. *Science* 1995;**269**:699–702.

25. Ellenberger T, Tomkinson AE. Eukaryotic DNA ligases: structural and functional insights. *Annu Rev Biochem* 2008;**77**:313–38.
26. Kuo CF, McRee DE, Fisher CL, O'Handley SF, Cunningham RP, Tainer JA. Atomic structure of the DNA repair [4Fe-4S] enzyme endonuclease III. *Science* 1992;**258**:434–40.
27. Thayer MM, Ahern H, Xing D, Cunningham RP, Tainer JA. Novel DNA binding motifs in the DNA repair enzyme endonuclease III crystal structure. *EMBO J* 1995;**14**:4108–20.
28. Das A, Rajagopalan L, Mathura VS, Rigby SJ, Mitra S, Hazra TK. Identification of a zinc finger domain in the human NEIL2 (Nei-like-2) protein. *J Biol Chem* 2004;**279**:47132–8.
29. Parker JB, Bianchet MA, Krosky DJ, Friedman JI, Amzel LM, Stivers JT. Enzymatic capture of an extrahelical thymine in the search for uracil in DNA. *Nature* 2007;**449**:433–7.
30. Slupphaug G, Mol CD, Kavli B, Arvai AS, Krokan HE, Tainer JA. A nucleotide-flipping mechanism from the structure of human uracil-DNA glycosylase bound to DNA. *Nature* 1996;**384**:87–92.
31. Krokan HE, Nilsen H, Skorpen F, Otterlei M, Slupphaug G. Base excision repair of DNA in mammalian cells. *FEBS Lett* 2000;**476**:73–7.
32. Das A, Wiederhold L, Leppard JB, Kedar P, Prasad R, Wang H, et al. NEIL2-initiated, APE-independent repair of oxidized bases in DNA: evidence for a repair complex in human cells. *DNA Repair (Amst)* 2006;**5**:1439–48.
33. Hazra TK, Izumi T, Boldogh I, Imhoff B, Kow YW, Jaruga P, et al. Identification and characterization of a human DNA glycosylase for repair of modified bases in oxidatively damaged DNA. *Proc Natl Acad Sci USA* 2002;**99**:3523–8.
34. Hazra TK, Kow YW, Hatahet Z, Imhoff B, Boldogh I, Mokkapati SK, et al. Identification and characterization of a novel human DNA glycosylase for repair of cytosine-derived lesions. *J Biol Chem* 2002;**277**:30417–20.
35. Izumi T, Wiederhold LR, Roy G, Roy R, Jaiswal A, Bhakat KK, et al. Mammalian DNA base excision repair proteins: their interactions and role in repair of oxidative DNA damage. *Toxicology* 2003;**193**:43–65.
36. Awada M, Dedon PC. Formation of the 1, N2-glyoxal adduct of deoxyguanosine by phosphoglycolaldehyde, a product of 3′-deoxyribose oxidation in DNA. *Chem Res Toxicol* 2001;**14**:1247–53.
37. Parsons JL, Dianova II, Dianov GL. APE1 is the major 3′-phosphoglycolate activity in human cell extracts. *Nucleic Acids Res* 2004;**32**:3531–6.
38. Zhou T, Akopiants K, Mohapatra S, Lin PS, Valerie K, Ramsden DA, et al. Tyrosyl-DNA phosphodiesterase and the repair of 3′-phosphoglycolate-terminated DNA double-strand breaks. *DNA Repair (Amst)* 2009;**8**:901–11.
39. El-Khamisy SF, Saifi GM, Weinfeld M, Johansson F, Helleday T, Lupski JR, et al. Defective DNA single-strand break repair in spinocerebellar ataxia with axonal neuropathy-1. *Nature* 2005;**434**:108–13.
40. Pouliot JJ, Yao KC, Robertson CA, Nash HA. Yeast gene for a Tyr-DNA phosphodiesterase that repairs topoisomerase I complexes. *Science* 1999;**286**:552–5.
41. Yang SW, Burgin Jr. AB, Huizenga BN, Robertson CA, Yao KC, Nash HA. A eukaryotic enzyme that can disjoin dead-end covalent complexes between DNA and type I topoisomerases. *Proc Natl Acad Sci USA* 1996;**93**:11534–9.
42. Weinfeld M, Mani RS, Abdou I, Aceytuno RD, Glover JN. Tidying up loose ends: the role of polynucleotide kinase/phosphatase in DNA strand break repair. *Trends Biochem Sci* 2011;**36**:262–71.
43. Ahel I, Rass U, El-Khamisy SF, Katyal S, Clements PM, McKinnon PJ, et al. The neurodegenerative disease protein aprataxin resolves abortive DNA ligation intermediates. *Nature* 2006;**443**:713–6.

44. Caldecott KW. Mammalian single-strand break repair: mechanisms and links with chromatin. *DNA Repair (Amst)* 2007;**6**:443–53.
45. Guzder SN, Torres-Ramos C, Johnson RE, Haracska L, Prakash L, Prakash S. Requirement of yeast Rad1-Rad10 nuclease for the removal of 3′-blocked termini from DNA strand breaks induced by reactive oxygen species. *Genes Dev* 2004;**18**:2283–91.
46. Masson M, Niedergang C, Schreiber V, Muller S, Menissier-de Murcia J, de Murcia G. XRCC1 is specifically associated with poly(ADP-ribose) polymerase and negatively regulates its activity following DNA damage. *Mol Cell Biol* 1998;**18**:3563–71.
47. Lan L, Nakajima S, Oohata Y, Takao M, Okano S, Masutani M, et al. In situ analysis of repair processes for oxidative DNA damage in mammalian cells. *Proc Natl Acad Sci USA* 2004;**101**: 13738–43.
48. Hassa PO, Hottiger MO. The diverse biological roles of mammalian PARPS, a small but powerful family of poly-ADP-ribose polymerases. *Front Biosci* 2008;**13**:3046–82.
49. Krishnakumar R, Kraus WL. The PARP side of the nucleus: molecular actions, physiological outcomes, and clinical targets. *Mol Cell* 2010;**39**:8–24.
50. Boehler C, Gauthier LR, Mortusewicz O, Biard DS, Saliou JM, Bresson A, et al. Poly(ADP-ribose) polymerase 3 (PARP3), a newcomer in cellular response to DNA damage and mitotic progression. *Proc Natl Acad Sci USA* 2011;**108**:2783–8.
51. Fisher AE, Hochegger H, Takeda S, Caldecott KW. Poly(ADP-ribose) polymerase 1 accelerates single-strand break repair in concert with poly(ADP-ribose) glycohydrolase. *Mol Cell Biol* 2007;**27**:5597–605.
52. Rouleau M, Patel A, Hendzel MJ, Kaufmann SH, Poirier GG. PARP inhibition: PARP1 and beyond. *Nat Rev Cancer* 2010;**10**:293–301.
53. Schreiber V, Dantzer F, Ame JC, de Murcia G. Poly(ADP-ribose): novel functions for an old molecule. *Nat Rev Mol Cell Biol* 2006;**7**:517–28.
54. Rulten SL, Fisher AE, Robert I, Zuma MC, Rouleau M, Ju L, et al. PARP-3 and APLF function together to accelerate nonhomologous end-joining. *Mol Cell* 2011;**41**:33–45.
55. Akbari M, Visnes T, Krokan HE, Otterlei M. Mitochondrial base excision repair of uracil and AP sites takes place by single-nucleotide insertion and long-patch DNA synthesis. *DNA Repair (Amst)* 2008;**7**:605–16.
56. Bohr VA. Repair of oxidative DNA damage in nuclear and mitochondrial DNA, and some changes with aging in mammalian cells. *Free Radic Biol Med* 2002;**32**:804–12.
57. Szczesny B, Tann AW, Longley MJ, Copeland WC, Mitra S. Long patch base excision repair in mammalian mitochondrial genomes. *J Biol Chem* 2008;**283**:26349–56.
58. Hashiguchi K, Stuart JA, de Souza-Pinto NC, Bohr VA. The C-terminal alphaO helix of human Ogg1 is essential for 8-oxoguanine DNA glycosylase activity: the mitochondrial beta-Ogg1 lacks this domain and does not have glycosylase activity. *Nucleic Acids Res* 2004;**32**:5596–608.
59. Hu J, de Souza-Pinto NC, Haraguchi K, Hogue BA, Jaruga P, Greenberg MM, et al. Repair of formamidopyrimidines in DNA involves different glycosylases: role of the OGG1, NTH1, and NEIL1 enzymes. *J Biol Chem* 2005;**280**:40544–51.
60. Karahalil B, de Souza-Pinto NC, Parsons JL, Elder RH, Bohr VA. Compromised incision of oxidized pyrimidines in liver mitochondria of mice deficient in NTH1 and OGG1 glycosylases. *J Biol Chem* 2003;**278**:33701–7.
61. Chattopadhyay R, Wiederhold L, Szczesny B, Boldogh I, Hazra TK, Izumi T, et al. Identification and characterization of mitochondrial abasic (AP)-endonuclease in mammalian cells. *Nucleic Acids Res* 2006;**34**:2067–76.
62. Das BB, Dexheimer TS, Maddali K, Pommier Y. Role of tyrosyl-DNA phosphodiesterase (TDP1) in mitochondria. *Proc Natl Acad Sci USA* 2010;**107**:19790–5.

63. Sykora P, Croteau DL, Bohr VA, Wilson 3rd. DM. Aprataxin localizes to mitochondria and preserves mitochondrial function. *Proc Natl Acad Sci USA* 2011;**108**:7437–42.
64. Longley MJ, Prasad R, Srivastava DK, Wilson SH, Copeland WC. Identification of 5′-deoxyribose phosphate lyase activity in human DNA polymerase gamma and its role in mitochondrial base excision repair in vitro. *Proc Natl Acad Sci USA* 1998;**95**:12244–8.
65. Liu P, Demple B. DNA repair in mammalian mitochondria: much more than we thought? *Environ Mol Mutagen* 2010;**51**:417–26.
66. Dou H, Mitra S, Hazra TK. Repair of oxidized bases in DNA bubble structures by human DNA glycosylases NEIL1 and NEIL2. *J Biol Chem* 2003;**278**:49679–84.
67. Kavli B, Sundheim O, Akbari M, Otterlei M, Nilsen H, Skorpen F, et al. hUNG2 is the major repair enzyme for removal of uracil from U:A matches, U:G mismatches, and U in single-stranded DNA, with hSMUG1 as a broad specificity backup. *J Biol Chem* 2002;**277**:39926–36.
68. Arai T, Kelly VP, Minowa O, Noda T, Nishimura S. The study using wild-type and Ogg1 knockout mice exposed to potassium bromate shows no tumor induction despite an extensive accumulation of 8-hydroxyguanine in kidney DNA. *Toxicology* 2006;**221**:179–86.
69. Chan MK, Ocampo-Hafalla MT, Vartanian V, Jaruga P, Kirkali G, Koenig KL, et al. Targeted deletion of the genes encoding NTH1 and NEIL1 DNA N-glycosylases reveals the existence of novel carcinogenic oxidative damage to DNA. *DNA Repair (Amst)* 2009;**8**:786–94.
70. Ocampo MT, Chaung W, Marenstein DR, Chan MK, Altamirano A, Basu AK, et al. Targeted deletion of mNth1 reveals a novel DNA repair enzyme activity. *Mol Cell Biol* 2002;**22**:6111–21.
71. Russo MT, De Luca G, Degan P, Parlanti E, Dogliotti E, Barnes DE, et al. Accumulation of the oxidative base lesion 8-hydroxyguanine in DNA of tumor-prone mice defective in both the Myh and Ogg1 DNA glycosylases. *Cancer Res* 2004;**64**:4411–4.
72. Takao M, Kanno S, Kobayashi K, Zhang QM, Yonei S, van der Horst GT, et al. A back-up glycosylase in Nth1 knock-out mice is a functional Nei (endonuclease VIII) homologue. *J Biol Chem* 2002;**277**:42205–13.
73. Vartanian V, Lowell B, Minko IG, Wood TG, Ceci JD, George S, et al. The metabolic syndrome resulting from a knockout of the NEIL1 DNA glycosylase. *Proc Natl Acad Sci USA* 2006;**103**:1864–9.
74. Banerjee D, Mandal SM, Das A, Hegde ML, Das S, Bhakat KK, et al. Preferential repair of oxidized base damage in the transcribed genes of mammalian cells. *J Biol Chem* 2011;**286**:6006–16.
75. Das A, Boldogh I, Lee JW, Harrigan JA, Hegde ML, Piotrowski J, et al. The human Werner syndrome protein stimulates repair of oxidative DNA base damage by the DNA glycosylase NEIL1. *J Biol Chem* 2007;**282**:26591–602.
76. Dou H, Theriot CA, Das A, Hegde ML, Matsumoto Y, Boldogh I, et al. Interaction of the human DNA glycosylase NEIL1 with proliferating cell nuclear antigen. The potential for replication-associated repair of oxidized bases in mammalian genomes. *J Biol Chem* 2008;**283**:3130–40.
77. Hegde ML, Theriot CA, Das A, Hegde PM, Guo Z, Gary RK, et al. Physical and functional interaction between human oxidized base-specific DNA glycosylase NEIL1 and flap endonuclease 1. *J Biol Chem* 2008;**283**:27028–37.
78. Theriot CA, Hegde ML, Hazra TK, Mitra S. RPA physically interacts with the human DNA glycosylase NEIL1 to regulate excision of oxidative DNA base damage in primer-template structures. *DNA Repair (Amst)* 2010;**9**:643–52.
79. Das S, Chattopadhyay R, Bhakat KK, Boldogh I, Kohno K, Prasad R, et al. Stimulation of NEIL2-mediated oxidized base excision repair via YB-1 interaction during oxidative stress. *J Biol Chem* 2007;**282**:28474–84.

80. Chattopadhyay R, Das S, Maiti AK, Boldogh I, Xie J, Hazra TK, et al. Regulatory role of human AP-endonuclease (APE1/Ref-1) in YB-1-mediated activation of the multidrug resistance gene MDR1. *Mol Cell Biol* 2008;**28**:7066–80.
81. Marenstein DR, Ocampo MT, Chan MK, Altamirano A, Basu AK, Boorstein RJ, et al. Stimulation of human endonuclease III by Y box-binding protein 1 (DNA-binding protein B). Interaction between a base excision repair enzyme and a transcription factor. *J Biol Chem* 2001;**276**:21242–9.
82. Liu Y, Prasad R, Wilson SH. HMGB1: roles in base excision repair and related function. *Biochim Biophys Acta* 2010;**1799**:119–30.
83. Zhou J, Ahn J, Wilson SH, Prives C. A role for p53 in base excision repair. *EMBO J* 2001;**20**:914–23.
84. Kassam SN, Rainbow AJ. UV-inducible base excision repair of oxidative damaged DNA in human cells. *Mutagenesis* 2009;**24**:75–83.
85. Jaiswal AS, Bloom LB, Narayan S. Long-patch base excision repair of apurinic/apyrimidinic site DNA is decreased in mouse embryonic fibroblast cell lines treated with plumbagin: involvement of cyclin-dependent kinase inhibitor p21Waf-1/Cip-1. *Oncogene* 2002;**21**:5912–22.
86. Jaiswal AS, Banerjee S, Panda H, Bulkin CD, Izumi T, Sarkar FH, et al. A novel inhibitor of DNA polymerase beta enhances the ability of temozolomide to impair the growth of colon cancer cells. *Mol Cancer* 2009;**7**:1973–83.
87. Wilson SH, Kunkel TA. Passing the baton in base excision repair. *Nat Struct Biol* 2000;**7**:176–8.
88. Mol CD, Izumi T, Mitra S, Tainer JA. DNA-bound structures and mutants reveal abasic DNA binding by APE1 and DNA repair coordination [corrected]. *Nature* 2000;**403**:451–6.
89. Otterlei M, Warbrick E, Nagelhus TA, Haug T, Slupphaug G, Akbari M, et al. Post-replicative base excision repair in replication foci. *EMBO J* 1999;**18**:3834–44.
90. Hegde ML, Hazra TK, Mitra S. Functions of disordered regions in mammalian early base excision repair proteins. *Cell Mol Life Sci* 2010;**67**:3573–87.
91. Ikeda S, Biswas T, Roy R, Izumi T, Boldogh I, Kurosky A, et al. Purification and characterization of human NTH1, a homolog of Escherichia coli endonuclease III. Direct identification of Lys-212 as the active nucleophilic residue. *J Biol Chem* 1998;**273**:21585–93.
92. Liu X, Roy R. Truncation of amino-terminal tail stimulates activity of human endonuclease III (hNTH1). *J Mol Biol* 2002;**321**:265–76.
93. Doublie S, Bandaru V, Bond JP, Wallace SS. The crystal structure of human endonuclease VIII-like 1 (NEIL1) reveals a zincless finger motif required for glycosylase activity. *Proc Natl Acad Sci USA* 2004;**101**:10284–9.
94. Lu AL, Li X, Gu Y, Wright PM, Chang DY. Repair of oxidative DNA damage: mechanisms and functions. *Cell Biochem Biophys* 2001;**35**:141–70.
95. Romero PR, Zaidi S, Fang YY, Uversky VN, Radivojac P, Oldfield CJ, et al. Alternative splicing in concert with protein intrinsic disorder enables increased functional diversity in multicellular organisms. *Proc Natl Acad Sci USA* 2006;**103**:8390–5.
96. Gunasekaran K, Tsai CJ, Kumar S, Zanuy D, Nussinov R. Extended disordered proteins: targeting function with less scaffold. *Trends Biochem Sci* 2003;**28**:81–5.
97. Haynes C, Oldfield CJ, Ji F, Klitgord N, Cusick ME, Radivojac P, et al. Intrinsic disorder is a common feature of hub proteins from four eukaryotic interactomes. *PLoS Comput Biol* 2006;**2**:e100.
98. Shimizu K, Toh H. Interaction between intrinsically disordered proteins frequently occurs in a human protein-protein interaction network. *J Mol Biol* 2009;**392**:1253–65.
99. Dunker AK, Cortese MS, Romero P, Iakoucheva LM, Uversky VN. Flexible nets. The roles of intrinsic disorder in protein interaction networks. *FEBS J* 2005;**272**:5129–48.

100. Dyson HJ, Wright PE. Intrinsically unstructured proteins and their functions. *Nat Rev Mol Cell Biol* 2005;**6**:197–208.
101. Mittag T, Kay LE, Forman-Kay JD. Protein dynamics and conformational disorder in molecular recognition. *J Mol Recognit* 2010;**23**:105–16.
102. Dingwall C, Laskey RA. Nuclear targeting sequences—a consensus? *Trends Biochem Sci* 1991;**16**:478–81.
103. Radivojac P, Iakoucheva LM, Oldfield CJ, Obradovic Z, Uversky VN, Dunker AK. Intrinsic disorder and functional proteomics. *Biophys J* 2007;**92**:1439–56.
104. Jackson EB, Theriot CA, Chattopadhyay R, Mitra S, Izumi T. Analysis of nuclear transport signals in the human apurinic/apyrimidinic endonuclease (APE1/Ref1). *Nucleic Acids Res* 2005;**33**:3303–12.
105. Ikeda S, Kohmoto T, Tabata R, Seki Y. Differential intracellular localization of the human and mouse endonuclease III homologs and analysis of the sorting signals. *DNA Repair (Amst)* 2002;**1**:847–54.
106. Otterlei M, Haug T, Nagelhus TA, Slupphaug G, Lindmo T, Krokan HE. Nuclear and mitochondrial splice forms of human uracil-DNA glycosylase contain a complex nuclear localisation signal and a strong classical mitochondrial localisation signal, respectively. *Nucleic Acids Res* 1998;**26**:4611–7.
107. Berg OG, Winter RB, von Hippel PH. Diffusion-driven mechanisms of protein translocation on nucleic acids. 1. Models and theory. *Biochemistry* 1981;**20**:6929–48.
108. Vuzman D, Azia A, Levy Y. Searching DNA via a "Monkey Bar" mechanism: the significance of disordered tails. *J Mol Biol* 2010;**396**:674–84.
109. Vuzman D, Levy Y. DNA search efficiency is modulated by charge composition and distribution in the intrinsically disordered tail. *Proc Natl Acad Sci USA* 2010;**107**:21004–9.
110. Vuzman D, Polonsky M, Levy Y. Facilitated DNA search by multidomain transcription factors: cross talk via a flexible linker. *Biophys J* 2010;**99**:1202–11.
111. Toth-Petroczy A, Simon I, Fuxreiter M, Levy Y. Disordered tails of homeodomains facilitate DNA recognition by providing a trade-off between folding and specific binding. *J Am Chem Soc* 2009;**131**:15084–5.
112. Iwahara J, Zweckstetter M, Clore GM. NMR structural and kinetic characterization of a homeodomain diffusing and hopping on nonspecific DNA. *Proc Natl Acad Sci USA* 2006;**103**:15062–7.
113. Dunn AR, Kad NM, Nelson SR, Warshaw DM, Wallace SS. Single Qdot-labeled glycosylase molecules use a wedge amino acid to probe for lesions while scanning along DNA. *Nucleic Acids Res* 2011;**17**:7487–98.
114. Krueger KE, Srivastava S. Posttranslational protein modifications: current implications for cancer detection, prevention, and therapeutics. *Mol Cell Proteomics* 2006;**5**:1799–810.
115. Seet BT, Dikic I, Zhou MM, Pawson T. Reading protein modifications with interaction domains. *Nat Rev Mol Cell Biol* 2006;**7**:473–83.
116. Bhakat KK, Izumi T, Yang SH, Hazra TK, Mitra S. Role of acetylated human AP-endonuclease (APE1/Ref-1) in regulation of the parathyroid hormone gene. *EMBO J* 2003;**22**:6299–309.
117. Busso CS, Iwakuma T, Izumi T. Ubiquitination of mammalian AP endonuclease (APE1) regulated by the p53-MDM2 signaling pathway. *Oncogene* 2009;**28**:1616–25.
118. Lee H, Mok KH, Muhandiram R, Park KH, Suk JE, Kim DH, et al. Local structural elements in the mostly unstructured transcriptional activation domain of human p53. *J Biol Chem* 2000;**275**:29426–32.
119. Fantini D, Vascotto C, Marasco D, D'Ambrosio C, Romanello M, Vitagliano L, et al. Critical lysine residues within the overlooked N-terminal domain of human APE1 regulate its biological functions. *Nucleic Acids Res* 2010;**38**:8239–56.

120. Bhattacharyya A, Chattopadhyay R, Burnette BR, Cross JV, Mitra S, Ernst PB, et al. Acetylation of apurinic/apyrimidinic endonuclease-1 regulates Helicobacter pylori-mediated gastric epithelial cell apoptosis. *Gastroenterology* 2009;**136**:2258–69.
121. Sengupta S, Mantha AK, Mitra S, Bhakat KK. Human AP endonuclease (APE1/Ref-1) and its acetylation regulate YB-1-p300 recruitment and RNA polymerase II loading in the drug-induced activation of multidrug resistance gene MDR1. *Oncogene* 2010;**30**:482–93.
122. Bhakat KK, Hazra TK, Mitra S. Acetylation of the human DNA glycosylase NEIL2 and inhibition of its activity. *Nucleic Acids Res* 2004;**32**:3033–9.
123. Tini M, Benecke A, Um SJ, Torchia J, Evans RM, Chambon P. Association of CBP/p300 acetylase and thymine DNA glycosylase links DNA repair and transcription. *Mol Cell* 2002;**9**:265–77.
124. Bhakat KK, Mokkapati SK, Boldogh I, Hazra TK, Mitra S. Acetylation of human 8-oxoguanine-DNA glycosylase by p300 and its role in 8-oxoguanine repair in vivo. *Mol Cell Biol* 2006;**26**:1654–65.
125. Hsieh MM, Hegde V, Kelley MR, Deutsch WA. Activation of APE/Ref-1 redox activity is mediated by reactive oxygen species and PKC phosphorylation. *Nucleic Acids Res* 2001;**29**:3116–22.
126. Yacoub A, Kelley MR, Deutsch WA. The DNA repair activity of human redox/repair protein APE/Ref-1 is inactivated by phosphorylation. *Cancer Res* 1997;**57**:5457–9.
127. Bhakat KK, Mantha AK, Mitra S. Transcriptional regulatory functions of mammalian AP-endonuclease (APE1/Ref-1), an essential multifunctional protein. *Antioxid Redox Signal* 2009;**11**:621–38.
128. Fritz G, Kaina B. Phosphorylation of the DNA repair protein APE/REF-1 by CKII affects redox regulation of AP-1. *Oncogene* 1999;**18**:1033–40.
129. Parsons JL, Dianova II, Finch D, Tait PS, Strom CE, Helleday T, et al. XRCC1 phosphorylation by CK2 is required for its stability and efficient DNA repair. *DNA Repair (Amst)* 2010;**9**:835–41.
130. Huang E, Qu D, Zhang Y, Venderova K, Haque ME, Rousseaux MW, et al. The role of Cdk5-mediated apurinic/apyrimidinic endonuclease 1 phosphorylation in neuronal death. *Nat Cell Biol* 2010;**12**:563–71.
131. Lu X, Bocangel D, Nannenga B, Yamaguchi H, Appella E, Donehower LA. The p53-induced oncogenic phosphatase PPM1D interacts with uracil DNA glycosylase and suppresses base excision repair. *Mol Cell* 2004;**15**:621–34.
132. Hagen L, Kavli B, Sousa MM, Torseth K, Liabakk NB, Sundheim O, et al. Cell cycle-specific UNG2 phosphorylations regulate protein turnover, activity and association with RPA. *EMBO J* 2008;**27**:51–61.
133. Gu Y, Lu AL. Differential DNA recognition and glycosylase activity of the native human MutY homolog (hMYH) and recombinant hMYH expressed in bacteria. *Nucleic Acids Res* 2001;**29**:2666–74.
134. Dantzer F, Luna L, Bjoras M, Seeberg E. Human OGG1 undergoes serine phosphorylation and associates with the nuclear matrix and mitotic chromatin in vivo. *Nucleic Acids Res* 2002;**30**:2349–57.
135. Parsons JL, Tait PS, Finch D, Dianova II, Edelmann MJ, Khoronenkova SV, et al. Ubiquitin ligase ARF-BP1/Mule modulates base excision repair. *EMBO J* 2009;**28**:3207–15.
136. Chau V, Tobias JW, Bachmair A, Marriott D, Ecker DJ, Gonda DK, et al. A multiubiquitin chain is confined to specific lysine in a targeted short-lived protein. *Science* 1989;**243**:1576–83.
137. Busso CS, Lake MW, Izumi T. Posttranslational modification of mammalian AP endonuclease (APE1). *Cell Mol Life Sci* 2010;**67**:3609–20.
138. Nijman SM, Luna-Vargas MP, Velds A, Brummelkamp TR, Dirac AM, Sixma TK, et al. A genomic and functional inventory of deubiquitinating enzymes. *Cell* 2005;**123**:773–86.

139. Busso CS, Wedgeworth CM, Izumi T. Ubiquitination of human AP-endonuclease 1 (APE1) enhanced by T233E substitution and by CDK5. *Nucleic Acids Res* 2011;**39**:8017–28.
140. Fischer JA, Muller-Weeks S, Caradonna S. Proteolytic degradation of the nuclear isoform of uracil-DNA glycosylase occurs during the S phase of the cell cycle. *DNA Repair (Amst)* 2004;**3**:505–13.
141. Chen J, Bozza W, Zhuang Z. Ubiquitination of PCNA and its essential role in eukaryotic translesion synthesis. *Cell Biochem Biophys* 2011;**60**:47–60.
142. Lehmann AR. Ubiquitin-family modifications in the replication of DNA damage. *FEBS Lett* 2011;**585**:2772–9.
143. Katyal S, McKinnon PJ. DNA strand breaks, neurodegeneration and aging in the brain. *Mech Ageing Dev* 2008;**129**:483–91.
144. Al-Tassan N, Chmiel NH, Maynard J, Fleming N, Livingston AL, Williams GT, et al. Inherited variants of MYH associated with somatic G:C→T:A mutations in colorectal tumors. *Nat Genet* 2002;**30**:227–32.
145. Goodenberger M, Lindor NM. Lynch syndrome and MYH-associated polyposis: review and testing strategy. *J Clin Gastroenterol* 2011;**45**:488–500.
146. Boiteux S, Radicella JP. The human OGG1 gene: structure, functions, and its implication in the process of carcinogenesis. *Arch Biochem Biophys* 2000;**377**:1–8.
147. Shinmura K, Yokota J. The OGG1 gene encodes a repair enzyme for oxidatively damaged DNA and is involved in human carcinogenesis. *Antioxid Redox Signal* 2001;**3**:597–609.
148. Dherin C, Radicella JP, Dizdaroglu M, Boiteux S. Excision of oxidatively damaged DNA bases by the human alpha-hOgg1 protein and the polymorphic alpha-hOgg1(Ser326Cys) protein which is frequently found in human populations. *Nucleic Acids Res* 1999;**27**:4001–7.
149. Hill JW, Evans MK. Dimerization and opposite base-dependent catalytic impairment of polymorphic S326C OGG1 glycosylase. *Nucleic Acids Res* 2006;**34**:1620–32.
150. Weiss JM, Goode EL, Ladiges WC, Ulrich CM. Polymorphic variation in hOGG1 and risk of cancer: a review of the functional and epidemiologic literature. *Mol Carcinog* 2005;**42**: 127–41.
151. Klungland A, Rosewell I, Hollenbach S, Larsen E, Daly G, Epe B, et al. Accumulation of premutagenic DNA lesions in mice defective in removal of oxidative base damage. *Proc Natl Acad Sci USA* 1999;**96**:13300–5.
152. Osterod M, Hollenbach S, Hengstler JG, Barnes DE, Lindahl T, Epe B. Age-related and tissue-specific accumulation of oxidative DNA base damage in 7,8-dihydro-8-oxoguanine-DNA glycosylase (Ogg1) deficient mice. *Carcinogenesis* 2001;**22**:1459–63.
153. Xie Y, Yang H, Cunanan C, Okamoto K, Shibata D, Pan J, et al. Deficiencies in mouse Myh and Ogg1 result in tumor predisposition and G to T mutations in codon 12 of the K-ras oncogene in lung tumors. *Cancer Res* 2004;**64**:3096–102.
154. Parsons JL, Elder RH. DNA N-glycosylase deficient mice: a tale of redundancy. *Mutat Res* 2003;**531**:165–75.
155. Wilson 3rd DM, Bohr VA. The mechanics of base excision repair, and its relationship to aging and disease. *DNA Repair (Amst)* 2007;**6**:544–59.
156. Friedberg EC, Walker GC, Siede W, Wood RD, Schultz RA, Ellenberger T. *DNA repair and mutagenesis*. 2nd ed. Washington: ASM Press; 2006.
157. Wilson 3rd DM, Thompson LH. Life without DNA repair. *Proc Natl Acad Sci USA* 1997;**94**:12754–7.
158. Zheng L, Dai H, Hegde ML, Zhou M, Guo Z, Wu X, et al. Fen1 mutations that specifically disrupt its interaction with PCNA cause aneuploidy-associated cancer. *Cell Res* 2011;**21**: 1052–67.
159. Zheng L, Dai H, Zhou M, Li M, Singh P, Qiu J, et al. Fen1 mutations result in autoimmunity, chronic inflammation and cancers. *Nat Med* 2007;**13**:812–9.

160. Starcevic D, Dalal S, Sweasy JB. Is there a link between DNA polymerase beta and cancer? *Cell Cycle* 2004;**3**:998–1001.
161. Lang T, Maitra M, Starcevic D, Li SX, Sweasy JB. A DNA polymerase beta mutant from colon cancer cells induces mutations. *Proc Natl Acad Sci USA* 2004;**101**:6074–9.
162. Hu JJ, Smith TR, Miller MS, Mohrenweiser HW, Golden A, Case LD. Amino acid substitution variants of APE1 and XRCC1 genes associated with ionizing radiation sensitivity. *Carcinogenesis* 2001;**22**:917–22.
163. Jiao L, Bondy ML, Hassan MM, Wolff RA, Evans DB, Abbruzzese JL, et al. Selected polymorphisms of DNA repair genes and risk of pancreatic cancer. *Cancer Detect Prev* 2006;**30**:284–91.
164. Goode EL, Ulrich CM, Potter JD. Polymorphisms in DNA repair genes and associations with cancer risk. *Cancer Epidemiol Biomarkers Prev* 2002;**11**:1513–30.
165. Hung RJ, Hall J, Brennan P, Boffetta P. Genetic polymorphisms in the base excision repair pathway and cancer risk: a HuGE review. *Am J Epidemiol* 2005;**162**:925–42.
166. Hanawalt PC. Emerging links between premature ageing and defective DNA repair. *Mech Ageing Dev* 2008;**129**:503–5.
167. Obulesu M, Rao DM. DNA damage and impairment of DNA repair in Alzheimer's disease. *Int J Neurosci* 2010;**120**:397–403.
168. Subba Rao K. Mechanisms of disease: DNA repair defects and neurological disease. *Nat Clin Pract Neurol* 2007;**3**:162–72.
169. Weissman L, Jo DG, Sorensen MM, de Souza-Pinto NC, Markesbery WR, Mattson MP, et al. Defective DNA base excision repair in brain from individuals with Alzheimer's disease and amnestic mild cognitive impairment. *Nucleic Acids Res* 2007;**35**:5545–55.
170. Dogru-Abbasoglu S, Aykac-Toker G, Hanagasi HA, Gurvit H, Emre M, Uysal M. The Arg194Trp polymorphism in DNA repair gene XRCC1 and the risk for sporadic late-onset Alzheimer's disease. *Neurol Sci* 2007;**28**:31–4.
171. Migliore L, Fontana I, Trippi F, Colognato R, Coppede F, Tognoni G, et al. Oxidative DNA damage in peripheral leukocytes of mild cognitive impairment and AD patients. *Neurobiol Aging* 2005;**26**:567–73.
172. Milutinovic S, Zhuang Q, Szyf M. Proliferating cell nuclear antigen associates with histone deacetylase activity, integrating DNA replication and chromatin modification. *J Biol Chem* 2002;**277**:20974–8.
173. el-Khamisy SF, Caldecott KW. DNA single-strand break repair and spinocerebellar ataxia with axonal neuropathy-1. *Neuroscience* 2007;**145**:1260–6.
174. Shen J, Gilmore EC, Marshall CA, Haddadin M, Reynolds JJ, Eyaid W, et al. Mutations in PNKP cause microcephaly, seizures and defects in DNA repair. *Nat Genet* 2010;**42**:245–9.
175. Coppede F, Ceravolo R, Migheli F, Fanucchi F, Frosini D, Siciliano G, et al. The hOGG1 Ser326Cys polymorphism is not associated with sporadic Parkinson's disease. *Neurosci Lett* 2010;**473**:248–51.
176. Coppede F, Mancuso M, Lo Gerfo A, Carlesi C, Piazza S, Rocchi A, et al. Association of the hOGG1 Ser326Cys polymorphism with sporadic amyotrophic lateral sclerosis. *Neurosci Lett* 2007;**420**:163–8.
177. Coppede F, Mancuso M, Lo Gerfo A, Manca ML, Petrozzi L, Migliore L, et al. A Ser326Cys polymorphism in the DNA repair gene hOGG1 is not associated with sporadic Alzheimer's disease. *Neurosci Lett* 2007;**414**:282–5.
178. Coppede F, Migliore L. DNA damage and repair in Alzheimer's disease. *Curr Alzheimer Res* 2009;**6**:36–47.
179. Hayward C, Colville S, Swingler RJ, Brock DJ. Molecular genetic analysis of the APEX nuclease gene in amyotrophic lateral sclerosis. *Neurology* 1999;**52**:1899–901.

180. Hirano R, Interthal H, Huang C, Nakamura T, Deguchi K, Choi K, et al. Spinocerebellar ataxia with axonal neuropathy: consequence of a Tdp1 recessive neomorphic mutation? *EMBO J* 2007;**26**:4732–43.
181. Rass U, Ahel I, West SC. Actions of aprataxin in multiple DNA repair pathways. *J Biol Chem* 2007;**282**:9469–74.
182. Hegde ML, Hegde PM, Holthauzen LM, Hazra TK, Rao KS, Mitra S. Specific Inhibition of NEIL-initiated repair of oxidized base damage in human genome by copper and iron: potential etiological linkage to neurodegenerative diseases. *J Biol Chem* 2010;**285**:28812–25.
183. Hegde ML, Hegde PM, Jagannatha Rao KS, Mitra S. Oxidative genome damage and its repair in neurodegenerative diseases: function of transition metals as a double-edged sword. *J Alzheimers Dis* 2011;**24**:183–98.
184. Whiteside JR, Box CL, McMillan TJ, Allinson SL. Cadmium and copper inhibit both DNA repair activities of polynucleotide kinase. *DNA Repair (Amst)* 2010;**9**:83–9.
185. Li H, Swiercz R, Englander EW. Elevated metals compromise repair of oxidative DNA damage via the base excision repair pathway: implications of pathologic iron overload in the brain on integrity of neuronal DNA. *J Neurochem* 2009;**110**:1774–83.
186. Bobola MS, Blank A, Berger MS, Stevens BA, Silber JR. Apurinic/apyrimidinic endonuclease activity is elevated in human adult gliomas. *Clin Cancer Res* 2001;**7**:3510–8.
187. Bobola MS, Emond MJ, Blank A, Meade EH, Kolstoe DD, Berger MS, et al. Apurinic endonuclease activity in adult gliomas and time to tumor progression after alkylating agent-based chemotherapy and after radiotherapy. *Clin Cancer Res* 2004;**10**:7875–83.
188. Bobola MS, Finn LS, Ellenbogen RG, Geyer JR, Berger MS, Braga JM, et al. Apurinic/apyrimidinic endonuclease activity is associated with response to radiation and chemotherapy in medulloblastoma and primitive neuroectodermal tumors. *Clin Cancer Res* 2005;**11**:7405–14.
189. Wilson 3rd DM, Simeonov A. Small molecule inhibitors of DNA repair nuclease activities of APE1. *Cell Mol Life Sci* 2010;**67**:3621–31.
190. Fishel ML, Jiang Y, Rajeshkumar NV, Scandura G, Sinn AL, He Y, et al. Impact of APE1/Ref-1 redox inhibition on pancreatic tumor growth. *Mol Cancer Ther* 2011;**10**:1698–708.
191. Bapat A, Glass LS, Luo M, Fishel ML, Long EC, Georgiadis MM, et al. Novel small-molecule inhibitor of apurinic/apyrimidinic endonuclease 1 blocks proliferation and reduces viability of glioblastoma cells. *J Pharmacol Exp Ther* 2010;**334**:988–98.
192. Ashworth A. A synthetic lethal therapeutic approach: poly(ADP) ribose polymerase inhibitors for the treatment of cancers deficient in DNA double-strand break repair. *J Clin Oncol* 2008;**26**:3785–90.
193. Farmer H, McCabe N, Lord CJ, Tutt AN, Johnson DA, Richardson TB, et al. Targeting the DNA repair defect in BRCA mutant cells as a therapeutic strategy. *Nature* 2005;**434**:917–21.
194. Loser DA, Shibata A, Shibata AK, Woodbine LJ, Jeggo PA, Chalmers AJ. Sensitization to radiation and alkylating agents by inhibitors of poly(ADP-ribose) polymerase is enhanced in cells deficient in DNA double-strand break repair. *Mol Cancer Ther* 2010;**9**:1775–87.
195. Rassool FV, Tomkinson AE. Targeting abnormal DNA double strand break repair in cancer. *Cell Mol Life Sci* 2010;**67**:3699–710.

Homologous Recombination in Eukaryotes

Ravindra Amunugama[*,†] and
Richard Fishel[†,‡]

[*]*Biophysics Graduate Program, The Ohio State University, Columbus, Ohio, USA*

[†]*Department of Molecular Virology, Immunology, and Medical Genetics, Human Cancer Genetics, The Ohio State University Medical Center and Comprehensive Cancer Center, The Ohio State University, Columbus, Ohio, USA*

[‡]*Physics Department, The Ohio State University, Columbus, Ohio USA*

Progress in Molecular Biology
and Translational Science, Vol. 110
DOI: 10.1016/B978-0-12-387665-2.00007-9

1877-1173/12 $35.00

Homologous recombination (HR) is a mechanistically conserved pathway that ensures maintenance of genomic integrity. During meiosis, HR results in DNA crossover events between homologous chromosomes that produce the genetic diversity inherent in germ cells. The physical connection established between homologs during the crossover event is essential to facilitate correct chromosome segregation. HR is also involved in maintenance of somatic cell genomic stability by restoring replication after a stalled replication fork has encountered a DNA lesion or strand break, as well as following exogenous stresses such as ionizing radiation that induce DNA double-strand breaks. The importance of HR can be gauged by the conservation of HR genes and functions from bacteria to man. Here we review the players and mechanics of eukaryotic HR.

DNA double-stranded breaks (DSBs) are generated spontaneously by radiation and chemical damage as well as intentionally as part of the chromosome-pairing process during meiosis. Genome instability resulting from DSB recombination repair (RR) defects has been linked to a variety of human cancers including hereditary breast cancer (BRCA1/2) as well as hematopoietic and other solid tumors (ataxia telangiectasia mutated (ATM), Nijmegen breakage syndrome (NBS), Fanconi anemia (FANC), and Bloom's syndrome (BLM)) among others.[1–5] Unlike many repair pathways, RR engenders a complex cascade of responses that include cellular signaling integrated with the physical processes of DSB repair.[6,7] The DSB repair reaction itself involves a complex cascade of enzymatic reactions that must manage the chromatin composite on the broken donor DNA in order to search and pair with the assembled chromatin of a homologous acceptor DNA. Deficiencies in any one of the multitudes of steps will affect the outcome of the RR process and ultimately affect genome stability. Understanding of the biophysical events associated with the DSB repair reaction, which rely on targeting redundant or overlapping repair pathways that ultimately result in a synergistic therapeutic response in cancer patients, is important as combinatorial chemical strategies are under development.[8]

I. Meiosis

All sexually reproducing organisms undergo meiosis—a process that reduces the cellular diploid content to produce haploid gametes. RR has been coopted and is essential for the completion of meiosis. Meiosis begins with replication that forms sister chromosomes (chromatids) and is followed by a pairing process that spatially associates chromosome homologs.[9] The segregation of chromosome homologs is completed in the first meiotic division (meiosis I) and the segregation of chromatids is completed in the second meiotic division (meiosis II), ultimately producing haploid gametes. The mechanism, regulation, and checkpoint functions of meiosis II chromatid segregation appear similar to the well-defined processes associated with mitosis.[10] In contrast, meiosis I requires suppression of the tendency to segregate sister chromatids and instead the homologous chromosomes are separated. More than 50% of all spontaneous miscarriages are due to errors in chromosome segregation (nondisjunction) at the first meiotic division.[11] Moreover, 90% of Down syndrome cases can be attributed to errors in maternal meiosis.[12] With few exceptions, the critical meiosis genes appear identical in all eukaryotes.[13,14]

The pairing of homologous chromosomes in meiosis I is a complex process fraught with many pitfalls that may ultimately result in infertility. Homologous chromosome pairing is initiated in Prophase I by the SPO11 gene product,[14,15] which *actively* introduces *hundreds* of DSBs into the sister chromatids.[16] The homologous recombination (HR) repair of these DSBs by the nearest sister is suppressed by the formation of meiosis-specific lateral elements between the chromatids.[17] This leaves the homologous chromosome as the only DNA sequences available for HR repair to restore the exact integrity of the genome (Fig. 1). The DSBs are first resected by a $5' \rightarrow 3'$ exonuclease.[18,19] The resulting $3'$ single-stranded DNA (ssDNA) end is then used in a classic homologous pairing and strand invasion reaction with the chromosome homolog to form a D-loop. Strand invasion requires RAD51 and/or the meiosis-specific DMC1, which are homologs to the prototypical bacterial recombination–initiation protein RecA.[20,21] The ssDNA-binding (SSB) protein, replication protein A (RPA), is an essential cofactor in this process.[22,23] Mutation of SPO11 or RAD51 results in a dramatic reduction of homologous chromosome pairing, a high frequency of meiosis I nondisjunction, and gamete inviability. In mice, more than 400 DSB sites are formed that contain RAD51 and RPA beginning in leptotene.[24] That the DSBs are almost always faithfully repaired is a testament to the accuracy and dependability of the process in the preservation of the many sexual species on earth.

Approximately 90% of the DSB sites are resolved following repair in a process that converts one parental homolog DNA sequence to the other parental homolog sequence with concurrent loss of that parental homolog DNA sequence (gene conversion; Ref. 25). These events leave the remaining chromosome of both parents intact.[25] The remaining 10% (40–50 in human) introduce visible chromosomal crossovers known as chiasmata, which exchange entire arms of genetic information reciprocally from one parental chromosome to the other.[26] Ultimately, there are two significant events associated with meiotic DSB repair: (1) genetic information is exchanged between chromosomes, which is the basis of modern genetics,[27] and (2) homologous chromosomes become linked via *chiasmata* that are essential for proper chromosome segregation.

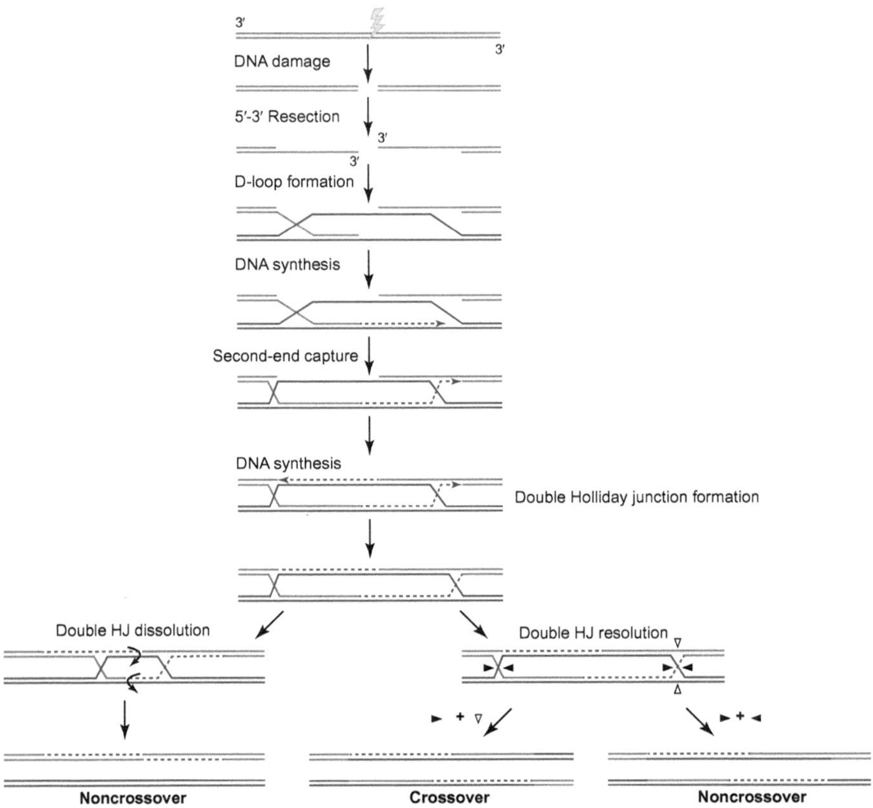

FIG. 1. DNA double-strand break repair (DSBR) by homologous recombination (HR). DSBR is initiated by D-loop formation (strand invasion) by the 3′ ssDNA overhang that results from strand resection. The invading DNA strand primes DNA synthesis. During double Holliday junction (dHJ) formation, the second-end is captured and the strands are ligated after DNA synthesis. Branch migration can either dissolve dHJs that result in noncrossover products or stabilize dHJs to undergo resolution. dHJ resolution can result in either crossover or noncrossover products. (See Color Insert.)

II. DSB Repair in Somatic Cells

Eukaryotes have four main pathways that repair DSBs generated spontaneously in somatic cells: HR, nonhomologous end-joining (NHEJ), alternative NHEJ (Alt-NHEJ, also known as microhomology-mediated end-joining, MMEJ), and single-strand annealing (SSA; Fig. 2; Ref. 28). The pathway of choice depends on the nature of DSB, the species, cell type, and the cell cycle stage where the DSB occurs.[29]

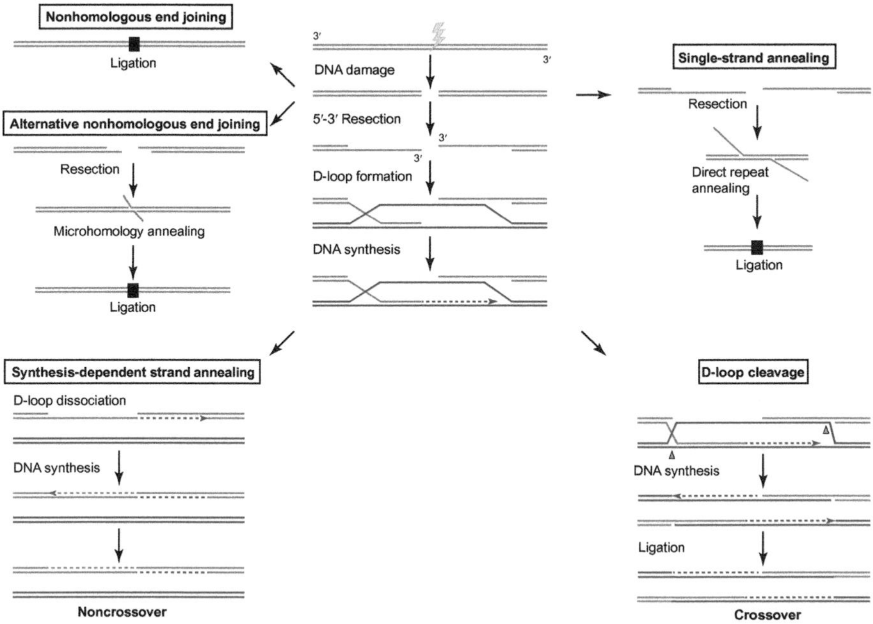

FIG. 2. Alternative double-strand break repair (DSBR) mechanisms. Nonhomologous end-joining (NHEJ) or alternative NHEJ (Alt-NHEJ) occurs by either direct ligation of the broken DNA strand or ligation after minimal processing. Both NHEJ and Alt-NHEJ are error-prone repair mechanisms. DSBR within direct repeat sequences could occur by single-strand annealing (SSA). SSA causes loss of a repeat sequence due to direct resection, annealing, and ligation. During DSBR through synthesis-dependent strand annealing (SDSA), synthesized nascent strand is displaced by D-loop dissociation, and anneals with the other 3′ ssDNA overhang to complete DNA synthesis. SDSA results in noncrossover products. After the initial D-loop formation and DNA synthesis, the D-loop can also be cleaved to produce crossover products. (See Color Insert.)

As discussed, HR is a template-dependent repair process and was long ago recognized to require the formation of a DNA crossover structure at the site of homology between chromosomes (termed Holliday junction; Ref. 30). In the 1980s, a DSB repair model that involves the formation of a double Holliday junction (dHJ) was developed based on transformation studies in budding yeast (*Saccharomyces cerevisiae*), where a linear plasmid was faithfully integrated into a homologous region of the host genome (Fig. 1; Refs. 31,32). Even though the DSB repair model also explains meiotic recombination products, during mitotic recombination very few crossover events are observed since the sister chromatid appears to be largely used as the template. The use of a sister chromatid strongly suggests that mitotic recombination mainly occurs during late S and G2 phases of the cell cycle.[29] Importantly, crossover events during mitotic recombination could lead to loss of heterozygosity (LOH), which is a common process during tumorigenesis.[29,33,34]

A synthesis-dependent strand annealing (SDSA) model was proposed to account for the low numbers of crossover recombination events (Fig. 2; Refs. 35–37). In SDSA, once DNA synthesis occurs on the invading strand of a D-loop, it is unwound and displaced such that it may anneal with the second end to prime DNA synthesis on the latter (Fig. 1). This pathway leads exclusively to noncrossover products. SDSA aside, dHJ-based recombination products have been observed between homologous chromosomes in yeast during mitotic DSB repair, but as a minor pathway.[38]

During NHEJ, the broken DSB ends are prevented from resection by the Ku70–Ku80 heterodimer.[39] The strong affinity of Ku70–Ku80 for DSB ends appears to recruit DNA ligase IV, which is capable of sealing the DSB (Fig. 2). The Alt-NHEJ pathway appears to resect a 5- to 25-nt region where microhomology may be used prior to ligation of the ends (Fig. 2; Refs. 28,40).

SSA occurs in regions flanked by direct repeat DNA sequences (Fig. 2; Ref. 41). Hence, this pathway is seen in higher eukaryotes where direct repeated sequences are prevalent.[41–44] In SSA, both 5′ flanking the DSB are resected by nucleases and the resulting 3′ overhangs annealed by RPA and RAD52.[29] As strand invasion and exchange are not involved, this process is independent of RAD51. 3′ Single-stranded overhangs are subsequently resected and ligated by nucleases and ligases, respectively.[41] NHEJ, Alt-NHEJ, and SSA are mutation-prone pathways due to the lack of fidelity and loss of genetic information during the repair process.

A. Collapsed or Stalled Replication Forks

When a replication fork collapses or if a telomere becomes uncapped, a single-ended DSB is formed.[45–47] This end may be processed to produce a 3′ overhang that can invade a homologous region of the sister chromatid, the homologous chromosome, or a homologous region of another chromosome to initiate DNA synthesis. This process is called break-induced replication (BIR; Fig. 3). Use of any other template other than the sister chromatid leads to LOH during BIR. During mitotic recombination, however, BIR is disfavored over SDSA apparently because of its slower kinetics.[48]

Stalled replication machinery or lesions on leading or lagging strands may lead to the formation of DNA gaps or DSBs that often result in replication fork collapse.[49,50] In vertebrate cells, replication fork collapse occurs in every cell cycle.[51,52] Translesion synthesis (TLS), template switching, or HR can restore replication. TLS is error-prone due to the low fidelity of the polymerase employed.[53] However, template switching and HR are error-free. Many replication mutants with defective checkpoint activation are dependent on HR gene

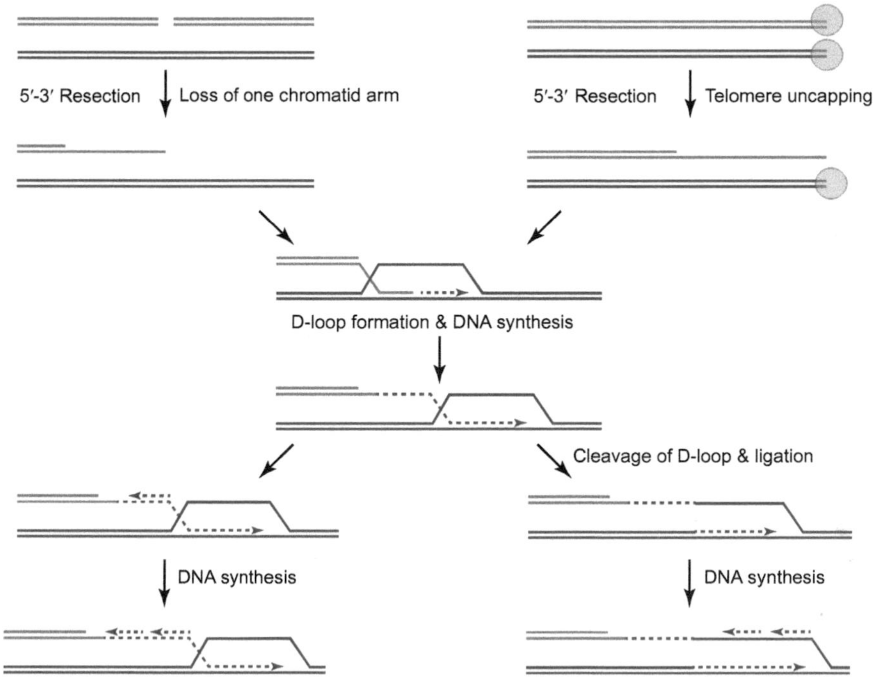

FIG. 3. Break-induced replication (BIR) initiates from a single-ended strand invasion. If one arm of the chromatid is lost after the double-strand break (DSB) or if a telomere is uncapped, a 3′ overhang is formed. DNA synthesis can continue to the end of the chromatid either by migration of the D-loop or after D-loop cleavage. (See Color Insert.)

products for viability.[54,55] Uncontrolled HR, however, during replication fork collapse can lead to gross genomic instability. These observations suggest that cell cycle checkpoints tightly regulate the HR pathway to ensure genomic integrity.[54,55]

If the nascent strand encounters a nick during replication, the fork may stall and the incomplete replicated strand may undergo resection, which can invade the sister chromatid once the latter is ligated (Fig. 4, left). After DNA synthesis, a resulting partial Holliday junction may be resolved to reinitiate DNA synthesis at the fork. If the leading strand stalls due to a lesion on its template, the newly synthesized strands may pair up via reverse branch migration to form a chicken-foot structure (a pseudo-Holliday junction; Fig. 4, middle; Ref. 56). Following a short DNA synthesis to fill in the chicken-foot ssDNA tail, the replication fork may be reinitiated by forward branch migration. A lesion on the lagging strand template would lead to a template switch mechanism

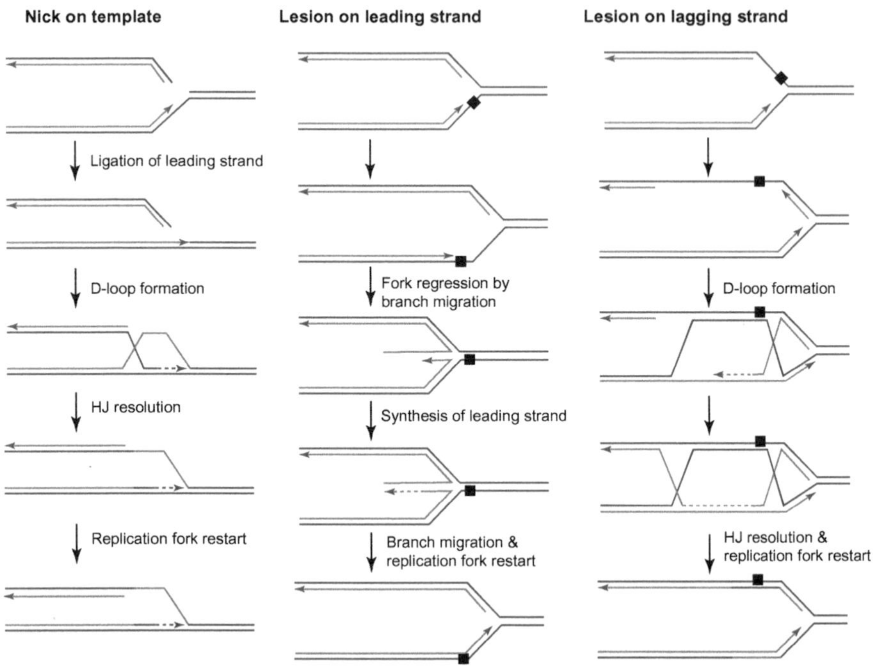

FIG. 4. Homologous recombination (HR) restores collapsed or stalled replication forks. Nicks of template strands lead to replication fork collapse, which can be repaired by D-loop formation and Holliday junction cleavage to restore replication. A lesion on a leading strand may result in replication fork regression and DNA synthesis on the leading strand. Subsequent branch migration restores the replication fork. (See Color Insert.)

where the blocked strand may invade the nascent complementary strand to bypass the lesion by DNA synthesis (Fig. 4, right). Resolution of the resulting Holliday junction may then restore replication.

B. DSB Recognition and End Resection

The substrate for HR is an ssDNA region with a 3′-end generated by resection of the 5′ strand of the duplex.[57] In the event of a DSB by ionizing radiation (IR) or chemical agents, the terminal nucleotides are often modified structurally or exist as protein-bound entities.[58–60] Such modifications pose an obstacle for downstream repair pathways and are effectively removed by nucleases and proteases or both.

The Mre11–Rad50–Xrs2 (MRX) complex in yeast and the homologous MRE11–RAD50–NBS1 (MRN) complex in higher eukaryotes recognize DSBs.[61–65] Even though MRE11 possesses an inherent $3' \rightarrow 5'$ nuclease activity, the initial $5' \rightarrow 3'$ resection is initiated by the Sae2 (CtIP, in mammals) endonuclease in complex with MRX/MRN. Sae2/CtIP appears to process possible adducts on DSB ends for other nucleases to act upon.[66–68] The cell cycle-mediated phosphorylation of Sae2/CtIP by cyclin-dependent kinases (CDKs) appears to determine the choice between HR and NHEJ.[69,70] In mammalian cells, CtIP is ubiquitinated by BRCA1 during S and G2 phases of the cell cycle, which appears to facilitate its association with DSB sites.[71,72]

For extensive resection of the 5′ strand, the Exo1 and Dna2 nucleases as well as the Sgs1–Top3–Rmi1 (STR) complex (BLM–TOP3α–RMI1–RMI2 in mammals; Ref. 73) are recruited.[74,75] The MRX complex is implicated in direct recruitment of Exo1 and Dna2.[75] Deletion of Exo1, Dna2, or Sgs1 leads to reduction of resection and the generation of poor HR substrates.[75,76] Furthermore, deletion of EXO1 in mammalian cells causes impaired recruitment of RPA and ATR at the DSB sites.[77] The initial resection complex that includes the MRX along with STR and Dna2 has been reconstituted *in vitro*.[78,79] Even though Top3 and Rmi1 stimulate resection by recruiting Sgs1, they are not required for the 5′ strand resection processes.[78,79] Stimulation of Sgs1 helicase activity by RPA occurs in a species-specific manner, as the bacterial SSB is unable to stimulate Sgs1.[78–80] RPA also suppresses the inherent 3′ endonuclease function of Dna2 while stimulating the $5' \rightarrow 3'$ exonuclease required for DSB resection.[78,79] Two functional human resection complexes have been reconstituted *in vitro*, one comprising MRN–EXO1–BLM–RPA and the other MRN–DNA2–BLM–RPA.[80] BLM exhibits direct protein–protein interactions with both EXO1 and DNA2.[80,81] Furthermore, the nuclease activity of EXO1 is stimulated by BLM, RPA, and MRN.

In the absence of a recombinase or a homologous sequence, resection could continue over several thousand nucleotides at a rate of approximately 4 kb/h in yeast.[75,82] During meiotic recombination, resection tracts average $\sim$850 nt.[83,84] However, the resection length required for recombination between sister chromatids during mitotic recombination has not been determined.[82]

III. RAD52 Epistasis Group

Many of the proteins involved in the RR pathway are genes of the *RAD52* epistasis group (Table I). The name was derived from radiation sensitivity genetic screening analysis of budding yeast.[85–88] Among eukaryotes, this group of genes is structurally and functionally conserved.

TABLE I

COMPARISON OF DOUBLE-STRAND BREAK REPAIR (DSBR) FACTORS IN BUDDING YEAST AND HUMANS

HR repair process	Budding yeast	Humans
5′ Resection	Mre11–**Rad50**–Xrs2 (MRX)	MRE11–RAD50–NBS1(MRN)
	Exo1	EXO1
	Dna2	DNA2
	Sae2	CtIP
	Sgs1–Top3–RmiI	BLM–TOP3α–RMII–RMI2
	RPA	RPA
Presynapsis and synapsis	**Rad52**	BRCA2
	Rad51	RAD51
	Rad55–Rad57	RAD51B, RAD51C, RAD51D, XRCC2, XRCC3
	Rad54	RAD54, RAD54B, RAD51AP1
	Dmc1	DMC1
DNA synthesis	DNA polymerase δ	DNA polymerase δ DNA polymerase η
Strand displacement	Srs2	BLM
	Mph1	RTEL1 FANCM
HJ[a] dissolution	Sgs1–Top3–Rmi1	BLM–TOP3–RMI1–RMI2
HJ[a] resolution	Yen1	ResA (GEN1)
	Mus81–Mms4	MUS81–EME1
	Slx1–Slx4	SLX1–SLX4
Chromatin remodeling	Ino80	INO80
	Swi/Snf	SWI/SNF
	Swr1	SWR1
	RSC	TIP60

Rad52 epistasis group in bold face.
[a]Holliday junction.

A. RAD51

RAD51 is unequivocally the central component in HR pathways (Table I). It preserves a high sequence homology to the prototypical bacterial recombinase RecA.[41,85,89] In eukaryotes, homologous pairing and strand exchange is primarily mediated by RAD51.[85,89] RAD51 exists as a heptamer in solution.[90] Yeast Rad51 and human RAD51 are 43 kDa and 37 kDa in size, respectively.[91,92] The main catalytic ATPase core region that includes the Walker A/B regions and SSB domains are conserved among the RecA/RAD51 recombinases.[91,93–95] However, RAD51 possesses an N-terminal extension that is absent in RecA, while in RecA a C-terminal extension is found that is not found in RAD51.[96] These extensions have been implicated as possible double-stranded DNA (dsDNA)-binding sites.[97,98]

Formation and maintenance of a stable nucleoprotein filament (NPF) is required for the DNA homology search and strand exchange by RecA/RAD51 recombinases.[85,99,100] RAD51 nucleates on both ssDNA and dsDNA *in vitro*, forming an extended NPF.[101,102] Biochemical studies have shown that incorporation of monovalent salts in the reaction buffer biases RAD51 nucleation on ssDNA.[103,104] This observation appears to presage the possibility that other recombination mediators might be responsible for efficient nucleation of RAD51 on ssDNA. A single RAD51 molecule binds to 3–4 nt or bp, extending the helical pitch by ~50% compared to canonical B-form DNA.[101,102] Interestingly, when calcium was substituted for magnesium, human RAD51 displayed enhanced strand exchange activity *in vitro*.[100] This effect has been attributed to the suppression of ATP hydrolysis by the calcium, which appears to enhance the lifetime of the ATP-bound active NPF.[100] This calcium-mediated stimulation is unique to human RAD51.[100]

Budding yeast containing a *rad51* deletion is viable.[86] However, knockout of RAD51 in vertebrates leads to chromosomal instability and embryonic lethality.[51,52] Although no mutations have been reported in the open reading frame, in many cancers and cancer cell lines the expression of RAD51 is increased, presumably providing a replicative advantage to the rapidly dividing cells via its role in the HR repair of collapsed forks.[105–107]

A catalytically conserved lysine residue in the Walker A box (yeast Rad51 (K191) and human RAD51(K133)) is essential for ATP binding and hydrolysis. Mutation of this conserved lysine to alanine (Rad51(K191A)) leads to a null phenotype and is a dominant negative phenotype in diploid cells.[41,108] Expression of Rad51(K191R) in *rad51* null yeast strains leads to resistance to DSB-causing agents, suggesting nucleotide binding is sufficient for HR repair *in vivo*.[109] Human RAD51(K133R) binds ATP but is unable to hydrolyze ATP, similar to the yeast mutation.[110,111] Overexpression of human RAD51 (K133R) in chicken DT40 *RAD51* knockout cells confers partial resistance to IR.[111] RAD51(K133R) forms a stable NPF on DNA and has enhanced recombinase activity *in vitro*.[110] However, generated mouse embryonic stem cells that express RAD51(K133R) display increased sensitivity to DSBs and reduced efficiency of spontaneous sister chromatid exchanges.[112]

ATP binds at the interface region of two adjacent RAD51 monomers within the NPF.[95,113] The bottom subunit provides the catalytic Walker A (P-loop) domain while the top subunit shields the nucleotide with an *ATP cap* containing a conserved proline residue.[113] For the homology search and strand exchange, ATP binding but not necessarily ATP hydrolysis is required.[99] When an ATP or a nonhydrolysable ATP analog binds at the subunit interface, the NPF adopts an active extended conformation. Several biochemical studies have shown that incorporation of monovalent cations such as ammonium and potassium as well as divalent cations such as calcium extends the NPF further by ~30%.[113,114] The NPF exists in the collapsed configuration with ADP.[115] The extended conformation is believed to facilitate homology sampling on duplex DNA. The ATP hydrolysis rate of RAD51 is severalfold slower compared to the bacterial RecA.[116] However, this hydrolysis is still important for the dissociation of recombinase from both newly formed heteroduplex DNA and fortuitously bound RAD51 on duplex DNA.[85]

The RAD51 NPF is not a static structure. Breast cancer susceptibility gene product 2 (BRCA2) is known to nucleate RAD51 NPF formation on ssDNA at the dsDNA–ssDNA junction.[117–120] Many helicases in yeast and humans have been identified both *in vivo* and *in vitro* that act as anti-recombinases capable of dissociating RAD51 from ssDNA. These include Srs2 in yeast, and BLM and RECQ5 in humans.[68,121–124] Biophysical studies also indicate that Srs2 augments the Rad51 ATPase activity within the NPF to facilitate rapid protein turnover.[121] Moreover, both ensemble and single-molecule experiments show a direct correlation of recombinase turnover from DNA and its ATPase activity.[99,100,110,125–127]

Homology search during the synapsis phase by RAD51/RecA family is by random collision that involves transient nonspecific interactions with dsDNA, presumably bound at the secondary DNA-binding site.[99,127] The transient triplex DNA formed during the homology search is paramenic and not topologically interwound.[99] Biophysical studies using fluorescence resonance energy transfer and selective substitution of guanine to inosine on both ssDNA and its identical strand on the duplex DNA showed that human RAD51 facilitated homology search by a rapid A:T base-flipping mechanism.[128] Once a homologous sequence is found, strand exchange occurs to produce a topologically interwound plectonemic heteroduplex DNA product.[99,127]

B. DMC1

Disrupted meiotic cDNA (DMC1) was first isolated from a budding yeast meiotic cDNA library screen.[129] DMC1 is only expressed during meiosis and displays considerable homology to the RecA/RAD51 family of recombinases (Table I; Ref. 129). During meiosis, both Rad51 and Dmc1 colocalize at DSBs in yeast.[130] Disruption of yeast Dmc1 leads to a number of abnormal meiotic phenotypes, including accumulation of DSBs, reduced reciprocal recombination, abnormal synaptonemal complex formation, and defective meiotic prophase arrest.[129] Dmc1 exhibits both overlapping and yet nonredundant functions to Rad51.[131] However, overexpression of Rad51 enables yeast cells to circumvent the defective meiotic phenotype of Dmc1 mutants.[131] Human DMC1 exists as an octamer in solution[132] and functions similar to RAD51 in assays for recombination and ATPase activity *in vitro*.[126,133,134]

IV. Recombination Mediators

A. RPA

RPA is a heterotrimeric (70 kD, 32 kD, 14 kD) SSB that binds to ssDNA with high affinity (Table I; Refs. 135,136). It was first shown to stimulate strand exchange *in vitro* with strand exchange protein 1 (SEP1).[137,138] A similar stimulatory effect was shown later with yeast Rad51.[139] RPA has a dual stimulatory role during RAD51-mediated strand exchange. During the presynaptic phase, it binds to ssDNA to prevent secondary structure formation that could potentially lead to inhibitory effects during RAD51 NPF formation.[140] During the isoenergetic strand exchange phase, RPA ensures unidirectional heteroduplex extension by binding to the displaced ssDNA.[141,142] However, during recombination assays *in vitro* where only RPA and RAD51 are present, if RPA is added to ssDNA prior to the addition of RAD51, strand exchange is inhibited due to the nanomolar affinity of RPA for ssDNA. This inhibition can be overcome by the addition of recombination mediators such as Rad52 or Rad55/Rad57 in yeast Rad51 recombinase reactions, and by the addition of BRCA2 or the RAD51 paralog heterodimer RAD51B–RAD51C in human RAD51 recombinase reactions[117,119,143–147] (R. Amunugama and R. Fishel, unpublished data).

B. RAD52

Rad52 plays an essential role in HR and SSA and its deletion leads to severe sensitivity to DSB-causing agents and defects during meiosis in budding yeast (Table I; Refs. 85,86). Electron microscopic (EM) evidence indicates that both yeast and human RAD52 form oligomeric ring structures.[148–150] The EM structure of human RAD52 indicated that the N terminus is responsible for the formation of a heptameric ring structure and the C terminus then self-assembles the heptameric rings into a higher ordered structure.[148,149] However, two independent X-ray crystallographic analyses revealed that the yeast Rad52 N-terminal residues 1–201 (1–209 of human RAD52) formed an undecameric (11 subunit) ring.[151,152] The overall structure resembles a mushroom top with positively charged residues lining a groove on the outside of the ring.[151] Even though no DNA-containing structures of RAD52 have been solved, the dimensions of the groove indicate that it is large enough to bind ssDNA in a sequence-independent manner that would position the bases away from the protein surface for possible annealing with complementary bases.[89,151] The ssDNA-binding property of purified yeast Rad52 was first demonstrated by Rothstein and colleagues in 1996 and found to reside in the N terminus.[153] A similar ssDNA-binding pattern was observed when ssDNA–RAD52 complexes were probed for hydroxyl radical hypersensitivity.[154]

During meiotic and mitotic DSB repair, RAD51 recruitment is dependent on RAD52.[64,155–157] Yeast Rad52 has been shown to stimulate Rad51-mediated strand exchange activity by mediating Rad51 NPF formation on RPA-coated ssDNA filaments.[144–146,158] This mediator activity is critical to overcome the inhibitory action of RPA, when RPA is added prior to Rad51. Several methods including yeast two-hybrids, co-immunoprecipitation of whole cell extracts, and direct protein–protein affinity pull-down techniques have shown a direct interaction between Rad52 and Rad51.[91,158–160] The interaction with Rad51 is mediated by the C-terminal 409–412 residues of Rad52.[159,161] Overexpression of Rad51 alleviates the defective RR phenotype of a *rad52* (D409–412) C-terminal deletion mutation.[159] *In vitro* and *in vivo* evidence suggests that human RAD52 interacts with the cognate RAD51 via residues 291–330, a region that does not share homology with yeast Rad51.[162] These results appear to imply a species-specific interaction between RAD51 and RAD52. Rad52 has also been implicated in yeast Rad51-independent events such as BIR and SSA.[41,86,163]

Both yeast and human RAD52 have been shown to interact with RPA.[150,164,165] Yeast two-hybrid assays indicated Rad52 interaction with all three subunits of yeast RPA.[165] However, human RAD52 was shown to bind directly to large (70 kD) and middle (32 kD) subunits of RPA.[164] Interestingly, RAD52 interaction with RPA inhibits higher-order self-association of RAD52.[164] The strand-annealing activity of both yeast and human RAD52 facilitates the second-end capture during HR repair.[166,167]

Even though Rad52 plays an essential role in HR in budding yeast, in vertebrate cells, or in cells with BRCA2 homologs, loss of *RAD52* gene leads to few phenotypic defects in RR. RAD52 knockdown in mouse embryonic cells and in chicken B-cell line DT40 cell lines does not cause an apparent sensitivity to IR or DSB-causing chemical agents.[168,169] In the corn smut *Ustilago maydis* (which contains the Brh2 BRCA2 homolog), no defects in HR were found in Rad52 mutants.[170] A recent study of human breast cancer cell line suggests that knockdown of *RAD52* acts as a synthetically lethal agent in the case of *BRCA2* deficiency.[171] This finding elevates RAD52 as a target for antitumorigenic therapy for breast cancer.[171,172] A model has been proposed where RAD52 functions in an alternative pathway to BRCA2.[172] Human RAD52 does not appear to possess any RR mediator activity *in vitro*.[85,117] However, chicken DT40 cells were nonviable and exhibited severe HR defects in a double knockdown of *RAD52* and the RAD51 paralog *XRCC3*.[173] In addition, *U. maydis* Rad52 mutants demonstrated an enhanced UV and IR sensitivity when their sole RAD51 paralog *rec2* was mutated.[170] Collectively, this implies that in human cells, RAD52 might function as a recombination mediator in conjunction with any one or a combination of the RAD51 paralogs, RAD51B, RAD51C, RAD51D, XRCC2, or XRCC3.[172]

C. BRCA2

Germline mutations of the BRCA2 gene predispose individuals to highly penetrant, autosomal-dominant breast and ovarian cancers as well as predisposition to other types of cancers.[5,174–176] Mutations of BRCA2 in metazoans lead to gross chromosomal rearrangements, accumulation of chromosomal breaks, developmental arrest, meiotic defects, and increased hypersensitivity of DSB and interstrand cross-linking (ICL)-causing agents.[177–182] Biallelic *BRCA2* mutations that lead to expression of truncated forms of BRCA2 predispose individuals to Fanconi anemia (FA) and the designation as FANCD1.[183] The similar radiation sensitivity and developmental defective phenotypes observed in *RAD51*- and *BRCA2*-deficient cell types suggest that BRCA2 is intricately involved in RAD51-mediated RR repair (Table I). These assertions have been solidified by an observed interaction between BRCA2 and RAD51 using the yeast two-hybrid system.[181,184]

BRCA2 possesses two spatially distinct RAD51-binding regions. The first region involves repeated sequences of BRC motifs and the second RAD51 interaction motif is located at the C terminus of the protein (C-terminal RAD51-binding domain (CTRB); Fig. 5A; Refs. 85,181,184–188). Each BRC repeat consists of about 35 amino acids and several of the residues within each motif are conserved. This conservation is seen among metazoan BRCA2 orthologs.[189,190] However, the number of repeats in each organism varies. For instance, humans, mice, and chicken have eight BRC repeats, *Drosophila* has four, *Caenorhabditis elegans* and *U. maydis* have a single one, and the plant species rice and *Arabidopsis thaliana* have eight and four repeats, respectively.[191] The BRC repeats are not functionally equivalent.[192] Mutations within BRC repeats lead to abrogation of RAD51 binding and thus manifest defective DNA repair.[193] Structural analysis of the human BRCA2 BRC-4 repeat with the core region of RAD51 revealed an interface on BRC-4 that mimics a binding motif of RAD51.[194] This surface was suggested to function as an oligomerization site for RAD51 to facilitate RAD51 NPF formation.[194] Studies with *U. maydis* Brh2 also suggested a similar recruitment mechanism of RAD51.[119] All BRC repeats of human BRCA2 bind to RAD51 with variable affinity but with a binding stoichiometry of 1:1.[195] For example, BRC-1, -2, -3, and -4 bind to free RAD51 with a higher affinity compared to BRC-5, -6, -7, and -8.[195] BRC repeats can also modulate the loading of RAD51 onto DNA.[118,195,196] RAD51 filament formation on dsDNA leads to a dead-end complex that is recombination-deficient both *in vivo* and *in vitro*.[197,198] BRC-1, -2, -3, and -4 are able to suppress the ATP turnover rate of RAD51 and facilitate nucleation on RPA-coated ssDNA, while suppressing RAD51 binding to dsDNA[117,195] (Fig. 5B). This in turn leads to enhanced recombinase activity.[195] The latter group, BRC-5, -6, -7, and -8, however, do not enhance

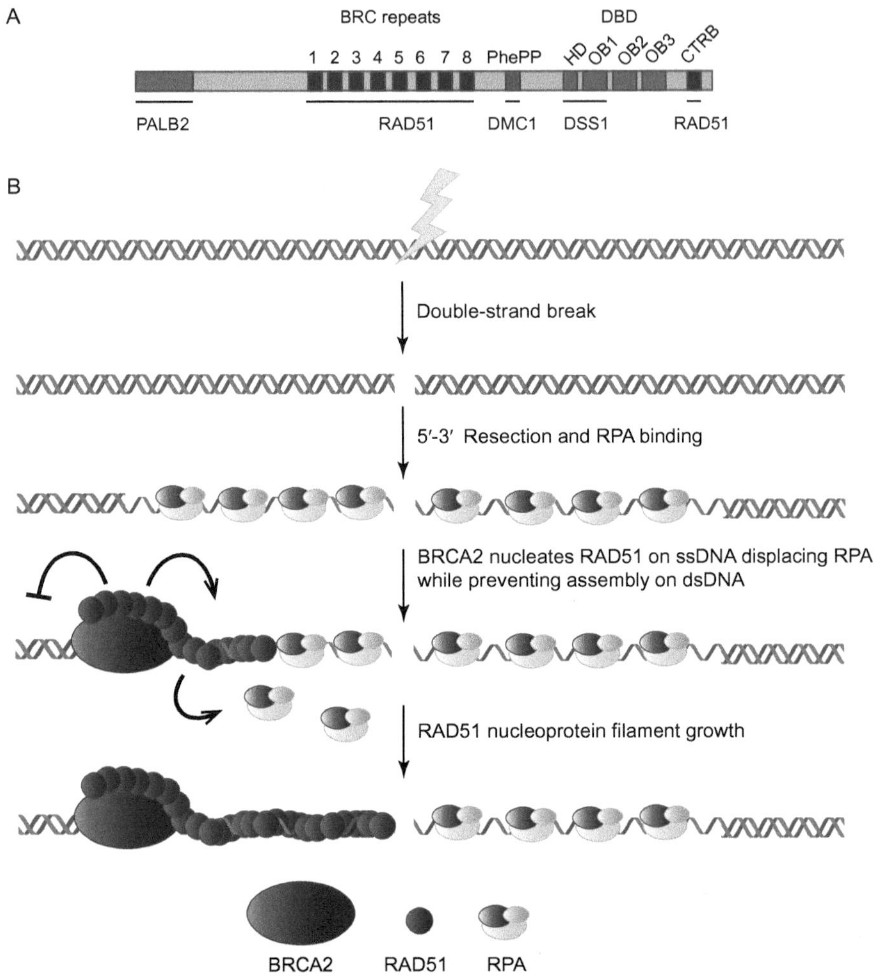

FIG. 5. BRCA2 and its proposed role in homologous recombination (HR). (A) Schematic representation of the functional and structural domains of human BRCA2. (B) Upon formation of a double-strand break (DSB), the 5′ strand is resected to leave a 3′ ssDNA. Replication protein A (RPA) binds and prevents secondary structure formation. BRCA2 binds at the dsDNA–ssDNA junction and initiates RAD51 nucleation on RPA-coated ssDNA while limiting nucleation on dsDNA. Continued RAD51 nucleoprotein filament growth results in a functional nucleoprotein filament that performs a homology search and strand exchange. (See Color Insert.)

preferential filament formation of RAD51.[195] The collective action of these two BRC groups can facilitate RAD51 nucleation and nascent filament formation on ssDNA prior to dissociation of BRCA2.[195] Similar attenuation of Rad51 ATPase activity is seen with *C. elegans* BRCA2 homolog BRC-2.[199]

The CTRB region that interacts with RAD51 is seen only in vertebrates and this region binds RAD51 filaments, but not free RAD51, likely to stabilize the NPF.[187,188] CDK phosphorylation of BRCA2(S3291) appears to inhibit its interaction with the RAD51 filament, leading to its disassembly.[188,200] As might be predicted, during the S phase when RR is highly active, there is very low BRCA2 (S3291) phosphorylation that gradually increases with the approach of the M phase.[200] These results suggest that the CTRB region functions as a cell cycle regulator of RR. In addition, the C terminus of BRCA2 is essential for the nuclear transport of RAD51 from the cytoplasm as RAD51 lacks a nuclear localization signal (NLS; Refs. 201,202). Thus, in human pancreatic cancer cell line CAPAN-1 that expresses a truncated version of BRCA2 (BRCA26174delT; Ref. 203), BRCA2 transportation into the nucleus is compromised and the levels of nuclear RAD51 are greatly diminished.[201] Finally, it has been recently shown that the CTRB region of BRCA2 is essential for protection of stalled replication forks against the MRE11 nuclease by stabilizing the RAD51 NPF.[204,205]

BRCA2 has also been shown to interact with DMC1 through a mammalian-specific 26 amino acid interaction motif containing BRCA2 residues 2386–2411.[206] This motif contains three critical amino acids, BRCA2 F2406, P2408, and P2409 (PhePP motif), that are essential for DMC1 interaction.[206] The N terminus of BRCA2 has also been shown to interact with RPA in a DNA-independent manner by co-immunoprecipitation[207] and the cancer-predisposing mutation BRCA2(Y42C) compromises this interaction.[207]

PALB2 (partner and localizer of BRCA2), also known as FANCN due to its involvement in FA,[208] interacts with the N terminus of BRCA2 and was shown to be essential for stable nuclear localization, recombination, and checkpoint functions of the latter.[209] The BRCA2(Y42C) also disrupts the interaction with PALB2.[209] In addition, PALB2 has been shown to stimulate RAD51-catalyzed D-loop formation by physically interacting with RAD51 and displays a cooperative effect in the presence of RAD51AP1, another stimulator of RAD51 recombinase activity.[210,211]

DSS1 (the *deleted in split hand/split foot* gene) is a small acidic protein first shown to interact with BRCA2 by yeast and mammalian two-hybrid assays.[212] Disruption of DSS1 leads to compromised RAD51 foci formation and DSB repair in both fungal and mammalian species.[213,214] DSS1 binds to the DNA-binding region of BRCA2,[215] and in *U. maydis*, Dss1 prevents dimerization of Brh2, allowing the formation of a monomeric functional protein.[216] These seemingly contradictory phenotypes will require some resolution in the coming years.

Other than the role of BRCA2 in DSB repair, it is also suggested to be involved in post-replication repair of ssDNA gaps by a template switch mechanism due to damage in the original template.[117] In fact, studies with Brh2 have confirmed the template switch mechanism *in vitro*.[217]

Recently, three independent groups were able to successfully express and purify the full-length BRCA2, which had initially been a challenge as a result of its sheer size (3418 amino acids; Refs. 117,218,219). The full-length BRCA2 was shown to bind six RAD51 molecules and enable their nucleation on RPA-coated ssDNA by binding to 3′ tailed structures.[117,218,219] Furthermore, as with BRC fragment analysis, the full-length BRCA2 suppressed RAD51 filament formation on dsDNA and stabilized RAD51–ssDNA NPF by suppressing the ATPase activity of the recombinase.[117,218] EM evidence of BRCA2-bound forked DNA structures illustrates its involvement in DNA replication coupled repair.[218]

V. RAD51 Paralogs

A. Yeast RAD51 Paralogs

RAD51 paralogs are products of gene duplication events of *RecA/RAD51* genes that function as accessory proteins in HR repair (Table I; Ref. 220). Budding yeast encodes two Rad51 paralogs, Rad55 and Rad57.[221,222] Mutants of Rad55 or Rad57 are cold-sensitive to DNA damage.[223] Overexpression of either Rad51 or Rad52 suppresses the DNA repair defect of these mutants, consistent with the notion that Rad55 and Rad57 function as accessory proteins in recombination.[223,224] In addition, Rad55 has been shown to interact with Rad57 and Rad51 both *in vivo* and *in vitro*.[223,224] Furthermore, inclusion of the stable heterodimer Rad55–Rad57 in substoichiometric amounts suppresses the inhibitory effects of RPA *in vitro*, suggesting an involvement during presynapsis.[147] Even though Rad55–Rad57 paralog heterodimer complex does not exhibit recombinase activity,[147] mutation of a Walker A box-conserved lysine to alanine in Rad55, but not in Rad57, results in defective meiotic recombination phenotypes.[224] In response to DSB-causing genotoxic stress cell cycle checkpoint, kinase-mediated phosphorylation of the Rad55 S2, S8, S14, and S378 residues has been shown essential for activating HR.[225–227] In the fission yeast *Schizosaccharomyces pombe*, mutants of *Rhp55* and *Rhp57* exhibited similar mutator phenotypes as budding yeast *Rad55* or *Rad57* mutants, which indicated structural and functional homology between Rad55 and Rad57 with Rhp55 and Rhp57, respectively.[228,229]

B. Vertebrate RAD51 Paralogs

Vertebrates encode five RAD51 paralogs, RAD51B (RAD51L1/hRec2/R51H2), RAD51C (RAD51L2), RAD51D (RAD51L3/R51H3), XRCC2, and XRCC3, that share 20–30% homology with RAD51 as well as with one other (Table I; Refs. 86,230). Similar to yeast Rad51 paralogs, these gene products appear to be the result of gene duplication events of an ancestral *RecA/RAD51* gene.[220] However, these paralogs show a high degree of evolutionary divergence from RAD51 as well as from each other.[230] XRCC2 and XRCC3 (X-ray repair cross complementing) were first identified in their ability to complement the extreme sensitivity of irs1 and irs1SF hamster cell lines.[231–233] Homology searches further identified RAD51B,[234–236] RAD51C,[237] and RAD51D.[236,238] Yeast two- and three-hybrid analyses, co-immunoprecipitation techniques, and biochemical studies have indicated interaction among the five RAD51 paralogs.[86,239] For example, RAD51B forms a stable heterodimer with RAD51C,[237,240,241] while RAD51 interacts with XRCC3 and weakly with RAD51C.[239,242] The latter interaction is improved in the presence of XRCC3.[239] RAD51D forms a stable complex with XRCC2.[243] In HeLa cells, two discrete complexes containing XRCC3 and RAD51C and the other containing RAD51B, RAD51C, RAD51D, and XRCC2 were found.[244,245] Knockout mutants of the RAD51 paralogs in chicken DT40 are viable yet exhibit increased sensitivity to cross-linking agents and IR as well as reduced RAD51 foci formation upon damage induction.[246,247] These phenotypes can be partially corrected by overexpression of RAD51.[246]

Compared to RAD51, the RAD51 paralogs display weaker DNA-stimulated ATPase activities.[143,241,243,248] RAD51B and RAD51C bind to ssDNA, dsDNA, and 3′-tailed dsDNA.[241] Moreover, RAD51C has an apparent strand exchange activity perhaps by destabilizing dsDNA.[241] RAD51D preferentially binds to ssDNA.[243] RAD51B–RAD51C heterodimer possesses *in vitro* recombination mediator activity for RAD51-catalyzed strand exchange by suppressing the inhibitory effect of RPA.[143] Furthermore, RAD51B–RAD51C is able to suppress the anti-recombinogenic activity of BLM during RAD51-mediated D-loop formation (R. Amunugama and R. Fishel, unpublished data). XRCC2 has been shown to enhance the ATP-processing activity of RAD51 by facilitating adenosine diphosphate (ADP) to adenosine triphosphate (ATP) exchange by reducing the affinity for ADP.[248] The *in vivo* complexes RAD51B, RAD51C, RAD51D, and XRCC2 and RAD51C–XRCC3 bind to ssDNA, 3′- and 5′-tailed dsDNA, forked DNA structures, and Holliday junctions.[244,245,249]

C. RAD54

RAD54 is a highly conserved eukaryotic gene of the *RAD52* epistasis group (Table I; Refs. 250–252). RAD54 homologs have been identified in a number of eukaryotes including yeast, Drosophila, plants, zebrafish, chicken, mice, and humans.[253–259] A RAD54 homolog has been identified in the archaebacterium *Sulfolobus solfataricus*, but not in eubacteria.[251,260] RAD54 homologs of budding yeast and humans share a 66% similarity and 48% homology.[253,259] Budding yeast *Rad54* was first discovered in a genetic screen to isolate mutants sensitive for IR.[88,261] Similar to *Rad51* and *Rad52* mutants, *Rad54* mutants were hypersensitive to IR as well as DNA cross-linking and alkylating agents that eventually cause DSBs.[41] *Rad54* mutants display only minor defects in meiotic recombination in yeast due to the presence of the meiotic homolog Rdh54/Tid1.[262] *RAD54* knockdown in mice causes hypersensitivity to IR at embryonic stages but not in adult stages due to rescue by NHEJ repair.[263] However, all developmental stages of *RAD54*-deficient mice are hypersensitive to DNA cross-linking agents.[263] Mutational analysis of Walker A box-conserved lysine indicated that in both mice and yeast, ATP hydrolysis of Rad54 is essential for its function *in vivo*.[264–266]

Rad54 expression levels increase during the late G1 phase of the cell cycle,[267,268] presumably to connect HR repair of DSBs during the late S and G2 phases.[269] *Rad54* expression levels are upregulated during DSB formation and Rad54 foci formation is dependent on Rad51.[64,270] However, Rad51 foci formation is not dependent on Rad54, indicating that Rad54 acts downstream of Rad51.[64] Similarly, RAD54 colocalizes with RAD51 foci following IR in mammalian cells.[271,272]

RAD54 belongs to the Swi2/Snf2 SF2 (superfamily 2) of proteins.[250–252] The Snf2/Swi2 proteins are commonly known for dsDNA-dependent ATPase, ATP-dependent chromatin remodeling, DNA translocase, and DNA-supercoiling activities.[250–252] Like other members of SF1 and SF2, members of RAD54 possess several signature helicase motifs I, Ia, II, III, IV, V, and VI that constitute the two tandem RecA-like lobes that utilize the energy of ATP binding and hydrolysis for functioning.[273,274] However, unlike helicases, the SF2 family of proteins does not unwind but translocates on dsDNA.[250,251] Also unlike helicases, the ATPase activity of RAD54 is not stimulated by ssDNA, nor does it translocate on ssDNA.[274,275] Both yeast and human RAD54 are strictly dsDNA-dependent ATPases, with a catalytic turnover rate ranging from 3000 to 6000 min^{-1}.[250] The binding affinity for branched DNA structures such as PX junctions (partial Holliday junctions) is approximately 200 times higher than for ssDNA or dsDNA.[276]

RAD54 function has been implicated in all three stages of recombination: presynapsis, synapsis, and postsynapsis.[251,252] These include interaction with RAD51 to stabilize the ssDNA NPF, stimulation of homology search and strand exchange catalyzed by RAD51, chromatin remodeling during the homology search, disruption of RAD51–dsDNA filaments, branch migration of Holliday junctions, and interaction with specific endonucleases to stimulate resolution of Holliday junctions.

RAD54 interacts with RAD51 in a species-specific manner through its N-terminal domain.[277–280] This interaction is seen with both free RAD51 and the RAD51 ssDNA NPF.[281] The ATPase activity of RAD54 is not required for RAD51 NPF stability. This was shown *in vivo* and *in vitro* using a RAD54 ATPase-deficient mutation where a Walker A box lysine to arginine substitution allows only ATP binding.[281,282]

The role of RAD54 in stimulating RAD51-mediated three-strand exchange activity and D-loop formation was first demonstrated with recombinant yeast proteins.[283] This stimulation is seen with many RAD54 orthologs and occurs in a species-specific manner.[277,284–286] Sub-stoichiometric amounts of RAD54 are sufficient to greatly stimulate the RAD51 recombinase activity *in vitro*,[283,284,286] indicating that the protein functions in a catalytic manner. In fact, for RAD51-mediated strand exchange stimulation, the ATPase activity of RAD54 is required.[264] Conversely, RAD51 improves the ATP hydrolysis and translocation ability of RAD54 on dsDNA.[284,287]

RAD54 facilitates dissociation of RAD51 from heteroduplex DNA following strand exchange in an ATP-dependent manner.[198,288] Overexpression of Rad51 in Rdh54-deficient background leads to arrest of cell growth in yeast due to accumulation of Rad51 on undamaged chromatin.[197] Furthermore, it was revealed that Rad54 is specialized for removal of Rad51 from damage-induced foci, while Rdh54 is involved in disassembly of Rad51 from undamaged toxic dead-end dsDNA complexes.[197]

During the postsynaptic phases, Rad54 enhances the heteroduplex extension of Rad51.[289] Human RAD54 has a relatively higher affinity for Holliday junctions and PX junctions compared to dsDNA, and it exhibits branch migration activity in a multimeric functional complex.[276,290] Furthermore, budding yeast Rad54 and human RAD54 have been shown to physically interact with the Holliday junction resolvase Mus81–Mms4 and MUS81–EME1, respectively, to stimulate Holliday junction resolution.[291,292] Even though the branch migration activity of RAD54 was not required for MUS81–EME1 stimulation, ATP was required.[292]

D. RAD51AP1

RAD51AP1 (RAD51-associated protein 1), previously known as PIR5, enhances RAD51-mediated joint molecule formation by physically interacting with both RAD51 and joint DNA structures and dsDNA molecules.[293–295] Knockdown of RAD51AP1 in human cells increases genotoxic stress to DSB-inducing agents.[293,294] However, the detailed role of RAD51AP in RR remains a mystery.

VI. DSB Repair in Chromatin

A. DSB-Induced Histone Modifications

In higher eukaryotes, DNA is compacted into chromatin and the basic unit is a nucleosome, which consists of 146 bp of DNA wrapped approximately 1.7 times in left-handed superhelical turns around a tetramer of histones H3 and H4 with two H2A–H2B dimers.[296] There are several levels of compaction of chromatin *in vivo*.[296,297] The basic nucleosome structure is arranged into an array of nucleosomes resembling "beads on a string".[298] This array is further compacted into a 30-nm filament solenoid through internucleosomal interactions and interactions with linker histones H1 or H5.[299]

In order to perform DNA replication, repair, and transcription, the chromatin must be dynamic. The dynamic character is achieved by posttranslational modifications (PMTs) of amino acid residues on both the solvent-exposed histone tails and the core regions.[300] The PMTs include phosphorylation, methylation, acetylation, ubiquitination, SUMOylation, ADP-ribosylation, and proline-isomerization.[300,301] Collectively, these modifications have been termed the "histone code".[301–303] Once the histones are modified, their affinity for DNA may change. For example, biochemical studies have shown that nucleosomes containing acetylated histone are more mobile than unmodified nucleosomes.[304,305] These nucleosomes can then be evicted or pushed from the region of DNA that needs to be replicated, repaired, or transcribed by chromatin remodelers using the free energy of ATP binding and/or hydrolysis.[306–312]

The initial chromatin-mediated response to a DSB is the phosphorylation of the C terminus of either major H2A variant in budding yeast on S129 (often referred to as H2AX for simplicity[313]) or the H2AX variant in mammals on S139.[314–316] H2AX constitutes approximately 10% of the nucleosomal H2A complement and the phosphorylated form is referred to as γH2AX.[315] The phosphorylation is ATM and MDC1 mediated, and in yeast it spans several kilobases from the DSB, while in mammals, γH2AX spreads to megabase distances flanking the DSB.[314,317,318] MDC1 directly interacts with γH2AX via its C-terminal BRCT motifs.[319] Upon DSB formation, MDC1 is phosphorylated by casein kinase 2 (CK2) on its N-terminal S-D-T triamino acid repeats and these phospho-domains interact with FHA and BRCT repeats of NBS1 in the MRN complex, which facilitates its recruitment to the DSB site.[320–322] Phosphorylated MDC1 once recruited to the DSB site functions as a positive feedback regulator by binding to γH2AX via its BRCT domain and to ATM through its FHA domain, respectively, to facilitate ATM-mediated additional phosphorylation of H2AX to amplify the DNA damage signal.[323] Sustaining γH2AX flanking, a DSB is critical for the recruitment of downstream repair factors.[324]

γH2AX functions as a molecular beacon to recruit the cohesion complex that is involved in linking sister chromatids during the postreplicative phase of the cell cycle.[325–327] This process prevents LOH during mitotic HR by allowing the broken strand to use the sister chromatid as the donor template. Studies on budding yeast indicate that γH2AX recruits ATP-dependent chromatin-remodeling factor INO80 through direct interaction.[313,328] Furthermore, ssDNA formation that functions as the substrate for RAD51 NPF assembly is compromised in *arp8* INO80 subunit mutants in yeast.[328] The histone acetyl transferase (HAT) NuA4 has also shown to interact with γH2AX and appears to acetylate histone H4 following DSB formation.[329]

H2AX(Y142) is also constitutively phosphorylated by the Williams–Beuren syndrome transcription factor (WSTF) kinase.[330] Following that advent of a DSB, H2AX(Y142) is dephosphorylated by the EYA1/EYA3 phosphatases.[331,332] Interestingly, MDC1 interaction with γH2AX depends on a dephosphorylation of H2AX(Y142), and the relative amount of dephosphorylation determines the recruitment of either proapoptotic or DNA repair factors to the damaged site.[331] This has suggested that the H2AX(Y142) phosphorylation–dephosphorylation is a "molecular switch" in response to DSB damage.[333]

An essential role for histone ubiquitination during the DSB repair response emerged after the discovery of the E3 ubiquitin ligase RNF8 at DSB foci.[334–336] RNF8 recruitment to DSB is mediated by the interaction between its FHA domain with the phosphorylated motifs of MDC1.[334–336] RNF8 was shown to catalyze the ubiquitinylation of histone H2A and H2AX upon DSB formation, and knockdown of RNF8 or disruption of FHA domains leads to failure to recruit the checkpoint-activating proteins BRCA1 and 53BP1 to the DSB.[334,335] Furthermore, depletion of the E2 ubiquitin adapter UBC13 compromises RNF8 function.[335,336] UBC13 is an essential ubiquitin adapter for HR that is also recruited to DSBs.[337,338] However, RNF8-mediated ubiquitination is not sufficient to sustain the damage signal at DSB foci and another ubiquitin E3 ligase, RNF168, that acts with UBC13 to amplify the ubiquitination signal via K63-linked ubiquitination of H2A and H2AX.[339,340] Another nonproteolytic E3 ligase, HERC2, is recruited to damage-induced foci and also forms a complex with RNF8 and RNF168 to extend their retention at the repair site.[341] The K63-linked polyubiquitinated histones function as substrates for BRCA1 A-complex binding that includes BRCA1/BARD1, ABRAXAS, RAP80, and BRCC36 through the ubiquitin-interaction motif (UIM) of RAP80.[342] ABRAXAS is thought to recruit RAP80 to the DSB site.[343] Just as ubiquitination of H2AX is critical for DSB repair factor recruitment, deubiquitination is also tightly regulated during the damage response. In fact, the deubiquitinating enzymes BRCC36 and OTUB1 are simultaneously recruited damage-induced foci.[344,345] BRCC36 is part of the BRCA1 A-complex and OTUB1 physically binds to UBC13 and inhibits RNF8- and RNF168-mediated polyubiquitination.[344,345]

SUMOylation also appears to be essential for effective DSB damage response. Recently it was shown that SUMO1, SUMO2, and SUMO3 are recruited to damage-induced foci along with the E3 ligases PIAS1 and PIAS4.[346,347] SUMOylation is critical for productive assembly of BRCA1, 53BP1, and RNF168 at damage-induced foci, and PIAS-mediated SUMOylation of BRCA1 leads to increased ubiquitin ligase activity of the BRCA1–BARD1 heterodimer *in vitro*.[346,347]

Histone H2B(K120) has recently been shown to be ubiquitinylated by the E3 ligase RNF20–RNF40 heterodimer following DSB damage. The H2B (K120ub) modification leads to the recruitment of chromatin-remodeling factor SNF2h to the DSB.[348,349]

Other histone modifications include constitutively methylated histone H3 (K79) (H3(K79me)), which has been shown to provide an interacting domain for 53BP1 upon relaxation of higher-order chromatin structure during a DSB.[350] Furthermore, 53BP1 also interacts with the dimethylated form of histone H4(K20) (H4(K20me2)) via its tandem tudor domains.[351] HAT acetylation of histones is common during DNA replication and repair.[300,352] Acetylation is found not only on the histone tails but also on the core regions. For example, histones H3(K9) as well as H3(K56) are acetylated by GCN5/KAT2A and p300 in response to DSB damage in human cells.[353,354] It is believed that H3(K56ac) influences the mobility of nucleosomes by neutralizing the positive charge on the lysine and modelizing the entry–exit region of the nucleosome.[352,355] Finally, TIP60/Esa1 has been found to acetylate H4 and H2AX during DSB repair.[356–359]

B. ATP-Dependent Chromatin Remodeling

A yeast system that employed a galactose-inducible *HO* endonuclease-provoked single DSB at a defined position in the "mating-type" (*MAT*) locus followed by chromatin immunoprecipitation allowed monitoring of chromatin remodeler and HR repair machinery recruitment at the recipient and donor sites in real time.[360,361] Once *HO* endonuclease cleaves the specific site within the MAT locus, which is otherwise protected by highly positioned nucleosomes, the break can be repaired by either NHEJ or HR.[362] The latter pathway is employed if the donor sequence *HMRa* or *HMLα* is present.[360,363] These studies have identified several chromatin remodelers recruited to the DSB site including INO80, SWI/SNF, RSC, SWR1, RAD54, and TIP60[155,282,313,328,356,358,364–377].

Inositol auxotroph 80 (INO80) is a multisubunit chromatin-remodeling complex which was first characterized in a budding yeast mutant strain that exhibited defective transcription activation following inositol depletion (Table I; Ref. 378). Relatively widely studied compared to other ATP-dependent chromatin remodelers, it is composed of several subunits that are shared by yeast and other eukaryotes.[362] These core subunits include INO80 ATPase, two AAA^+ ATPases (Rvb1 and Rvb2 in yeast, RuvB-like 1 and RuvB-like 2 in humans), actin and actin-related proteins Arp4, Arp5, and Arp8, and INO80 subunits (Ies2 and Ies6).[379] In addition, the yeast Ino80 complex contains unique polypeptides Taf14, HMG, Nhp10, and Ies1, Ies3, Ies4, and Ies5, whereas the human homolog contains YY1, Uch37, and NFRKB.[379] The Ino80ATPase subunit of INO80 appears to be essential for its cellular function and for ATP-dependent chromatin remodeling *in vitro*.[380–382] Ino80 is recruited to the HO endonuclease DSB within an hour.[313,328,364] In yeast, deletion of Ino80ATPase, Arp5, or Arp8 subunit leads to increased sensitivity to DSB-inducing agents.[313,328,364] Recruitment of Ino80 to the DSB is dependent on a specific interaction between the

Nhp10 (an HMG-like subunit of the Ino80 complex) and γ-H2AX, and the loss of either component leads to compromised DSB repair.[313] However, two conflicting observations were reported with respect to recruitment of repair mediators at the DSB. One group observed that resection of 5′ strands at the break was compromised in Arp8 and H2A mutants,[328] while the other reported that even though the resection occurred regularly in the *Ino80* mutant, Rad51 recruitment was delayed.[364]

The SWI/SNF chromatin-remodeling complex was first identified in two genetic screens in budding yeast (Table I). The first gene regulates the mating type switch (SWI) and the other regulates sucrose nonfermenting (SNF) phenotypes.[383–385] The multisubunit complex consists of 9–12 subunits and has shown to possess ATP-dependent chromatin-remodeling activity *in vitro*.[386,387] Several homologs of SWI/SWF have been identified in metazoans; for example, in *Drosophila*, BRM (Brahma) and BAPs (BRM-associated proteins) were characterized as SWI/SNF homologs and in humans, BRM, BRG1 (Brahma-related gene 1) and BAFs (BRM- or BRG1-associated factors) are found as SWI/SNF complexes.[73,388–390] SWI/SNF remodels nucleosomes both by nucleosome sliding and nucleosome ejection.[391] Its activity is implicated both in transcription and DSB repair.[362,366,367] SWI/SNF involvement in DSB repair is extensively studied in yeast. Although Rad51 NPF formation does not require chromatin remodelers for homology searches and capture even on positioned nucleosomal surfaces *in vivo* or *in vitro*,[156,282,373,392] Swi/Snf remodelers are essential for recombinational repair within heterochromatin.[365] When nucleosomal donor sequences are constrained by Sir2, Sir3, and Sir4 structural proteins that are found at telomeres and silent *MAT* loci, remodeling activity of Swi/Snf was required for efficient joint molecule formation.[365] Interestingly, Rad54, Ino80, RSC, and Swr1 were incapable of promoting joint molecule formation within heterochromatin.[365] Several mammalian SWI/SNF complexes have also been shown to interact with DSB repair response proteins such as BRCA1, p53, and with FA pathway proteins.[393–395]

Remodel structure of chromatin (RSC) is another multisubunit ATP-dependent chromatin remodeler that is rapidly recruited to the DSB site upon damage.[366] It is homologous to the SWI/SNF complex.[372] Its rapid recruitment to DSB that coincides with MRX recruitment suggests that initial nucleosome remodeling at the DSB might facilitate DNA end-processing by the MRX complex.[362,368,369] RSC is also required for loading of cohesins during DSB repair to ensure that repair occurs between sister chromatids during mitotic HR repair.[368,370] However, there is also evidence that suggests that RSC might be involved in the latter stages of HR repair pathway, particularly during the DNA ligation step after DNA synthesis.[366]

SWR1 is closely related to INO80 that is recruited to the DSB site in a γH2AX-dependent manner.[329,396] However, its involvement in DSB repair is different than the role of INO80. SWR1 is known for exchanging histone H2A with the H2AZ variant.[374,397] SWR1 is also implicated in regulating γH2AX levels. Inactivation of SWR1 ceases H2AZ incorporation at nucleosomes surrounding the DSB; however, this restores γH2AX levels and checkpoint adaptation, which functions antagonistically to INO80.[373]

Other than the acetylation of histones H4 and H2AX by TIP60 following DNA damage,[356–359] the *Drosophila* TIP60 has been shown to exchange acetylated histone H2Av (H2AX homolog) with unmodified H2Av using the domino/p400 ATPase.[358,398] The conjunctional action of TIP60/p400 that leads to nucleosomal destabilization at the DSB is required for efficient RNF8-mediated ubiquitination of histones and recruitment of BRCA1 and 53BP1 for DSB response.[398] Furthermore, the exchange of acetylated H2Av with unmodified H2Av might be critical for attenuation of DSB signal propagation.[362]

RAD54 appears to have the ability to remodel chromatin.[399] RAD54 from budding yeast, *Drosophila*, and humans has been shown to possess ATP-driven chromatin-remodeling activity *in vitro* using assembled mononucleosomes and nucleosomal arrays.[285,400–402] The RAD51 NPF stimulates chromatin-remodeling activity as well as the dsDNA-dependent ATPase activity of RAD54.[250–252] Strand exchange by RAD51 is greatly enhanced by RAD54 in chromatinized substrates,[285,400] even though for the initial homology, capture by RAD51 NPF on a chromatin substrate RAD54 is not required.[392] Interestingly, a specific protein–protein interaction between the amino terminus of histone H3 and RAD54 has also been reported, implying a specific role in RAD54-mediated chromatin remodeling.[403]

VII. Postsynaptic Removal of RAD51

Postsynaptic RAD51 turnover plays a critical role in regulating the HR repair pathway (Table I). The deproteinization step in conventional strand exchange studies *in vitro* obliterates the possible analysis of RAD51 dissociation from dsDNA.[404] Furthermore, unlike RecA that dissociates from dsDNA upon ATP hydrolysis,[99] RAD51 remains bound to heteroduplex DNA, probably due to its intrinsic slow ATP hydrolysis. These results suggest that eukaryotes accessory proteins have evolved to facilitate RAD51 removal from the nascent heteroduplex to allow the 3′-end to prime DNA synthesis.[85,404] RAD54 has been shown to dissociate RAD51 from dsDNA.[198,251] Recent studies on the nematode *C. elegans* identified two more gene products, a helicase HELQ-1 (homologous to human HEL308) and the single RAD51 paralog RFS-1, that are essential for postsynaptic RAD51 turnover during meiotic HR.[405] Both these gene products have been shown to have synthetic lethal interactions.[405] It is suggested that once RAD54 removes the RAD51 from the 3′-end of the invading strand, HELQ-1 and RFS-1 might be involved in removing the remaining RAD51 from the heteroduplex DNA.[405,406]

VIII. Second-End Capture

When the RAD51 NPF forms a D-loop, the displaced strand can potentially pair with the second 3′ overhang that is produced. This process is called second-end capture. In yeast and mammalian cells, this process is mediated by RAD52 through its inherent ability of pair with RPA-coated ssDNA.[166,407–409] In *U. maydis*, Brh2 is also capable of catalyzing second-end capture in conditions where annealing by RAD52 is inhibited.[217]

IX. dHJ Dissolution

In the classical DSB repair model, after the second-end capture, DNA synthesis and ligation result in a dHJ, which can be either dissolved or resolved. During mitotic recombinational repair, the former is the preferred pathway.[29,34,410] dHJ dissolution is mainly mediated by RecQ helicases (Table I; Ref. 410). The budding yeast RecQ homolog Sgs1 (fission yeast Rqh1) interacts strongly with the Top3 topoisomerase.[411–414] In human cells, the homologous RecQ helicase BLM strongly interacts with TOP3α.[415,416] This helicase–topoisomerase complex has been shown to interact with yeast Rmi1/ Nce4 (RMI1/ BLAP75 in human cells) DNA-binding protein. In budding yeast, dHJ dissolution is mediated by Sgs1–Top3–Rmi1.[417,418] In human cells, the dHJ dissolution complex is comprised of BLM–TOP3α–RMI1–RMI2.[410,419]

X. Holliday Junction Resolution

Weisberg and colleagues reported the first biochemical evidence for an enzyme that has Holliday junction resolution activity in 1982. They identified the T4-endonuclease activity of bacteriophage T4 that was capable of cutting the branched structures of the phage genome before it was packaged into new phage particles.[420–423] Soon afterwards, Holliday junction resolvases were identified in several organisms including budding yeast mitochondrial Cce1 (from cell-free extracts), fission yeast Ydc2, and endonuclease I of the bacteriophage T7.[424–429] The first prokaryotic resolvase to be identified was *Escherichia coli* RuvC.[430,431] Biochemical characterization of RuvC revealed that the resolvase binds the

Holliday junction as a dimer and unfolds the antiparallel stacked-X structure (observed in the presence of divalent cation Mg^{++}) into an open-planar structure (observed in the absence of divalent ions) and nicks strands of the same polarity.[430–436] RuvC activity in bacterial cells are closely associated with the RuvA, the tetrameric protein that binds to Holliday junctions and unfolds it into an open-planar structure and RuvB, an hexameric ATP hydrolysis-driven translocase which promotes branch migration.[89,437–441]

Holliday junction resolvase activity in mammalian systems was first observed in extracts prepared from homogenized calf thymus tissues.[442] Similar resolvase activity was later observed in extracts prepared from cultured cells.[443,444] Given the complexity of eukaryotic genomes and the stringency of maintaining its integrity, eukaryotic cells have evolved multiple pathways and resolvases to process Holliday junctions.[445] This in turn made identification of single mutants defective in Holliday junction resolution challenging in eukaryotic model organisms.[446] In 2001, the Mus81–Eme1 (Mus81–Mms4 in budding yeast) heterodimer was identified in fission yeast as an endonuclease capable of cleaving Holliday junctions as well as branched DNA structures (Table I; Refs. 447–449). Mus81 is homologous to the XPF subunit of the ERCC1–XPF nucleotide excision repair endonuclease.[448,449] Eme1 is a noncatalytic subunit.[420,448,449] Mus81–Eme1 depletion caused some meiotic defects and stalled replication forks.[448,449] Meiotic defective cells could be rescued by ectopic expression of bacterial Holliday junction resolvases.[449] In 2003, human MUS81–EME1 was characterized as a replication fork/flap endonuclease that is essential to maintain the integrity of replication, even though it possessed inefficient Holliday junction resolvases *in vitro*.[450] In the case of budding yeast, *Mus81* deletion only exhibited modest decrease in crossover formation during meiosis, implying that Mus81–Eme1 is not the sole resolvase complex of Holliday junctions.[451,452] Similar minor meiotic defects were observed in *MUS81* knockout studies in mice.[453,454] Search for a RuvC-type Holliday junction resolvase in eukaryotic cells had been quite challenging by conventional sequence homology-based queries due to the absence of conservation of primary amino acid sequences.[421] However, tertiary structure-level conservation was seen among most of the identified Holliday junction resolvases, categorizing them into two main superfamilies of integrase and nuclease.[421] Identification of a Holliday junction resolvase in eukaryotes that resembled bacterial RuvC was inadvertently assigned to the RAD51C–XRCC3 heterodimer, which was isolated from mammalian cells.[446,455] However, the absence of an apparent nuclease domain and the inability of recombinant RAD51C–XRCC3 to recapitulate the Holliday junction resolvase activity led to questions regarding the assignment of the Holliday junction resolvase to the RAD51C–XRCC3 heterodimer.[420,456] In 2008, after a tedious effort by the West laboratory (the group that suggested that RAD51C–XRCC3 was a

Holliday junction resolvase), the eukaryotic ResA Holliday junction resolvase was identified.[457] It appears that ResA copurifies with RAD51C–XRCC3 isolated from human cells.[446,457]

Recently, another group of Holliday junction resolvases, namely, the SLX1–SLX4 complex, were identified (Table I; Refs. 458–461). Even though Slx1 was conserved, conventional homology searches did not indicate a conservation of Slx4 outside the yeast genome.[420,445,462] By more refined in silico analyses, human SLX4 was identified as BTBD12, which was the ortholog *Drosophila* MUS312 and fungal Slx4.[458–460,463] In a separate study, BTBD12 was identified as a phosphorylation substrate of ATM/ATR kinases.[459,464] SLX1 possesses the endonuclease domain for Holliday junction cleavage, while SLX4 acts as a protein-interacting scaffold that interacts with multiple nucleases that cleave Holliday junctions both symmetrically and asymmetrically.[420,445] In fact, SLX4 has been implicated in multiple genome maintenance pathways including replication and repair.[445,458–460,463] Because SLX4 interacts with other Holliday junction resolvases such as MUS81–EME1, when the SLX1–SLX4 complex was isolated from human cells, symmetrical cleavage of Holliday junctions was not observed.[458,459,464] However, bacterially expressed, the SLX1–SLX4 heterodimer with a truncated SLX4 region that does not interact with MUS81–EME1 did possess symmetrical cleavage ability of Holliday junctions.[445,458,459,464] Among the many interacting partners of mammalian SLX4 are proteins of diverse functions. These include the endonucleases SLX1, ERCC4–ERCC1, and MUS81–EME1; mismatch repair heterodimer MSH2–MSH3; telomere proteins TRF2/RAP1; and polo-like kinase PLK1.[445,458–460,464]

XI. Homeologous Recombination: The Interplay Between Mismatch Repair and HR

Mismatch repair (MMR) is a conserved process that plays an important role in maintaining genome integrity by correcting DNA mismatches formed during replication and recombination.[465–467] MMR ensures that HR occurs between perfectly homologous sequences and suppresses recombination between sequences that contain partial homology (homeologous recombination).[465] Genetic studies in budding yeast indicate that even a single mismatch reduces the recombination rates by at least fourfold compared to recombination between substrates of perfect homology.[468,469] In yeast, Msh2–Msh6, Msh2–Msh3, and Mlh1–Pms1 MMR heterodimers as well as Rad1–Rad10 and Exo1 nucleases and helicases Sgs1 and Srs2 have been implicated in suppressing homeologous recombination.[29,470–473] In mice and human cells, a BLM (Sgs1 homolog) deficiency still suppresses homeologous recombination. However, in combination with an Msh2 deficiency, the amount of homeologous recombination events increased.[474] RecQ helicase Werner syndrome (WRN) has also been implicated in suppressing homeologous recombination by exhibiting strong interaction with the MSH2–MSH6, MSH2–MSH3, and MLH1–PMS2 heterodimers.[475] To date, a reaction that suppresses homeologous recombination *in vitro* has not been developed.

XII. Conclusion

HR is mechanistically conserved in prokaryotes and eukaryotes. However, because of the genome complexity in eukaryotes, additional mediators are required for the successful repair of DSBs. Defective HR leads to genomic instability and tumorigenesis. Paradoxically, unregulated HR also leads to the same outcome. Therefore, eukaryotic cells have evolved elegant mechanisms to regulate each step of HR to ultimately produce an accurate repair outcome.

References

1. D'Andrea AD, Grompe M. The Fanconi anaemia/BRCA pathway. *Nat Rev Cancer* 2003;**3**:23–34.
2. Ellis NA, Groden J, Ye TZ, et al. The Bloom's syndrome gene product is homologous to RecQ helicases. *Cell* 1995;**83**:655–66.
3. Miki Y, Swensen J, Shattuck Eidens D, et al. A strong candidate for the breast and ovarian cancer susceptibility gene BRCA1. *Science* 1994;**266**:66–71.
4. Savitsky K, Bar Shira A, Gilad S, et al. A single ataxia telangiectasia gene with a product similar to PI-3 kinase. [See comments]. *Science* 1995;**268**:1749–53.
5. Tavtigian SV, Simard J, Rommens J, et al. The complete BRCA2 gene and mutations in chromosome 13q-linked kindreds. [See comments]. *Nat Genet* 1996;**12**:333–7.
6. Hoeijmakers JH. Genome maintenance mechanisms for preventing cancer. *Nature* 2001;**411**:366–74.
7. Lobrich M, Jeggo PA. The impact of a negligent G2/M checkpoint on genomic instability and cancer induction. *Nat Rev Cancer* 2007;**7**:861–9.
8. Farmer H, McCabe N, Lord CJ, et al. Targeting the DNA repair defect in BRCA mutant cells as a therapeutic strategy. *Nature* 2005;**434**:917–21.
9. Zickler D, Kleckner N. Meiotic chromosomes: integrating structure and function. *Annu Rev Genet* 1999;**33**:603–754.
10. Marston AL, Amon A. Meiosis: cell-cycle controls shuffle and deal. *Nat Rev Mol Cell Biol* 2004;**5**:983–97.
11. Hassold T, Hall H, Hunt P. The origin of human aneuploidy: where we have been, where we are going. *Hum Mol Genet* 2007;**16**(Spec No. 2):R203–8.
12. Hassold T, Sherman S, Hunt PA. The origin of trisomy in humans. *Prog Clin Biol Res* 1995;**393**:1–12.
13. Heyting C. Synaptonemal complexes: structure and function. *Curr Biol* 1996;**8**:389–96.

14. Keeney S. Mechanism and control of meiotic recombination initiation. *Curr Top Dev Biol* 2001;**52**:1–53.
15. Klapholz S, Waddell CS, Esposito RE. The role of the SPO11 gene in meiotic recombination in yeast. *Genetics* 1985;**110**:187–216.
16. Kleckner N. Meiosis: how could it work? *Proc Natl Acad Sci U S A* 1996;**93**:8167–74.
17. Cromie GA, Smith GR. Branching out: meiotic recombination and its regulation. *Trends Cell Biol* 2007;**17**:448–55.
18. Alani E, Padmore R, Kleckner N. Analysis of wild-type and rad50 mutants of yeast suggests an intimate relationship between meiotic chromosome synapsis and recombination. *Cell* 1990;**61**:419–36.
19. Liu J, Wu TC, Lichten M. The location and structure of double-strand DNA breaks induced during yeast meiosis: evidence for a covalently linked DNA-protein intermediate. *EMBO J* 1995;**14**:4599–608.
20. Masson JY, Davies AA, Hajibagheri N, et al. The meiosis-specific recombinase hDmc1 forms ring structures and interacts with hRad51. *EMBO J* 1999;**18**:6552–60.
21. Masson JY, West SC. The Rad51 and Dmc1 recombinases: a non-identical twin relationship. *Trends Biochem Sci* 2001;**26**:131–6.
22. Kantake N, Sugiyama T, Kolodner RD, Kowalczykowski SC. The recombination-deficient mutant RPA (rfa1-t11) is displaced slowly from single-stranded DNA by Rad51 protein. *J Biol Chem* 2003;**278**:23410–7.
23. Wang X, Haber JE. Role of Saccharomyces single-stranded DNA-binding protein RPA in the strand invasion step of double-strand break repair. *PLoS Biol* 2004;**2**:E21.
24. Moens PB, Kolas NK, Tarsounas M, Marcon E, Cohen PE, Spyropoulos B. The time course and chromosomal localization of recombination-related proteins at meiosis in the mouse are compatible with models that can resolve the early DNA–DNA interactions without reciprocal recombination. *J Cell Sci* 2002;**115**:1611–22.
25. Houswerth EA, Stahl FW. Crossover interference in humans. *Am J Hum Genet* 2003;**73**:188–97.
26. Kneitz B, Cohen PE, Avdievich E, et al. MutS homolog 4 localization to meiotic chromosomes is required for chromosome pairing during meiosis in male and female mice. *Genes Dev* 2000;**14**:1085–97.
27. Stahl FW. *Genetic recombination.* San Francisco: W.H. Freeman and Co. 1979.
28. Ciccia A, Elledge SJ. The DNA damage response: making it safe to play with knives. *Mol Cell* 2010;**40**:179–204.
29. Heyer WD, Ehmsen KT, Liu J. Regulation of homologous recombination in eukaryotes. *Annu Rev Genet* 2010;**44**:113–39.
30. Holliday RA. A mechanism for gene conversion in fungi. *Genet Res* 1964;**5**:282–304.
31. Orr-Weaver TL, Szostak JW. Yeast recombination: the association between double-strand gap repair and crossing-over. *Proc Natl Acad Sci U S A* 1983;**80**:4417–21.
32. Orr-Weaver TL, Szostak JW. Multiple, tandem plasmid integration in Saccharomyces cerevisiae. *Mol Cell Biol* 1983;**3**:747–9.
33. Moynahan ME, Jasin M. Mitotic homologous recombination maintains genomic stability and suppresses tumorigenesis. *Nat Rev Mol Cell Biol* 2010;**11**:196–207.
34. Wu L, Hickson ID. DNA helicases required for homologous recombination and repair of damaged replication forks. *Annu Rev Genet* 2006;**40**:279–306.
35. Ferguson DO, Holloman WK. Recombinational repair of gaps in DNA is asymmetric in Ustilago maydis and can be explained by a migrating D-loop model. *Proc Natl Acad Sci* 1996;**93**:5419–24.
36. Nassif N, Penney J, Pal S, Engels WR, Gloor GB. Efficient copying of nonhomologous sequences from ectopic sites via P-element-induced gap repair. *Mol Cell Biol* 1994;**14**:1613–25.

37. Strathern JN, Klar AJS, Hicks JB, et al. Homothallic switching of yeast mating type cassettes is initiated by a double-stranded cut in the MAT locus. *Cell* 1982;**31**:183–92.
38. Bzymek M, Thayer NH, Oh SD, Kleckner N, Hunter N. Double Holliday junctions are intermediates of DNA break repair. *Nature* 2010;**464**:937–41.
39. Mahaney BL, Meek K, Lees-miller SP. Repair of ionizing radiation-induced DNA double-strand breaks by non-homologous end-joining. *Biochem J* 2009;**417**:639–50.
40. Hartlerode AJ, Scully R. Mechanisms of double-strand break repair in somatic mammalian cells. *Biochem J* 2009;**423**:157–68.
41. Krogh BO, Symington LS. Recombination proteins in yeast. *Annu Rev Genet* 2004;**38**:233–71.
42. Fishman-Lobell J, Rudin N, Haber JE. Two alternative pathways of double-strand break repair that are kinetically separable and independently modulated. *Mol Cell Biol* 1992;**12**:1292–303.
43. Lin FL, Sperle K, Sternberg N. Model for homologous recombination during transfer of DNA into mouse L cells: role for DNA ends in the recombination process. *Mol Cell Biol* 1984;**4**:1020–34.
44. Maryon E, Carroll D. Involvement of single-stranded tails in homologous recombination of DNA injected into Xenopus laevis oocyte nuclei. *Mol Cell Biol* 1991;**11**:3268–77.
45. Lundblad V, Blackburn EH. An alternative pathway for yeast telomere maintenance rescues est1-senescence. *Cell* 1993;**73**:347–60.
46. Garvik B, Carson M, Hartwell L. Single-stranded DNA arising at telomeres in cdc13 mutants may constitute a specific signal for the RAD9 checkpoint. [Published erratum appears in Mol Cell Biol 1996;16(1):457.]. *Mol Cell Biol* 1995;**15**:6128–38 .
47. Hackett JA, Greider CW. End resection initiates genomic instability in the absence of telomerase. *Mol Cell Biol* 2003;**23**:8450–61.
48. Malkova A, Naylor ML, Yamaguchi M, Ira G, Haber JE. RAD51-dependent break-induced replication differs in kinetics and checkpoint responses from RAD51-mediated gene conversion. *Mol Cell Biol* 2005;**25**:933–44.
49. Branzei D, Foiani M. Regulation of DNA repair throughout the cell cycle. *Nat Rev Mol Cell Biol* 2008;**9**:297–308.
50. Heller RC, Marians KJ. Replication fork reactivation downstream of a blocked nascent leading strand. *Nature* 2006;**439**:557–62.
51. Sonoda E, Sasaki MS, Buerstedde JM, et al. Rad51-deficient vertebrate cells accumulate chromosomal breaks prior to cell death. *EMBO J* 1998;**17**:598–608.
52. Tsuzuki T, Fujii Y, Sakumi K, et al. Targeted disruption of the Rad51 gene leads to lethality in embryonic mice. *Proc Natl Acad Sci U S A* 1996;**93**:6236–40.
53. Prakash S, Johnson RE, Prakash L. Eukaryotic translesion synthesis dna polymerases: specificity of structure and function. *Annu Rev Biochem* 2005;**74**:317–53.
54. Andreassen PR, Ho GPH, D'Andrea AD. DNA damage responses and their many interactions with the replication fork. *Carcinogenesis* 2006;**27**:883–92.
55. Branzei D, Foiani M. Interplay of replication checkpoints and repair proteins at stalled replication forks. *DNA Repair (Amst)* 2007;**6**:994–1003.
56. Higgins NP, Kato K, Strauss B. A model for replication repair in mammalian cells. *J Mol Biol* 1976;**101**:417–25.
57. Mimitou EP, Symington LS. DNA end resection: many nucleases make light work. *DNA Repair (Amst)* 2009;**8**:983–95.
58. Lobachev K, Vitriol E, Stemple J, Resnick MA, Bloom K. Chromosome fragmentation after induction of a double-strand break is an active process prevented by the RMX repair complex. *Curr Biol* 2004;**14**:2107–12.
59. McKee AH, Kleckner N. Mutations in Saccharomyces cerevisiae that block meiotic prophase chromosome metabolism and confer cell cycle arrest at pachytene identify two new meiosis-specific genes SAE1 and SAE3. *Genetics* 1997;**146**:817–34.

60. Mimitou EP, Symington LS. Nucleases and helicases take center stage in homologous recombination. *Trends Biochem Sci* 2009;**34**:264–72.
61. Lisby M, Rothstein R. Choreography of recombination proteins during the DNA damage response. *DNA Repair (Amst)* 2009;**8**:1068–76.
62. Wu D, Topper LM, Wilson TE. Recruitment and dissociation of nonhomologous end joining proteins at a DNA double-strand break in Saccharomyces cerevisiae. *Genetics* 2008;**178**:1237–49.
63. Dupaigne P, Le Breton C, Fabre F, Gangloff S, Le Cam E, Veaute X. The Srs2 helicase activity is stimulated by Rad51 filaments on dsDNA: implications for crossover incidence during mitotic recombination. *Mol Cell* 2008;**29**:243–54.
64. Lisby M, Barlow JH, Burgess RC, Rothstein R. Choreography of the DNA damage response: spatiotemporal relationships among checkpoint and repair proteins. *Cell* 2004;**118**:699–713.
65. Martin SG, Laroche T, Suka N, Grunstein M, Gasser SM. Relocalization of telomeric Ku and SIR proteins in response to DNA strand breaks in yeast. *Cell* 1999;**97**:621–33.
66. Lengsfeld BM, Rattray AJ, Bhaskara V, Ghirlando R, Paull TT. Sae2 is an endonuclease that processes hairpin DNA cooperatively with the Mre11/Rad50/Xrs2 complex. *Mol Cell* 2007;**28**:638–51.
67. Clerici M, Mantiero D, Lucchini G, Longhese MP. The Saccharomyces cerevisiae Sae2 protein promotes resection and bridging of double strand break ends. *J Biol Chem* 2005;**280**:38631–8.
68. Krejci L, Van Komen S, Li Y, et al. DNA helicase Srs2 disrupts the Rad51 presynaptic filament. *Nature* 2003;**423**:305–9.
69. Huertas P, Jackson SP. Human CtIP mediates cell cycle control of DNA end resection and double strand break repair. *J Biol Chem* 2009;**284**:9558–65.
70. Huertas P, Cortes-Ledesma F, Sartori AA, Aguilera A, Jackson SP. CDK targets Sae2 to control DNA-end resection and homologous recombination. *Nature* 2008;**455**:689–92.
71. Huen MS, Chen J. Assembly of checkpoint and repair machineries at DNA damage sites. *Trends Biochem Sci* 2010;**35**:101–8.
72. Huen MS, Sy SM, Chen J. BRCA1 and its toolbox for the maintenance of genome integrity. *Nat Rev Mol Cell Biol* 2010;**11**:138–48.
73. Khavari PA, Peterson CL, Tamkun JW, Mendel DB, Crabtree GR. BRG1 contains a conserved domain of the SWI2/SNF2 family necessary for normal mitotic growth and transcription. *Nature* 1993;**366**:170–4.
74. Shim EY, Chung W-H, Nicolette ML, et al. Saccharomyces cerevisiae Mre11/Rad50/Xrs2 and Ku proteins regulate association of Exo1 and Dna2 with DNA breaks. *EMBO J* 2010;**29**:3370–80.
75. Zhu Z, Chung W-H, Shim EY, Lee SE, Ira G. Sgs1 Helicase and Two Nucleases Dna2 and Exo1 Resect DNA Double-Strand Break Ends. *Cell* 2008;**134**:981–94.
76. Mimitou EP, Symington LS. Sae2, Exo1 and Sgs1 collaborate in DNA double-strand break processing. *Nature* 2008;**455**:770–4.
77. Schaetzlein S, Kodandaramireddy NR, Ju Z, et al. Exonuclease-1 deletion impairs DNA damage signaling and prolongs lifespan of telomere-dysfunctional mice. *Cell* 2007; **130**:863–77.
78. Cejka P, Cannavo E, Polaczek P, et al. DNA end resection by Dna2-Sgs1-RPA and its stimulation by Top3-Rmi1 and Mre11-Rad50-Xrs2. *Nature* 2010;**467**:112–6.
79. Niu H, Chung W-H, Zhu Z, et al. Mechanism of the ATP-dependent DNA end-resection machinery from Saccharomyces cerevisiae. *Nature* 2010;**467**:108–11.
80. Nimonkar AV, Genschel J, Kinoshita E, et al. BLM-DNA2-RPA-MRN and EXO1-BLM-RPA-MRN constitute two DNA end resection machineries for human DNA break repair. *Genes Dev* 2011;**25**:350–62.

81. Nimonkar AV, Ozsoy AZ, Genschel J, Modrich P, Kowalczykowski SC. Human exonuclease 1 and BLM helicase interact to resect DNA and initiate DNA repair. *Proc Natl Acad Sci* 2008;**105**:16906–11.
82. Symington LS, Gautier J. Double-strand break end resection and repair pathway choice. *Annu Rev Genet* 2011;**45**:247–71.
83. Zakharyevich K, Ma Y, Tang S, Hwang PY-H, Boiteux S, Hunter N. Temporally and biochemically distinct activities of Exo1 during meiosis: double-strand break resection and resolution of double holliday junctions. *Mol Cell* 2010;**40**:1001–15.
84. Chung W-H, Zhu Z, Papusha A, Malkova A, Ira G. Defective resection at DNA double-strand breaks leads to *De Novo* telomere formation and enhances gene targeting. *PLoS Genet* 2010;**6**: e1000948.
85. San Filippo J, Sung P, Klein H. Mechanism of eukaryotic homologous recombination. *Annu Rev Biochem* 2008;**77**:229–57.
86. Symington LS. Role of RAD52 epistasis group genes in homologous recombination and double-strand break repair. *Microbiol Mol Biol Rev* 2002;**66**:630–70.
87. McKee RA, Lawrence CW. Genetic analysis of [gamma]-mutagenesis in yeast III. Double-mutant strains. *Mutat Res* 1980;**70**:37–48.
88. Game JC, Mortimer RK. A genetic study of X-ray sensitive mutants in yeast. *Mutat Res* 1974;**24**:281–92.
89. West SC. Molecular views of recombination proteins and their control. *Nat Rev Mol Cell Biol* 2003;**4**:435–45.
90. Shin DS, Pellegrini L, Daniels DS, et al. Full-length archaeal Rad51 structure and mutants: mechanisms for RAD51 assembly and control by BRCA2. *EMBO J* 2003;**22**:4566–76.
91. Shinohara A, Ogawa H, Ogawa T. Rad51 protein involved in repair and recombination in S. cerevisiae is a RecA-like protein. *Cell* 1992;**69**:457–70.
92. Benson FE, Stasiak A, West SC. Purification and characterization of the human Rad51 protein, an analogue of E. coli RecA. *EMBO J* 1994;**13**:5764–71.
93. Aboussekhra A, Chanet R, Adjiri A, Fabre F. Semidominant suppressors of Srs2 helicase mutations of Saccharomyces cerevisiae map in the RAD51 gene, whose sequence predicts a protein with similarities to procaryotic RecA proteins. *Mol Cell Biol* 1992;**12**:3224–34.
94. Basile G, Aker M, Mortimer RK. Nucleotide sequence and transcriptional regulation of the yeast recombinational repair gene RAD51. *Mol Cell Biol* 1992;**12**:3235–46.
95. Wu Y, He Y, Moya IA, Qian X, Luo Y. Crystal structure of archaeal recombinase RADA: a snapshot of its extended conformation. *Mol Cell* 2004;**15**:423–35.
96. Baumann P, West SC. Role of the human RAD51 protein in homologous recombination and double-stranded-break repair. *Trends Biochem Sci* 1998;**23**:247–51.
97. Aihara H, Ito Y, Kurumizaka H, Terada T, Yokoyama S, Shibata T. An interaction between a specified surface of the C-terminal domain of RecA protein and double-stranded DNA for homologous pairing. *J Mol Biol* 1997;**274**:213–21.
98. Aihara H, Ito Y, Kurumizaka H, Yokoyama S, Shibata T. The N-terminal domain of the human Rad51 protein binds DNA: structure and a DNA binding surface as revealed by NMR. *J Mol Biol* 1999;**290**:495–504.
99. Kowalczykowski SC. Biochemistry of genetic recombination: energetics and mechanism of DNA strand exchange. *Annu Rev Biophys Biophys Chem* 1991;**20**:539–75.
100. Bugreev DV, Mazin AV. Ca2+ activates human homologous recombination protein Rad51 by modulating its ATPase activity. *Proc Natl Acad Sci U S A* 2004;**101**:9988–93.
101. Ogawa T, Yu X, Shinohara A, Egelman E. Similarity of the yeast RAD51 filament to the bacterial RecA filament. *Science* 1993;**259**:1896–9.
102. Sung P, Robberson DL. DNA strand exchange mediated by a RAD51-ssDNA nucleoprotein filament with polarity opposite to that of RecA. *Cell* 1995;**82**:453–61.

103. Shim KS, Schmutte C, Yoder K, Fishel R. Defining the salt effect on human RAD51 activities. *DNA Repair (Amst)* 2006;**5**:718–30.
104. Liu Y, Stasiak AZ, Masson JY, McIlwraith MJ, Stasiak A, West SC. Conformational changes modulate the activity of human RAD51 protein. *J Mol Biol* 2004;**337**:817–27.
105. Schild D, Wiese C. Overexpression of RAD51 suppresses recombination defects: a possible mechanism to reverse genomic instability. *Nucleic Acids Res* 2010;**38**:1061–70.
106. Klein HL. The consequences of Rad51 overexpression for normal and tumor cells. *DNA Repair (Amst)* 2008;**7**:686–93.
107. Richardson C, Stark JM, Ommundsen M, Jasin M. Rad51 overexpression promotes alternative double-strand break repair pathways and genome instability. *Oncogene* 2004;**23**:546–53.
108. Morgan EA, Shah N, Symington LS. The requirement for ATP hydrolysis by Saccharomyces cerevisiae Rad51 is bypassed by mating-type heterozygosity or RAD54 in high copy. *Mol Cell Biol* 2002;**22**:6336–43.
109. Sung P, Stratton SA. Yeast Rad51 recombinase mediates polar DNA strand exchange in the absence of ATP hydrolysis. *J Biol Chem* 1996;**271**:27983–6.
110. Chi P, Van Komen S, Sehorn MG, Sigurdsson S, Sung P. Roles of ATP binding and ATP hydrolysis in human Rad51 recombinase function. *DNA Repair (Amst)* 2006;**5**:381–91.
111. Morrison C, Shinohara A, Sonoda E, et al. The essential functions of human Rad51 are independent of ATP hydrolysis. *Mol Cell Biol* 1999;**19**:6891–7.
112. Stark JM, Hu P, Pierce AJ, Moynahan ME, Ellis N, Jasin M. ATP hydrolysis by mammalian RAD51 has a key role during homology-directed DNA repair. *J Biol Chem* 2002;**277**:20185–94.
113. Wu Y, Qian X, He Y, Moya IA, Luo Y. Crystal structure of an ATPase-active form of Rad51 homolog from Methanococcus voltae. Insights into potassium dependence. *J Biol Chem* 2005;**280**:722–8.
114. Qian X, He Y, Ma X, Fodje MN, Grochulski P, Luo Y. Calcium stiffens archaeal Rad51 recombinase from Methanococcus voltae for homologous recombination. *J Biol Chem* 2006;**281**:39380–7.
115. Qian X, Wu Y, He Y, Luo Y. Crystal structure of Methanococcus voltae RadA in complex with ADP: hydrolysis-induced conformational change. *Biochemistry* 2005;**44**:13753–61.
116. Tombline G, Fishel R. Biochemical characterization of the human RAD51 protein. I. ATP hydrolysis. *J Biol Chem* 2002;**277**:14417–25.
117. Jensen RB, Carreira A, Kowalczykowski SC. Purified human BRCA2 stimulates RAD51-mediated recombination. *Nature* 2010;**467**:678–83.
118. Carreira A, Hilario J, Amitani I, et al. The BRC repeats of BRCA2 modulate the DNA-binding selectivity of RAD51. *Cell* 2009;**136**:1032–43.
119. Yang H, Li Q, Fan J, Holloman WK, Pavletich NP. The BRCA2 homologue Brh2 nucleates RAD51 filament formation at a dsDNA-ssDNA junction. *Nature* 2005;**433**:653–7.
120. Galkin VE, Esashi F, Yu X, Yang S, West SC, Egelman EH. BRCA2 BRC motifs bind RAD51-DNA filaments. *Proc Natl Acad Sci U S A* 2005;**102**:8537–42.
121. Antony E, Tomko EJ, Xiao Q, Krejci L, Lohman TM, Ellenberger T. Srs2 disassembles Rad51 filaments by a protein–protein interaction triggering ATP turnover and dissociation of Rad51 from DNA. *Mol Cell* 2009;**35**:105–15.
122. Hu Y, Raynard S, Sehorn MG, et al. RECQL5/Recql5 helicase regulates homologous recombination and suppresses tumor formation via disruption of Rad51 presynaptic filaments. *Genes Dev* 2007;**21**:3073–84.
123. Bugreev DV, Yu X, Egelman EH, Mazin AV. Novel pro- and anti-recombination activities of the Bloom's syndrome helicase. *Genes Dev* 2007;**21**:3085–94.
124. Veaute X, Jeusset J, Soustelle C, Kowalczykowski SC, Le Cam E, Fabre F. The Srs2 helicase prevents recombination by disrupting Rad51 nucleoprotein filaments. *Nature* 2003;**423**:309–12.

125. Hilario J, Amitani I, Baskin RJ, Kowalczykowski SC. Direct imaging of human Rad51 nucleoprotein dynamics on individual DNA molecules. *Proc Natl Acad Sci U S A* 2009;**106**:361–8.
126. Bugreev DV, Golub EI, Stasiak AZ, Stasiak A, Mazin AV. Activation of human meiosis-specific recombinase Dmc1 by Ca2+. *J Biol Chem* 2005;**280**:26886–95.
127. Bianco PR, Tracy RB, Kowalczykowski SC. DNA strand exchange proteins: a biochemical and physical comparison. *Front Biosci* 1998;**3**:D570–603.
128. Gupta RC, Folta-Stogniew E, O'Malley S, Takahashi M, Radding CM. Rapid exchange of A: T base pairs is essential for recognition of DNA homology by human Rad51 recombination protein. *Mol Cell* 1999;**4**:705–14.
129. Bishop DK, Park D, Xu L, Kleckner N. DMC1: a meiosis-specific yeast homolog of E. coli recA required for recombination, synaptonemal complex formation, and cell cycle progression. *Cell* 1992;**69**:439–56.
130. Bishop DK. RecA homologs Dmc1 and Rad51 interact to form multiple nuclear complexes prior to meiotic chromosome synapsis. *Cell* 1994;**79**:1081–92.
131. Tsubouchi H, Roeder GS. The importance of genetic recombination for fidelity of chromosome pairing in meiosis. *Dev Cell* 2003;**5**:915–25.
132. Passy SI, Yu X, Li Z, et al. Human Dmc1 protein binds DNA as an octameric ring. *Proc Natl Acad Sci* 1999;**96**:10684–8.
133. Sauvageau S, Stasiak AZ, Banville I, Ploquin M, Stasiak A, Masson J-Y. Fission yeast Rad51 and Dmc1, two efficient DNA recombinases forming helical nucleoprotein filaments. *Mol Cell Biol* 2005;**25**:4377–87.
134. Sehorn MG, Sigurdsson S, Bussen W, Unger VM, Sung P. Human meiotic recombinase Dmc1 promotes ATP-dependent homologous DNA strand exchange. *Nature* 2004;**429**:433–7.
135. Wold MS, Kelly T. Purification and characterization of replication protein A, a cellular protein required for in vitro replication of simian virus 40 DNA. *Proc Natl Acad Sci* 1988;**85**:2523–7.
136. Erdile LF, Wold MS, Kelly TJ. The primary structure of the 32-kDa subunit of human replication protein A. *J Biol Chem* 1990;**265**:3177–82.
137. Alani E, Thresher R, Griffith JD, Kolodner RD. Characterization of DNA-binding and strand-exchange stimulation properties of y-RPA, a yeast single-strand-DNA-binding protein. *J Mol Biol* 1992;**227**:54–71.
138. Heyer WD, Rao MR, Erdile LF, Kelly TJ, Kolodner RD. An essential Saccharomyces cerevisiae single-stranded DNA binding protein is homologous to the large subunit of human RP-A. *EMBO J* 1990;**9**:2321–9.
139. Sung P. Catalysis of ATP-dependent homologous DNA pairing and strand exchange by yeast RAD51 protein. *Science* 1994;**265**:1241–3.
140. Sugiyama T, Zaitseva EM, Kowalczykowski SC. A single-stranded DNA-binding protein is needed for efficient presynaptic complex formation by the Saccharomyces cerevisiae Rad51 protein. *J Biol Chem* 1997;**272**:7940–5.
141. Van Komen S, Petukhova G, Sigurdsson S, Sung P. Functional cross-talk among Rad51, Rad54, and replication protein A in heteroduplex DNA joint formation. *J Biol Chem* 2002;**277**:43578–87.
142. Eggler AL, Inman RB, Cox MM. The Rad51-dependent pairing of long DNA substrates is stabilized by replication protein A. *J Biol Chem* 2002;**277**:39280–8.
143. Sigurdsson S, Van Komen S, Bussen W, Schild D, Albala JS, Sung P. Mediator function of the human Rad51B-Rad51C complex in Rad51/RPA-catalyzed DNA strand exchange. *Genes Dev* 2001;**15**:3308–18.
144. Shinohara A, Ogawa T. Stimulation by Rad52 of yeast Rad51-mediated recombination. *Nature* 1998;**391**:404–7.

145. New JH, Sugiyama T, Zaitseva E, Kowalczykowski SC. Rad52 protein stimulates DNA strand exchange by Rad51 and replication protein A. *Nature* 1998;**391**:407–10.
146. Kanaar R, Hoeijmakers JHJ. Genetic recombination: from competition to collaboration. *Nature* 1998;**391**:335–8.
147. Sung P. Yeast Rad55 and Rad57 proteins form a heterodimer that functions with replication protein A to promote DNA strand exchange by Rad51 recombinase. *Genes Dev* 1997;**11**:1111–21.
148. Ranatunga W, Jackson D, Lloyd JA, Forget AL, Knight KL, Borgstahl GEO. Human RAD52 exhibits two modes of self-association. *J Biol Chem* 2001;**276**:15876–80.
149. Stasiak AZ, Larquet E, Stasiak A, et al. The human Rad52 protein exists as a heptameric ring. *Curr Biol* 2000;**10**:337–40.
150. Shinohara A, Shinohara M, Ohta T, Matsuda S, Ogawa T. Rad52 forms ring structures and co-operates with RPA in single-strand DNA annealing. *Genes Cells* 1998;**3**:145–56.
151. Singleton MR, Wentzell LM, Liu Y, West SC, Wigley DB. Structure of the single-strand annealing domain of human RAD52 protein. *Proc Natl Acad Sci* 2002;**99**:13492–7.
152. Kagawa W, Kurumizaka H, Ishitani R, et al. Crystal structure of the homologous-pairing domain from the human Rad52 recombinase in the undecameric form. *Mol Cell* 2002;**10**:359–71.
153. Mortensen UH, Bendixen C, Sunjevaric I, Rothstein R. DNA strand annealing is promoted by the yeast Rad52 protein. *Proc Natl Acad Sci* 1996;**93**:10729–34.
154. Parsons CA, Baumann P, Van Dyck E, West SC. Precise binding of single-stranded DNA termini by human RAD52 protein. *EMBO J* 2000;**19**:4175–81.
155. Wolner B, van Komen S, Sung P, Peterson CL. Recruitment of the recombinational repair machinery to a DNA double-strand break in yeast. *Mol Cell* 2003;**12**:221–32.
156. Sugawara N, Wang X, Haber JE. In vivo roles of Rad52, Rad54, and Rad55 proteins in Rad51-mediated recombination. *Mol Cell* 2003;**12**:209–19.
157. Gasior SL, Wong AK, Kora Y, Shinohara A, Bishop DK. Rad52 associates with RPA and functions with Rad55 and Rad57 to assemble meiotic recombination complexes. *Genes Dev* 1998;**12**:2208–21.
158. Sung P. Function of yeast Rad52 protein as a mediator between replication protein A and the Rad51 recombinase. *J Biol Chem* 1997;**272**:28194–7.
159. Krejci L, Song B, Bussen W, Rothstein R, Mortensen UH, Sung P. Interaction with Rad51 is indispensable for recombination mediator function of Rad52. *J Biol Chem* 2002;**277**:40132–41.
160. Milne GT, Weaver DT. Dominant negative alleles of RAD52 reveal a DNA repair/recombination complex including Rad51 and Rad52. *Genes Dev* 1993;**7**:1755–65.
161. Boundy-Mills KL, Livingston DM. A Saccharomyces cerevisiae RAD52 allele expressing a C-terminal truncation protein: activities and intragenic complementation of missense mutations. *Genetics* 1993;**133**:39–49.
162. Shen Z, Cloud KG, Chen DJ, Park MS. Specific interactions between the human RAD51 and RAD52 proteins. *J Biol Chem* 1996;**271**:148–52.
163. McEachern MJ, Haber JE. Break-induced replication and recombinational telomere elongation in yeast. *Annu Rev Biochem* 2006;**75**:111–35.
164. Jackson D, Dhar K, Wahl JK, Wold MS, Borgstahl GEO. Analysis of the human replication protein A:Rad52 complex: evidence for crosstalk between RPA32, RPA70, Rad52 and DNA. *J Mol Biol* 2002;**321**:133–48.
165. Hays SL, Firmenich AA, Massey P, Banerjee R, Berg P. Studies of the interaction between Rad52 protein and the yeast single-stranded DNA binding protein RPA. *Mol Cell Biol* 1998;**18**:4400–6.
166. McIlwraith MJ, West SC. DNA repair synthesis facilitates RAD52-mediated second-end capture during DSB repair. *Mol Cell* 2008;**29**:510–6.

167. Wu Y, Kantake N, Sugiyama T, Kowalczykowski SC. Rad51 protein controls Rad52-mediated DNA annealing. *J Biol Chem* 2008;**283**:14883–92.
168. Yamaguchi-Iwai Y, Sonoda E, Buerstedde J-M, et al. Homologous recombination, but not DNA repair, is reduced in vertebrate cells deficient in RAD52. *Mol Cell Biol* 1998;**18**:6430–5.
169. Rijkers T, Van Den Ouweland J, Morolli B, et al. Targeted inactivation of mouse RAD52 reduces homologous recombination but not resistance to ionizing radiation. *Mol Cell Biol* 1998;**18**:6423–9.
170. Kojic M, Mao N, Zhou Q, Lisby M, Holloman WK. Compensatory role for Rad52 during recombinational repair in Ustilago maydis. *Mol Microbiol* 2008;**67**:1156–68.
171. Feng Z, Scott SP, Bussen W, et al. Rad52 inactivation is synthetically lethal with BRCA2 deficiency. *Proc Natl Acad Sci* 2011;**108**:686–91.
172. Liu J, Heyer W-D. Who's who in human recombination: BRCA2 and RAD52. *Proc Natl Acad Sci* 2011;**108**:441–2.
173. Fujimori A, Tachiiri S, Sonoda E, et al. Rad52 partially substitutes for the Rad51 paralog XRCC3 in maintaining chromosomal integrity in vertebrate cells. *EMBO J* 2001;**20**:5513–20.
174. Jasin M. Homologous repair of DNA damage and tumorigenesis: the BRCA connection. *Oncogene* 2002;**21**:8981–93.
175. Wooster R, Bignell G, Lancaster J, et al. Identification of the breast cancer susceptibility gene BRCA2. *Nature* 1995;**378**:789–92.
176. Wooster R, Neuhausen S, Mangion J, et al. Localization of a breast cancer susceptibility gene, BRCA2, to chromosome 13q12-13. *Science* 1994;**265**:2088–90.
177. Siaud N, Dray E, Gy I, Gerard E, Takvorian N, Doutriaux M-P. Brca2 is involved in meiosis in Arabidopsis thaliana as suggested by its interaction with Dmc1. *EMBO J* 2004;**23**:1392–401.
178. Yu VPCC, Koehler M, Steinlein C, et al. Gross chromosomal rearrangements and genetic exchange between nonhomologous chromosomes following BRCA2 inactivation. *Genes Dev* 2000;**14**:1400–6.
179. Martin JS, Winkelmann N, Petalcorin MIR, McIlwraith MJ, Boulton SJ. RAD-51-dependent and -independent roles of a Caenorhabditis elegans BRCA2-related protein during DNA double-strand break repair. *Mol Cell Biol* 2005;**25**:3127–39.
180. Moynahan ME, Pierce AJ, Jasin M. BRCA2 is required for homology-directed repair of chromosomal breaks. *Mol Cell* 2001;**7**:263–72.
181. Sharan SK, Morimatsu M, Albrecht U, et al. Embryonic lethality and radiation hypersensitivity mediated by Rad51 in mice lacking Brca2. *Nature* 1997;**386**:804–10.
182. Xia F, Taghian DG, DeFrank JS, et al. Deficiency of human BRCA2 leads to impaired homologous recombination but maintains normal nonhomologous end joining. *Proc Natl Acad Sci* 2001;**98**:8644–9.
183. Howlett NG, Taniguchi T, Olson S, et al. Biallelic inactivation of BRCA2 in Fanconi anemia. *Science* 2002;**297**:606–9.
184. Mizuta R, LaSalle JM, Cheng H-L, et al. RAB22 and RAB163/mouse BRCA2: proteins that specifically interact with the RAD51 protein. *Proc Natl Acad Sci* 1997;**94**:6927–32.
185. Chen P-L, Chen C-F, Chen Y, Xiao J, Sharp ZD, Lee W-H. The BRC repeats in BRCA2 are critical for RAD51 binding and resistance to methyl methanesulfonate treatment. *Proc Natl Acad Sci* 1998;**95**:5287–92.
186. Wong AKC, Pero R, Ormonde PA, Tavtigian SV, Bartel PL. RAD51 interacts with the evolutionarily conserved BRC motifs in the human breast cancer susceptibility Gene brca2. *J Biol Chem* 1997;**272**:31941–4.
187. Esashi F, Galkin VE, Yu X, Egelman EH, West SC. Stabilization of RAD51 nucleoprotein filaments by the C-terminal region of BRCA2. *Nat Struct Mol Biol* 2007;**14**:468–74.
188. Davies OR, Pellegrini L. Interaction with the BRCA2 C terminus protects RAD51-DNA filaments from disassembly by BRC repeats. *Nat Struct Mol Biol* 2007;**14**:475–83.

189. Bork P, Blomberg N, Nilges M. Internal repeats in the BRCA2 protein sequence. *Nat Genet* 1996;**13**:22–3.
190. Bignell G, Micklem G, Stratton MR, Ashworth A, Wooster R. The BRC repeats are conserved in mammalian BRCA2 proteins. *Hum Mol Genet* 1997;**6**:53–8.
191. Lo T, Pellegrini L, Venkitaraman AR, Blundell TL. Sequence fingerprints in BRCA2 and RAD51: implications for DNA repair and cancer. *DNA Repair (Amst)* 2003;**2**:1015–28.
192. Holloman WK. Unraveling the mechanism of BRCA2 in homologous recombination. *Nat Struct Mol Biol* 2011;**18**:748–54.
193. Chen C-F, Chen P-L, Zhong Q, Sharp ZD, Lee W-H. Expression of BRC repeats in breast cancer cells disrupts the BRCA2-Rad51 complex and leads to radiation hypersensitivity and loss of G2/M checkpoint control. *J Biol Chem* 1999;**274**:32931–5.
194. Pellegrini L, Yu DS, Lo T, et al. Insights into DNA recombination from the structure of a RAD51-BRCA2 complex. *Nature* 2002;**420**:287–93.
195. Carreira A, Kowalczykowski SC. Two classes of BRC repeats in BRCA2 promote RAD51 nucleoprotein filament function by distinct mechanisms. *Proc Natl Acad Sci* 2011; **108**:10448–53.
196. Shivji MKK, Mukund SR, Rajendra E, et al. The BRC repeats of human BRCA2 differentially regulate RAD51 binding on single- versus double-stranded DNA to stimulate strand exchange. *Proc Natl Acad Sci* 2009;**106**:13254–9.
197. Shah PP, Zheng X, Epshtein A, Carey JN, Bishop DK, Klein HL. Swi2/Snf2-related translocases prevent accumulation of toxic Rad51 complexes during mitotic growth. *Mol Cell* 2010;**39**:862–72.
198. Solinger JA, Kiianitsa K, Heyer WD. Rad54, a Swi2/Snf2-like recombinational repair protein, disassembles Rad51:dsDNA filaments. *Mol Cell* 2002;**10**:1175–88.
199. Petalcorin MIR, Sandall J, Wigley DB, Boulton SJ. CeBRC-2 stimulates D-loop formation by RAD-51 and promotes DNA single-strand annealing. *J Mol Biol* 2006;**361**:231–42.
200. Esashi F, Christ N, Gannon J, et al. CDK-dependent phosphorylation of BRCA2 as a regulatory mechanism for recombinational repair. *Nature* 2005;**434**:598–604.
201. Davies AA, Masson J-Y, McIlwraith MJ, et al. Role of BRCA2 in control of the RAD51 recombination and DNA repair protein. *Mol Cell* 2001;**7**:273–82.
202. Gildemeister OS, Sage JM, Knight KL. Cellular redistribution of Rad51 in response to DNA damage. *J Biol Chem* 2009;**284**:31945–52.
203. Goggins M, Schutte M, Lu J, et al. Germline BRCA2 gene mutations in patients with apparently sporadic pancreatic carcinomas. *Cancer Res* 1996;**56**:5360–4.
204. Schlacher K, Christ N, Siaud N, Egashira A, Wu H, Jasin M. Double-strand break repair-independent role for BRCA2 in blocking stalled replication fork degradation by MRE11. *Cell* 2011;**145**:529–42.
205. Hashimoto Y, Chaudhuri AR, Lopes M, Costanzo V. Rad51 protects nascent DNA from Mre11-dependent degradation and promotes continuous DNA synthesis. *Nat Struct Mol Biol* 2010;**17**:1305–11.
206. Thorslund T, Esashi F, West SC. Interactions between human BRCA2 protein and the meiosis-specific recombinase DMC1. *EMBO J* 2007;**26**:2915–22.
207. Wong JMS, Ionescu D, Ingles CJ. Interaction between BRCA2 and replication protein A is compromised by a cancer-predisposing mutation in BRCA2. *Oncogene* 2003;**22**:28–33.
208. Reid S, Schindler D, Hanenberg H, et al. Biallelic mutations in PALB2 cause Fanconi anemia subtype FA-N and predispose to childhood cancer. *Nat Genet* 2007;**39**:162–4.
209. Xia B, Sheng Q, Nakanishi K, et al. Control of BRCA2 cellular and clinical functions by a nuclear partner, PALB2. *Mol Cell* 2006;**22**:719–29.
210. Dray E, Etchin J, Wiese C, et al. Enhancement of RAD51 recombinase activity by the tumor suppressor PALB2. *Nat Struct Mol Biol* 2010;**17**:1255–9.

211. Buisson R, Dion-Cote A-M, Coulombe Y, et al. Cooperation of breast cancer proteins PALB2 and piccolo BRCA2 in stimulating homologous recombination. *Nat Struct Mol Biol* 2010;**17**:1247–54.
212. Marston NJ, Richards WJ, Hughes D, Bertwistle D, Marshall CJ, Ashworth A. Interaction between the product of the breast cancer susceptibility gene BRCA2 and DSS1, a protein functionally conserved from yeast to mammals. *Mol Cell Biol* 1999;**19**:4633–42.
213. Gudmundsdottir K, Lord CJ, Witt E, Tutt ANJ, Ashworth A. DSS1 is required for RAD51 focus formation and genomic stability in mammalian cells. *EMBO Rep* 2004;**5**:989–93.
214. Kojic M, Yang H, Kostrub CF, Pavletich NP, Holloman WK. The BRCA2-interacting protein DSS1 is vital for DNA repair, recombination, and genome stability in Ustilago maydis. *Mol Cell* 2003;**12**:1043–9.
215. Yang H, Jeffrey PD, Miller J, et al. BRCA2 function in DNA binding and recombination from a BRCA2-DSS1-ssDNA structure. *Science* 2002;**297**:1837–48.
216. Zhou Q, Kojic M, Cao Z, Lisby M, Mazloum NA, Holloman WK. Dss1 interaction with Brh2 as a regulatory mechanism for recombinational repair. *Mol Cell Biol* 2007;**27**:2512–26.
217. Mazloum N, Holloman WK. Second-end capture in DNA double-strand break repair promoted by Brh2 protein of Ustilago maydis. *Mol Cell* 2009;**33**:160–70.
218. Thorslund T, McIlwraith MJ, Compton SA, et al. The breast cancer tumor suppressor BRCA2 promotes the specific targeting of RAD51 to single-stranded DNA. *Nat Struct Mol Biol* 2010;**17**:1263–5.
219. Liu J, Doty T, Gibson B, Heyer W-D. Human BRCA2 protein promotes RAD51 filament formation on RPA-covered single-stranded DNA. *Nat Struct Mol Biol* 2010;**17**:1260–2.
220. Lin Z, Kong H, Nei M, Ma H. Origins and evolution of the recA/RAD51 gene family: evidence for ancient gene duplication and endosymbiotic gene transfer. *Proc Natl Acad Sci U S A* 2006;**103**:10328–33.
221. Lovett ST. Sequence of the RAD55 gene of Saccharomyces cerevisiae: similarity of RAD55 to prokaryotic RecA and other RecA-like proteins. *Gene* 1994;**142**:103–6.
222. Kans JA, Mortimer RK. Nucleotide sequence of the RAD57 gene of Saccharomyces cerevisiae. *Gene* 1991;**105**:139–40.
223. Hays SL, Firmenich AA, Berg P. Complex formation in yeast double-strand break repair: participation of Rad51, Rad52, Rad55, and Rad57 proteins. *Proc Natl Acad Sci* 1995;**92**:6925–9.
224. Johnson R, Symington L. Functional differences and interactions among the putative RecA homologs Rad51, Rad55, and Rad57. *Mol Cell Biol* 1995;**15**:4843–50.
225. Janke R, Herzberg K, Rolfsmeier M, et al. A truncated DNA-damage-signaling response is activated after DSB formation in the G1 phase of Saccharomyces cerevisiae. *Nucleic Acids Res* 2010;**38**:2302–13.
226. Herzberg K, Bashkirov VI, Rolfsmeier M, et al. Phosphorylation of Rad55 on serines 2, 8, and 14 is required for efficient homologous recombination in the recovery of stalled replication forks. *Mol Cell Biol* 2006;**26**:8396–409.
227. Bashkirov VI, King JS, Bashkirova EV, Schmuckli-Maurer J, Heyer W-D. DNA repair protein Rad55 is a terminal substrate of the DNA damage checkpoints. *Mol Cell Biol* 2000;**20**:4393–404.
228. Tsutsui Y, Morishita T, Iwasaki H, Toh H, Shinagawa H. A recombination repair gene of Schizosaccharomyces pombe, rhp57, is a functional homolog of the Saccharomyces cerevisiae RAD57 gene and is phylogenetically related to the human XRCC3 gene. *Genetics* 2000;**154**:1451–61.
229. Khasanov FK, Savchenko GV, Bashkirova EV, Korolev VG, Heyer W-D, Bashkirov VI. A new recombinational DNA repair gene from Schizosaccharomyces pombe with homology to Escherichia coli RecA. *Genetics* 1999;**152**:1557–72.

230. Thompson LH, Schild D. Homologous recombinational repair of DNA ensures mammalian chromosome stability. *Mutat Res* 2001;**477**:131–53.
231. Jones NJ, Zhao Y, Siciliano MJ, Thompson LH. Assignment of the XRCC2 human DNA repair gene to chromosome 7q36 by complementation analysis. *Genomics* 1995;**26**:619–22.
232. Thacker J, Tambini CE, Simpson PJ, Tsui LC, Scherer SW. Localization to chromosome 7q36.1 of the human XRCC2 gene, determining sensitivity to DNA-damaging agents. *Hum Mol Genet* 1995;**4**:113–20.
233. Tebbs RS, Zhao Y, Tucker JD, et al. Correction of chromosomal instability and sensitivity to diverse mutagens by a cloned cDNA of the XRCC3 DNA repair gene. *Proc Natl Acad Sci* 1995;**92**:6354–8.
234. Rice MC, Smith ST, Bullrich F, Havre P, Kmiec EB. Isolation of human and mouse genes based on homology to REC2, a recombinational repair gene from the fungus Ustilago maydis. *Proc Natl Acad Sci* 1997;**94**:7417–22.
235. Albala JS, Thelen MP, Prange C, et al. Identification of a novel human RAD51 homolog, RAD51B. *Genomics* 1997;**46**:476–9.
236. Cartwright R, Dunn AM, Simpson PJ, Tambini CE, Thacker J. Isolation of novel human and mouse genes of the recA/RAD51 recombination-repair gene family. *Nucleic Acids Res* 1998;**26**:1653–9.
237. Dosanjh MK, Collins DW, Schild D, et al. Isolation and characterization of RAD51C, a new human member of the RAD51 family of related genes. *Nucleic Acids Res* 1998;**26**:1179–84.
238. Pittman DL, Weinberg LR, Schimenti JC. Identification, characterization, and genetic mapping of Rad51d, a new mouse and human RAD51/RecA-related gene. *Genomics* 1998;**49**:103–11.
239. Schild D, Lio Y-c, Collins DW, Tsomondo T, Chen DJ. Evidence for simultaneous protein interactions between human Rad51 paralogs. *J Biol Chem* 2000;**275**:16443–9.
240. Miller KA, Yoshikawa DM, McConnell IR, Clark R, Schild D, Albala JS. RAD51C interacts with RAD51B and is central to a larger protein complex in vivo exclusive of RAD51. *J Biol Chem* 2002;**277**:8406–11.
241. Lio YC, Mazin AV, Kowalczykowski SC, Chen DJ. Complex formation by the human Rad51B and Rad51C DNA repair proteins and their activities in vitro. *J Biol Chem* 2003;**278**:2469–78.
242. Liu N, Lamerdin JE, Tebbs RS, et al. XRCC2 and XRCC3, new human Rad51-family members, promote chromosome stability and protect against DNA cross-links and other damages. *Mol Cell* 1998;**1**:783–93.
243. Braybrooke JP, Spink KG, Thacker J, Hickson ID. The RAD51 family member, RAD51L3, is a DNA-stimulated ATPase that forms a complex with XRCC2. *J Biol Chem* 2000;**275**:29100–6.
244. Masson JY, Tarsounas MC, Stasiak AZ, et al. Identification and purification of two distinct complexes containing the five RAD51 paralogs. *Genes Dev* 2001;**15**:3296–307.
245. Masson JY, Stasiak AZ, Stasiak A, Benson FE, West SC. Complex formation by the human RAD51C and XRCC3 recombination repair proteins. *Proc Natl Acad Sci U S A* 2001;**98**:8440–6.
246. Takata M, Sasaki MS, Tachiiri S, et al. Chromosome instability and defective recombinational repair in knockout mutants of the five Rad51 paralogs. *Mol Cell Biol* 2001;**21**:2858–66.
247. Takata M, Sasaki MS, Sonoda E, et al. The Rad51 paralog Rad51B promotes homologous recombinational repair. *Mol Cell Biol* 2000;**20**:6476–82.
248. Shim KS, Schmutte C, Tombline G, Heinen CD, Fishel R. hXRCC2 enhances ADP/ATP processing and strand exchange by hRAD51. *J Biol Chem* 2004;**279**:30385–94.
249. Compton SA, Özgür S, Griffith JD. Ring-shaped Rad51 paralog protein complexes bind holliday junctions and replication forks as visualized by electron microscopy. *J Biol Chem* 2010;**285**:13349–56.

250. Mazin AV, Mazina OM, Bugreev DV, Rossi MJ. Rad54, the motor of homologous recombination. *DNA Repair (Amst)* 2010;**9**:286–302.
251. Heyer W-D, Li X, Rolfsmeier M, Zhang X-P. Rad54: the Swiss Army knife of homologous recombination? *Nucleic Acids Res* 2006;**34**:4115–25.
252. Tan TL, Kanaar R, Wyman C. Rad54, a Jack of all trades in homologous recombination. *DNA Repair (Amst)* 2003;**2**:787–94.
253. Petrini JHJ, Bressan DA, Yao MS. TheRAD52 epistasis group in mammalian double strand break repair. *Semin Immunol* 1997;**9**:181–8.
254. Muris DFR, Vreeken K, Carr AM, et al. Isolation of the Schizosaccharomyces pombe RAD54 homologue, rhp54+, a gene involved in the repair of radiation damage and replication fidelity. *J Cell Sci* 1996;**109**:73–81.
255. Kooistra R, Vreeken K, Zonneveld JBM, et al. The Drosophila melanogaster RAD54 homolog, DmRAD54, is involved in the repair of radiation damage and recombination. *Mol Cell Biol* 1997;**17**:6097–104.
256. Shaked H, Avivi-Ragolsky N, Levy AA. Involvement of the Arabidopsis SWI2/SNF2 chromatin remodeling gene family in DNA damage response and recombination. *Genetics* 2006;**173**:985–94.
257. Thoma NH, Czyzewski BK, Alexeev AA, Mazin AV, Kowalczykowski SC, Pavletich NP. Structure of the SWI2/SNF2 chromatin-remodeling domain of eukaryotic Rad54. *Nat Struct Mol Biol* 2005;**12**:350–6.
258. Bezzubova O, Silbergleit A, Yamaguchi-Iwai Y, Takeda S, Buerstedde J-M. Reduced X-ray resistance and homologous recombination frequencies in a RAD54-/- mutant of the chicken DT40 cell line. *Cell* 1997;**89**:185–93.
259. Kanaar R, Troelstra C, Swagemakers SMA, et al. Human and mouse homologs of the Saccharomyces cerevisiae RAD54 DNA repair gene: evidence for functional conservation. *Curr Biol* 1996;**6**:828–38.
260. Seitz EM, Haseltine CA, Kowalczykowski SC. DNA recombination and repair in the Archaea. *Adv Appl Microbiol* 2001;**50**:101–69.
261. Snow R. Mutants of yeast sensitive to ultraviolet light. *J Bacteriol* 1967;**94**:571–5.
262. Klein HL. RDH54, a RAD54 homologue in Saccharomyces cerevisiae, is required for mitotic diploid-specific recombination and repair and for meiosis. *Genetics* 1997;**147**:1533–43.
263. Essers J, van Steeg H, de Wit J, et al. Homologous and non-homologous recombination differentially affect DNA damage repair in mice. *EMBO J* 2000;**19**:1703–10.
264. Petukhova G, Van Komen S, Vergano S, Klein H, Sung P. Yeast Rad54 promotes Rad51-dependent homologous DNA pairing via ATP hydrolysis-driven change in DNA double helix conformation. *J Biol Chem* 1999;**274**:29453–62.
265. Clever B, Schmuckli-Maurer J, Sigrist M, Glassner BJ, Heyer W-D. Specific negative effects resulting from elevated levels of the recombinational repair protein Rad54p in Saccharomyces cerevisiae. *Yeast* 1999;**15**:721–40.
266. Agarwal S, van Cappellen WA, Guenole A, et al. ATP-dependent and independent functions of Rad54 in genome maintenance. *J Cell Biol* 2011;**192**:735–50.
267. Johnston LH, Johnson AL. The DNA repair genes RAD54 and UNG1 are cell cycle regulated in budding yeast but MCB promoter elements have no essential role in the DNA damage response. *Nucleic Acids Res* 1995;**23**:2147–52.
268. Cole GM, Schild D, Mortimer RK. Two DNA repair and recombination genes in Saccharomyces cerevisiae, RAD52 and RAD54, are induced during meiosis. *Mol Cell Biol* 1989;**9**:3101–4.
269. Takata M, Sasaki MS, Sonoda E, et al. Homologous recombination and non-homologous end-joining pathways of DNA double-strand break repair have overlapping roles in the maintenance of chromosomal integrity in vertebrate cells. *EMBO J* 1998;**17**:5497–508.

270. Cole GM, Schild D, Lovett ST, Mortimer RK. Regulation of RAD54- and RAD52-lacZ gene fusions in Saccharomyces cerevisiae in response to DNA damage. *Mol Cell Biol* 1987;**7**:1078–84.
271. Essers J, Houtsmuller AB, van Veelen L, et al. Nuclear dynamics of RAD52 group homologous recombination proteins in response to DNA damage. *EMBO J* 2002;**21**:2030–7.
272. Tan TLR, Essers J, Citterio E, et al. Mouse Rad54 affects DNA conformation and DNA-damage-induced Rad51 foci formation. *Curr Biol* 1999;**9**:325–8.
273. Gorbalenya AE, Koonin EV. Helicases: amino acid sequence comparisons and structure-function relationships. *Curr Opin Struct Biol* 1993;**3**:419–29.
274. Singleton MR, Dillingham MS, Wigley DB. Structure and mechanism of helicases and nucleic acid translocases. *Annu Rev Biochem* 2007;**76**:23–50.
275. Swagemakers SMA, Essers J, de Wit J, Hoeijmakers JHJ, Kanaar R. The human Rad54 recombinational DNA repair protein is a double-stranded DNA-dependent ATPase. *J Biol Chem* 1998;**273**:28292–7.
276. Bugreev DV, Mazina OM, Mazin AV. Rad54 protein promotes branch migration of Holliday junctions. *Nature* 2006;**442**:590–3.
277. Haseltine CA, Kowalczykowski SC. An archaeal Rad54 protein remodels DNA and stimulates DNA strand exchange by RadA. *Nucleic Acids Res* 2009;**37**:2757–70.
278. Clever B, Interthal H, Schmuckli-Maurer J, King J, Sigrist M, Heyer W-D. Recombinational repair in yeast: functional interactions between Rad51 and Rad54 proteins. *EMBO J* 1997;**16**:2535–44.
279. Golub EI, Kovalenko OV, Gupta RC, Ward DC, Radding CM. Interaction of human recombination proteins Rad51 and Rad54. *Nucleic Acids Res* 1997;**25**:4106–10.
280. Jiang H, Xie Y, Houston P, et al. Direct association between the yeast Rad51 and Rad54 recombination proteins. *J Biol Chem* 1996;**271**:33181–6.
281. Mazin AV, Alexeev AA, Kowalczykowski SC. A novel function of Rad54 protein. *J Biol Chem* 2003;**278**:14029–36.
282. Wolner B, Peterson CL. ATP-dependent and ATP-independent roles for the Rad54 chromatin remodeling enzyme during recombinational repair of a DNA double strand break. *J Biol Chem* 2005;**280**:10855–60.
283. Petukhova G, Stratton S, Sung P. Catalysis of homologous DNA pairing by yeast Rad51 and Rad54 proteins. *Nature* 1998;**393**:91–4.
284. Mazina OM, Mazin AV. Human Rad54 protein stimulates DNA strand exchange activity of hRad51 protein in the presence of Ca2+. *J Biol Chem* 2004;**279**:52042–51.
285. Alexiadis V, Kadonaga JT. Strand pairing by Rad54 and Rad51 is enhanced by chromatin. *Genes Dev* 2002;**16**:2767–71.
286. Sigurdsson S, Van Komen S, Petukhova G, Sung P. Homologous DNA pairing by human recombination factors Rad51 and Rad54. *J Biol Chem* 2002;**277**:42790–4.
287. Van Komen S, Petukhova G, Sigurdsson S, Stratton S, Sung P. Superhelicity-driven homologous DNA pairing by yeast recombination factors Rad51 and Rad54. *Mol Cell* 2000;**6**:563–72.
288. Li X, Zhang X-P, Solinger JA, et al. Rad51 and Rad54 ATPase activities are both required to modulate Rad51-dsDNA filament dynamics. *Nucleic Acids Res* 2007;**35**:4124–40.
289. Solinger JA, Heyer W-D. Rad54 protein stimulates the postsynaptic phase of Rad51 protein-mediated DNA strand exchange. *Proc Natl Acad Sci* 2001;**98**:8447–53.
290. Mazina OM, Rossi MJ, Thomaa NH, Mazin AV. Interactions of human Rad54 protein with branched DNA molecules. *J Biol Chem* 2007;**282**:21068–80.
291. Matulova P, Marini V, Burgess RC, et al. Cooperativity of Mus81·Mms4 with Rad54 in the resolution of recombination and replication intermediates. *J Biol Chem* 2009;**284**:7733–45.
292. Mazina OM, Mazin AV. Human Rad54 protein stimulates human Mus81-Eme1 endonuclease. *Proc Natl Acad Sci* 2008;**105**:18249–54.

293. Wiese C, Dray E, Groesser T, et al. Promotion of homologous recombination and genomic stability by RAD51AP1 via RAD51 recombinase enhancement. *Mol Cell* 2007;**28**:482–90.
294. Modesti M, Budzowska M, Baldeyron Cl, Demmers JAA, Ghirlando R, Kanaar R. RAD51AP1 is a structure-specific DNA binding protein that stimulates joint molecule formation during RAD51-mediated homologous recombination. *Mol Cell* 2007;**28**:468–81.
295. Kovalenko OV, Golub EI, Bray-Ward P, Ward DC, Radding CM. A novel nucleic acid-binding protein that interacts with human Rad51 recombinase. *Nucleic Acids Res* 1997;**25**:4946–53.
296. Luger K, Mader AW, Richmond RK, Sargent DF, Richmond TJ. Crystal structure of the nucleosome core particle at 2.8 A resolution. *Nature* 1997;**389**:251–60.
297. Kornberg RD. Structure of chromatin. *Annu Rev Biochem* 1977;**46**:931–54.
298. Thoma F, Koller T. Influence of histone H1 on chromatin structure. *Cell* 1977;**12**:101–7.
299. Horn PJ, Peterson CL. Chromatin higher order folding—wrapping up transcription. *Science* 2002;**297**:1824–7.
300. Kouzarides T. Chromatin modifications and their function. *Cell* 2007;**128**:693–705.
301. Strahl BD, Allis CD. The language of covalent histone modifications. *Nature* 2000;**403**:41–5.
302. Jenuwein T, Allis CD. Translating the histone code. *Science* 2001;**293**:1074–80.
303. Rice JC, Allis CD. Code of silence. *Nature* 2001;**414**:258–61.
304. Manohar M, Mooney AM, North JA, et al. Acetylation of histone H3 at the nucleosome dyad alters DNA-histone binding. *J Biol Chem* 2009;**284**:23312–21.
305. Simon M, North JA, Shimko JC, et al. Histone fold modifications control nucleosome unwrapping and disassembly. *Proc Natl Acad Sci* 2011;**108**:12711–6.
306. Eberharter A, Becker PB. ATP-dependent nucleosome remodelling: factors and functions. *J Cell Sci* 2004;**117**:3707–11.
307. Flaus A, Owen-Hughes T. Dynamic properties of nucleosomes during thermal and ATP-driven mobilization. *Mol Cell Biol* 2003;**23**:7767–79.
308. Gavin I, Horn PJ, Peterson CL. SWI/SNF chromatin remodeling requires changes in DNA topology. *Mol Cell* 2001;**7**:97–104.
309. Kassabov SR, Zhang B, Persinger J, Bartholomew B. SWI/SNF unwraps, slides, and rewraps the nucleosome. *Mol Cell* 2003;**11**:391–403.
310. Lorch Y, Maier-Davis B, Kornberg RD. Chromatin remodeling by nucleosome disassembly in vitro. *Proc Natl Acad Sci U S A* 2006;**103**:3090–3.
311. Saha A, Wittmeyer J, Cairns BR. Chromatin remodeling by RSC involves ATP-dependent DNA translocation. *Genes Dev* 2002;**16**:2120–34.
312. Ulyanova NP, Schnitzler GR. Human SWI/SNF generates abundant, structurally altered dinucleosomes on polynucleosomal templates. *Mol Cell Biol* 2005;**25**:11156–70.
313. Morrison AJ, Highland J, Krogan NJ, et al. INO80 and gamma-H2AX interaction links ATP-dependent chromatin remodeling to DNA damage repair. *Cell* 2004;**119**:767–75.
314. Downs JA, Lowndes NF, Jackson SP. A role for Saccharomyces cerevisiae histone H2A in DNA repair. *Nature* 2000;**408**:1001–4.
315. Rogakou EP, Pilch DR, Orr AH, Ivanova VS, Bonner WM. DNA double-stranded breaks induce histone H2AX phosphorylation on serine 139. *J Biol Chem* 1998;**273**:5858–68.
316. Rogakou EP, Nieves-Neira W, Boon C, Pommier Y, Bonner WM. Initiation of DNA fragmentation during apoptosis induces phosphorylation of H2AX histone at serine 139. *J Biol Chem* 2000;**275**:9390–5.
317. Rogakou EP, Boon C, Redon C, Bonner WM. Megabase chromatin domains involved in DNA double-strand breaks in vivo. *J Cell Biol* 1999;**146**:905–16.
318. Harper JW, Elledge SJ. The DNA damage response: ten years after. *Mol Cell* 2007;**28**:739–45.
319. Stucki M, Clapperton JA, Mohammad D, Yaffe MB, Smerdon SJ, Jackson SP. MDC1 directly binds phosphorylated histone H2AX to regulate cellular responses to DNA double-strand breaks. *Cell* 2005;**123**:1213–26.

320. Chapman JR, Jackson SP. Phospho-dependent interactions between NBS1 and MDC1 mediate chromatin retention of the MRN complex at sites of DNA damage. *EMBO Rep* 2008;**9**:795–801.
321. Melander F, Bekker-Jensen S, Falck J, Bartek J, Mailand N, Lukas J. Phosphorylation of SDT repeats in the MDC1 N terminus triggers retention of NBS1 at the DNA damage-modified chromatin. *J Cell Biol* 2008;**181**:213–26.
322. Spycher C, Miller ES, Townsend K, et al. Constitutive phosphorylation of MDC1 physically links the MRE11-RAD50-NBS1 complex to damaged chromatin. *J Cell Biol* 2008;**181**:227–40.
323. Lou Z, Minter-Dykhouse K, Franco S, et al. MDC1 maintains genomic stability by participating in the amplification of ATM-dependent DNA damage signals. *Mol Cell* 2006;**21**:187–200.
324. Celeste A, Petersen S, Romanienko PJ, et al. Genomic instability in mice lacking histone H2AX. *Science* 2002;**296**:922–7.
325. Strom L, Karlsson C, Lindroos HB, et al. Postreplicative formation of cohesion is required for repair and induced by a single DNA break. *Science* 2007;**317**:242–5.
326. Strom L, Lindroos HB, Shirahige K, Sjogren C. Postreplicative recruitment of cohesin to double-strand breaks is required for DNA repair. *Mol Cell* 2004;**16**:1003–15.
327. Unal E, Arbel-Eden A, Sattler U, et al. DNA damage response pathway uses histone modification to assemble a double-strand break-specific cohesin domain. *Mol Cell* 2004;**16**:991–1002.
328. van Attikum H, Fritsch O, Hohn B, Gasser SM. Recruitment of the INO80 complex by H2A phosphorylation links ATP-dependent chromatin remodeling with DNA double-strand break repair. *Cell* 2004;**119**:777–88.
329. Downs JA, Allard S, Jobin-Robitaille O, et al. Binding of chromatin-modifying activities to phosphorylated histone H2A at DNA damage sites. *Mol Cell* 2004;**16**:979–90.
330. Xiao A, Li H, Shechter D, et al. WSTF regulates the H2A.X DNA damage response via a novel tyrosine kinase activity. *Nature* 2009;**457**:57–62.
331. Cook PJ, Ju BG, Telese F, Wang X, Glass CK, Rosenfeld MG. Tyrosine dephosphorylation of H2AX modulates apoptosis and survival decisions. *Nature* 2009;**458**:591–6.
332. Krishnan N, Jeong DG, Jung SK, et al. Dephosphorylation of the C-terminal tyrosyl residue of the DNA damage-related histone H2A.X is mediated by the protein phosphatase eyes absent. *J Biol Chem* 2009;**284**:16066–70.
333. Stucki M. Histone H2A.X Tyr142 phosphorylation: a novel sWItCH for apoptosis? *DNA Repair (Amst)* 2009;**8**:873–6.
334. Mailand N, Bekker-Jensen S, Faustrup H, et al. RNF8 ubiquitylates histones at DNA double-strand breaks and promotes assembly of repair proteins. *Cell* 2007;**131**:887–900.
335. Kolas NK, Chapman JR, Nakada S, et al. Orchestration of the DNA-damage response by the RNF8 ubiquitin ligase. *Science* 2007;**318**:1637–40.
336. Huen MS, Grant R, Manke I, et al. RNF8 transduces the DNA-damage signal via histone ubiquitylation and checkpoint protein assembly. *Cell* 2007;**131**:901–14.
337. Zhao GY, Sonoda E, Barber LJ, et al. A critical role for the ubiquitin-conjugating enzyme Ubc13 in initiating homologous recombination. *Mol Cell* 2007;**25**:663–75.
338. Ikura T, Tashiro S, Kakino A, et al. DNA damage-dependent acetylation and ubiquitination of H2AX enhances chromatin dynamics. *Mol Cell Biol* 2007;**27**:7028–40.
339. Stewart GS, Panier S, Townsend K, et al. The RIDDLE syndrome protein mediates a ubiquitin-dependent signaling cascade at sites of DNA damage. *Cell* 2009;**136**:420–34.
340. Doil C, Mailand N, Bekker-Jensen S, et al. RNF168 binds and amplifies ubiquitin conjugates on damaged chromosomes to allow accumulation of repair proteins. *Cell* 2009;**136**:435–46.

341. Bekker-Jensen S, Rendtlew Danielsen J, Fugger K, et al. HERC2 coordinates ubiquitin-dependent assembly of DNA repair factors on damaged chromosomes. *Nat Cell Biol* 2010;**12**:80–6 (sup pp 1–12).
342. Wang B, Elledge SJ. Ubc13/Rnf8 ubiquitin ligases control foci formation of the Rap80/Abraxas/Brca1/Brcc36 complex in response to DNA damage. *Proc Natl Acad Sci U S A* 2007;**104**:20759–63.
343. Wang B, Matsuoka S, Ballif BA, et al. Abraxas and RAP80 form a BRCA1 protein complex required for the DNA damage response. *Science* 2007;**316**:1194–8.
344. Nakada S, Tai I, Panier S, et al. Non-canonical inhibition of DNA damage-dependent ubiquitination by OTUB1. *Nature* 2010;**466**:941–6.
345. Sobhian B, Shao G, Lilli DR, et al. RAP80 targets BRCA1 to specific Ubiquitin structures at DNA damage sites. *Science* 2007;**316**:1198–202.
346. Galanty Y, Belotserkovskaya R, Coates J, Polo S, Miller KM, Jackson SP. Mammalian SUMO E3-ligases PIAS1 and PIAS4 promote responses to DNA double-strand breaks. *Nature* 2009;**462**:935–9.
347. Morris JR, Boutell C, Keppler M, et al. The SUMO modification pathway is involved in the BRCA1 response to genotoxic stress. *Nature* 2009;**462**:886–90.
348. Nakamura K, Kato A, Kobayashi J, et al. Regulation of homologous recombination by RNF20-dependent H2B ubiquitination. *Mol Cell* 2011;**41**:515–28.
349. Moyal L, Lerenthal Y, Gana-Weisz M, et al. Requirement of ATM-dependent monoubiquitylation of histone H2B for timely repair of DNA double-strand breaks. *Mol Cell* 2011;**41**:529–42.
350. Huyen Y, Zgheib O, Ditullio Jr. RA, et al. Methylated lysine 79 of histone H3 targets 53BP1 to DNA double-strand breaks. *Nature* 2004;**432**:406–11.
351. Botuyan MV, Lee J, Ward IM, et al. Structural basis for the methylation state-specific recognition of histone H4-K20 by 53BP1 and Crb2 in DNA repair. *Cell* 2006;**127**:1361–73.
352. Henikoff S. Nucleosome destabilization in the epigenetic regulation of gene expression. *Nat Rev Genet* 2008;**9**:15–26.
353. Tjeertes JV, Miller KM, Jackson SP. Screen for DNA-damage-responsive histone modifications identifies H3K9Ac and H3K56Ac in human cells. *EMBO J* 2009;**28**:1878–89.
354. Das C, Lucia MS, Hansen KC, Tyler JK. CBP/p300-mediated acetylation of histone H3 on lysine 56. *Nature* 2009;**459**:113–7.
355. Xu F, Zhang K, Grunstein M. Acetylation in histone H3 globular domain regulates gene expression in yeast. *Cell* 2005;**121**:375–85.
356. Ikura T, Ogryzko VV, Grigoriev M, et al. Involvement of the TIP60 histone acetylase complex in DNA repair and apoptosis. *Cell* 2000;**102**:463–73.
357. Murr R, Loizou JI, Yang YG, et al. Histone acetylation by Trrap-Tip60 modulates loading of repair proteins and repair of DNA double-strand breaks. *Nat Cell Biol* 2006;**8**:91–9.
358. Kusch T, Florens L, Macdonald WH, et al. Acetylation by Tip60 is required for selective histone variant exchange at DNA lesions. *Science* 2004;**306**:2084–7.
359. Bird AW, Yu DY, Pray-Grant MG, et al. Acetylation of histone H4 by Esa1 is required for DNA double-strand break repair. *Nature* 2002;**419**:411–5.
360. Hicks WM, Yamaguchi M, Haber JE. Real-time analysis of double-strand DNA break repair by homologous recombination. *Proc Natl Acad Sci U S A* 2011;**108**:3108–15.
361. Azar QnAs B, James Haber E. QnAs with James E. Haber. *Proc Natl Acad Sci* 2011;**108**:5479.
362. Osley MA, Tsukuda T, Nickoloff JA. ATP-dependent chromatin remodeling factors and DNA damage repair. *Mutat Res* 2007;**618**:65–80.
363. Lee SE, Moore JK, Holmes A, Umezu K, Kolodner RD, Haber JE. Saccharomyces Ku70, mre11/rad50 and RPA proteins regulate adaptation to G2/M arrest after DNA damage. *Cell* 1998;**94**:399–409.

364. Tsukuda T, Fleming AB, Nickoloff JA, Osley MA. Chromatin remodelling at a DNA double-strand break site in Saccharomyces cerevisiae. *Nature* 2005;**438**:379–83.
365. Sinha M, Watanabe S, Johnson A, Moazed D, Peterson CL. Recombinational repair within heterochromatin requires ATP-dependent chromatin remodeling. *Cell* 2009;**138**:1109–21.
366. Chai B, Huang J, Cairns BR, Laurent BC. Distinct roles for the RSC and Swi/Snf ATP-dependent chromatin remodelers in DNA double-strand break repair. *Genes Dev* 2005;**19**:1656–61.
367. Martens JA, Winston F. Recent advances in understanding chromatin remodeling by Swi/Snf complexes. *Curr Opin Genet Dev* 2003;**13**:136–42.
368. Shim EY, Hong SJ, Oum JH, Yanez Y, Zhang Y, Lee SE. RSC mobilizes nucleosomes to improve accessibility of repair machinery to the damaged chromatin. *Mol Cell Biol* 2007;**27**:1602–13.
369. Kent NA, Chambers AL, Downs JA. Dual chromatin remodeling roles for RSC during DNA double strand break induction and repair at the yeast MAT locus. *J Biol Chem* 2007;**282**:27693–701.
370. Shim EY, Ma JL, Oum JH, Yanez Y, Lee SE. The yeast chromatin remodeler RSC complex facilitates end joining repair of DNA double-strand breaks. *Mol Cell Biol* 2005;**25**:3934–44.
371. Koyama H, Itoh M, Miyahara K, Tsuchiya E. Abundance of the RSC nucleosome-remodeling complex is important for the cells to tolerate DNA damage in Saccharomyces cerevisiae. *FEBS Lett* 2002;**531**:215–21.
372. Cairns BR, Lorch Y, Li Y, et al. RSC, an essential, abundant chromatin-remodeling complex. *Cell* 1996;**87**:1249–60.
373. Papamichos-Chronakis M, Krebs JE, Peterson CL. Interplay between Ino80 and Swr1 chromatin remodeling enzymes regulates cell cycle checkpoint adaptation in response to DNA damage. *Genes Dev* 2006;**20**:2437–49.
374. Mizuguchi G, Shen X, Landry J, Wu W-H, Sen S, Wu C. ATP-driven exchange of histone H2AZ variant catalyzed by SWR1 chromatin remodeling complex. *Science* 2004;**303**:343–8.
375. Kruhlak MJ, Celeste A, Dellaire G, et al. Changes in chromatin structure and mobility in living cells at sites of DNA double-strand breaks. *J Cell Biol* 2006;**172**:823–34.
376. Doyon Y, Selleck W, Lane WS, Tan S, Cote J. Structural and functional conservation of the NuA4 histone acetyltransferase complex from yeast to humans. *Mol Cell Biol* 2004;**24**:1884–96.
377. Cai Y, Jin J, Tomomori-Sato C, et al. Identification of new subunits of the multiprotein mammalian TRRAP/TIP60-containing histone acetyltransferase complex. *J Biol Chem* 2003;**278**:42733–6.
378. Ebbert R, Birkmann A, Schuller HJ. The product of the SNF2/SWI2 paralogue INO80 of Saccharomyces cerevisiae required for efficient expression of various yeast structural genes is part of a high-molecular-weight protein complex. *Mol Microbiol* 1999;**32**:741–51.
379. Conaway RC, Conaway JW. The INO80 chromatin remodeling complex in transcription, replication and repair. *Trends Biochem Sci* 2009;**34**:71–7.
380. Shen X, Mizuguchi G, Hamiche A, Wu C. A chromatin remodelling complex involved in transcription and DNA processing. *Nature* 2000;**406**:541–4.
381. Shen X, Ranallo R, Choi E, Wu C. Involvement of actin-related proteins in ATP-dependent chromatin remodeling. *Mol Cell* 2003;**12**:147–55.
382. Jonsson ZO, Jha S, Wohlschlegel JA, Dutta A. Rvb1p/Rvb2p recruit Arp5p and assemble a functional Ino80 chromatin remodeling complex. *Mol Cell* 2004;**16**:465–77.
383. Breeden L, Nasmyth K. Cell cycle control of the yeast HO gene: Cis- and Trans-acting regulators. *Cell* 1987;**48**:389–97.
384. Stern M, Jensen R, Herskowitz I. Five SWI genes are required for expression of the HO gene in yeast. *J Mol Biol* 1984;**178**:853–68.

385. Neigeborn L, Carlson M. Genes affecting the regulation of SUC2 gene expression by glucose repression in Saccharomyces cerevisiae. *Genetics* 1984;**108**:845–58.
386. Cote J, Quinn J, Workman J, Peterson C. Stimulation of GAL4 derivative binding to nucleosomal DNA by the yeast SWI/SNF complex. *Science* 1994;**265**:53–60.
387. Cairns BR, Kim YJ, Sayre MH, Laurent BC, Kornberg RD. A multisubunit complex containing the SWI1/ADR6, SWI2/SNF2, SWI3, SNF5, and SNF6 gene products isolated from yeast. *Proc Natl Acad Sci* 1994;**91**:1950–4.
388. Tamkun JW. The role of brahma and related proteins in transcription and development. *Curr Opin Genet Dev* 1995;**5**:473–7.
389. Dingwall AK, Beek SJ, McCallum CM, et al. The Drosophila snr1 and brm proteins are related to yeast SWI/SNF proteins and are components of a large protein complex. *Mol Biol Cell* 1995;**6**:777–91.
390. Reisman D, Glaros S, Thompson EA. The SWI/SNF complex and cancer. *Oncogene* 2009;**28**:1653–68.
391. Saha A, Wittmeyer J, Cairns BR. Chromatin remodelling: the industrial revolution of DNA around histones. *Nat Rev Mol Cell Biol* 2006;**7**:437–47.
392. Sinha M, Peterson CL. A Rad51 presynaptic filament is sufficient to capture nucleosomal homology during recombinational repair of a DNA double-strand break. *Mol Cell* 2008;**30**:803–10.
393. Bochar DA, Wang L, Beniya H, et al. BRCA1 is associated with a human SWI/SNF-related complex: linking chromatin remodeling to breast cancer. *Cell* 2000;**102**:257–65.
394. Lee D, Kim JW, Seo T, Hwang SG, Choi E-J, Choe J. SWI/SNF complex interacts with tumor suppressor p53 and is necessary for the activation of p53-mediated transcription. *J Biol Chem* 2002;**277**:22330–7.
395. Otsuki T, Furukawa Y, Ikeda K, et al. Fanconi anemia protein, FANCA, associates with BRG1, a component of the human SWI/SNF complex. *Hum Mol Genet* 2001;**10**:2651–60.
396. van Attikum H, Fritsch O, Gasser SM. Distinct roles for SWR1 and INO80 chromatin remodeling complexes at chromosomal double-strand breaks. *EMBO J* 2007;**26**:4113–25.
397. Krogan NJ, Keogh MC, Datta N, et al. A Snf2 family ATPase complex required for recruitment of the histone H2A variant Htz1. *Mol Cell* 2003;**12**:1565–76.
398. Xu Y, Sun Y, Jiang X, et al. The p400 ATPase regulates nucleosome stability and chromatin ubiquitination during DNA repair. *J Cell Biol* 2010;**191**:31–43.
399. Haushalter KA, Kadonaga JT. Chromatin assembly by DNA-translocating motors. *Nat Rev Mol Cell Biol* 2003;**4**:613–20.
400. Zhang Z, Fan H-Y, Goldman JA, Kingston RE. Homology-driven chromatin remodeling by human RAD54. *Nat Struct Mol Biol* 2007;**14**:397–405.
401. Jaskelioff M, Van Komen S, Krebs JE, Sung P, Peterson CL. Rad54p is a chromatin remodeling enzyme required for heteroduplex DNA joint formation with chromatin. *J Biol Chem* 2003;**278**:9212–8.
402. Alexeev A, Mazin A, Kowalczykowski SC. Rad54 protein possesses chromatin-remodeling activity stimulated by the Rad51-ssDNA nucleoprotein filament. *Nat Struct Mol Biol* 2003;**10**:182–6.
403. Kwon Y, Chi P, Roh DH, Klein H, Sung P. Synergistic action of the Saccharomyces cerevisiae homologous recombination factors Rad54 and Rad51 in chromatin remodeling. *DNA Repair (Amst)* 2007;**6**:1496–506.
404. Symington LS, Heyer WD. Some disassembly required: role of DNA translocases in the disruption of recombination intermediates and dead-end complexes. *Genes Dev* 2006;**20**:2479–86.
405. Ward JD, Muzzini DM, Petalcorin MI, et al. Overlapping mechanisms promote postsynaptic RAD-51 filament disassembly during meiotic double-strand break repair. *Mol Cell* 2010;**37**:259–72.

406. Williams AB, Michael WM. Eviction notice: new insights into Rad51 removal from DNA during homologous recombination. *Mol Cell* 2010;**37**:157–8.
407. Shi I, Hallwyl SC, Seong C, Mortensen U, Rothstein R, Sung P. Role of the Rad52 amino-terminal DNA binding activity in DNA strand capture in homologous recombination. *J Biol Chem* 2009;**284**:33275–84.
408. Sugiyama T, Kantake N. Dynamic regulatory interactions of rad51, rad52, and replication protein-a in recombination intermediates. *J Mol Biol* 2009;**390**:45–55.
409. Nimonkar AV, Sica RA, Kowalczykowski SC. Rad52 promotes second-end DNA capture in double-stranded break repair to form complement-stabilized joint molecules. *Proc Natl Acad Sci U S A* 2009;**106**:3077–82.
410. Bernstein KA, Gangloff S, Rothstein R. The RecQ DNA helicases in DNA repair. *Annu Rev Genet* 2010;**44**:393–417.
411. Ahmad F, Stewart E. The N-terminal region of the Schizosaccharomyces pombe RecQ helicase, Rqh1p, physically interacts with Topoisomerase III and is required for Rqh1p function. *Mol Genet Genomics* 2005;**273**:102–14.
412. Bennett RJ, Noirot-Gros MF, Wang JC. Interaction between yeast sgs1 helicase and DNA topoisomerase III. *J Biol Chem* 2000;**275**:26898–905.
413. Fricke WM, Kaliraman V, Brill SJ. Mapping the DNA topoisomerase III binding domain of the Sgs1 DNA helicase. *J Biol Chem* 2001;**276**:8848–55.
414. Gangloff S, McDonald JP, Bendixen C, Arthur L, Rothstein R. The yeast type I topoisomerase Top3 interacts with Sgs1, a DNA helicase homolog: a potential eukaryotic reverse gyrase. *Mol Cell Biol* 1994;**14**:8391–8.
415. Johnson FB, Lombard DB, Neff NF, et al. Association of the Bloom syndrome protein with topoisomerase IIIalpha in somatic and meiotic cells. *Cancer Res* 2000;**60**:1162–7.
416. Wu L, Davies SL, North PS, et al. The Bloom's syndrome gene product interacts with topoisomerase III. *J Biol Chem* 2000;**275**:9636–44.
417. Chang M, Bellaoui M, Zhang C, et al. RMI1/NCE4, a suppressor of genome instability, encodes a member of the RecQ helicase/Topo III complex. *EMBO J* 2005;**24**:2024–33.
418. Mullen JR, Nallaseth FS, Lan YQ, Slagle CE, Brill SJ. Yeast Rmi1/Nce4 controls genome stability as a subunit of the Sgs1-Top3 complex. *Mol Cell Biol* 2005;**25**:4476–87.
419. Yin J, Sobeck A, Xu C, et al. BLAP75, an essential component of Bloom's syndrome protein complexes that maintain genome integrity. *EMBO J* 2005;**24**:1465–76.
420. Svendsen JM, Harper JW. GEN1/Yen1 and the SLX4 complex: solutions to the problem of Holliday junction resolution. *Genes Dev* 2010;**24**:521–36.
421. West SC. The search for a human Holliday junction resolvase. *Biochem Soc Trans* 2009;**37**:519–26.
422. Liu Y, West SC. Happy Hollidays: 40th anniversary of the Holliday junction. *Nat Rev Mol Cell Biol* 2004;**5**:937–44.
423. Mizuuchi K, Kemper B, Hays J, Weisberg RA. T4 endonuclease VII cleaves holliday structures. *Cell* 1982;**29**:357–65.
424. Oram M, Keeley A, Tsaneva I. Holliday junction resolvase in Schizosaccharomyces pombe has identical endonuclease activity to the CCE1 homologue YDC2. *Nucleic Acids Res* 1998;**26**:594–601.
425. Whitby MC, Dixon J. A new Holliday junction resolving enzyme from Schizosaccharomyces pombe that is homologous to CCE1 from Saccharomyces cerevisiae. *J Mol Biol* 1997;**272**:509–22.
426. Kleff S, Kemper B, Sternglanz R. Identification and characterization of yeast mutants and the gene for a cruciform cutting endonuclease. *EMBO J* 1992;**11**:699–704.
427. West SC, Korner A. Cleavage of cruciform DNA structures by an activity from Saccharomyces cerevisiae. *Proc Natl Acad Sci U S A* 1985;**82**:6445–9.

428. Symington LS, Kolodner R. Partial purification of an enzyme from Saccharomyces cerevisiae that cleaves Holliday junctions. *Proc Natl Acad Sci U S A* 1985;**82**:7247–51.
429. de Massy B, Studier FW, Dorgai L, Appelbaum E, Weisberg RA. Enzymes and sites of genetic recombination: studies with gene-3 endonuclease of phage T7 and with site-affinity mutants of phage lambda. *Cold Spring Harb Symp Quant Biol* 1984;**49**:715–26.
430. Sharples GJ, Lloyd RG. Resolution of Holliday junctions in Escherichia coli: identification of the ruvC gene product as a 19-kilodalton protein. *J Bacteriol* 1991;**173**:7711–5.
431. Connolly B, Parsons CA, Benson FE, et al. Resolution of Holliday junctions in vitro requires the Escherichia coli ruvC gene product. *Proc Natl Acad Sci U S A* 1991;**88**:6063–7.
432. West SC. Processing of recombination intermediates by the RuvABC proteins. *Annu Rev Genet* 1997;**31**:213–44.
433. Takahagi M, Iwasaki H, Nakata A, Shinagawa H. Molecular analysis of the Escherichia coli ruvC gene, which encodes a Holliday junction-specific endonuclease. *J Bacteriol* 1991;**173**:5747–53.
434. Iwasaki H, Takahagi M, Shiba T, Nakata A, Shinagawa H. Escherichia coli RuvC protein is an endonuclease that resolves the Holliday structure. *EMBO J* 1991;**10**:4381–9.
435. Dunderdale HJ, Benson FE, Parsons CA, Sharples GJ, Lloyd RG, West SC. Formation and resolution of recombination intermediates by E. coli RecA and RuvC proteins. *Nature* 1991;**354**:506–10.
436. Connolly B, West SC. Genetic recombination in Escherichia coli: Holliday junctions made by RecA protein are resolved by fractionated cell-free extracts. *Proc Natl Acad Sci U S A* 1990;**87**:8476–80.
437. Yamada K, Miyata T, Tsuchiya D, et al. Crystal structure of the RuvA-RuvB complex: a structural basis for the Holliday junction migrating motor machinery. *Mol Cell* 2002;**10**:671–81.
438. Ariyoshi M, Nishino T, Iwasaki H, Shinagawa H, Morikawa K. Crystal structure of the holliday junction DNA in complex with a single RuvA tetramer. *Proc Natl Acad Sci U S A* 2000;**97**:8257–62.
439. Hargreaves D, Rice DW, Sedelnikova SE, Artymiuk PJ, Lloyd RG, Rafferty JB. Crystal structure of E. coli RuvA with bound DNA Holliday junction at 6 A resolution. *Nat Struct Biol* 1998;**5**:441–6.
440. Parsons CA, Stasiak A, West SC. The E. coli RuvAB proteins branch migrate Holliday junctions through heterologous DNA sequences in a reaction facilitated by SSB. *EMBO J* 1995;**14**:5736–44.
441. Parsons CA, Stasiak A, Bennett RJ, West SC. Structure of a multisubunit complex that promotes DNA branch migration. *Nature* 1995;**374**:375–8.
442. Elborough KM, West SC. Resolution of synthetic Holliday junctions in DNA by an endonuclease activity from calf thymus. *EMBO J* 1990;**9**:2931–6.
443. Constantinou A, Davies AA, West SC. Branch migration and Holliday junction resolution catalyzed by activities from mammalian cells. *Cell* 2001;**104**:259–68.
444. Hyde H, Davies AA, Benson FE, West SC. Resolution of recombination intermediates by a mammalian activity functionally analogous to Escherichia coli RuvC resolvase. *J Biol Chem* 1994;**269**:5202–9.
445. Klein HL, Symington LS. Breaking up just got easier to do. *Cell* 2009;**138**:20–2.
446. Symington LS, Holloman WK. Resolving resolvases: the final act? *Mol Cell* 2008;**32**:603–4.
447. Hollingsworth NM, Brill SJ. The Mus81 solution to resolution: generating meiotic crossovers without Holliday junctions. *Genes Dev* 2004;**18**:117–25.
448. Chen XB, Melchionna R, Denis CM, et al. Human Mus81-associated endonuclease cleaves Holliday junctions in vitro. *Mol Cell* 2001;**8**:1117–27.

449. Boddy MN, Gaillard PH, McDonald WH, Shanahan P, Yates 3rd JR, Russell P. Mus81-Eme1 are essential components of a Holliday junction resolvase. *Cell* 2001;**107**:537–48.
450. Ciccia A, Constantinou A, West SC. Identification and characterization of the human mus81-eme1 endonuclease. *J Biol Chem* 2003;**278**:25172–8.
451. de los Santos T, Loidl J, Larkin B, Hollingsworth NM, et al. A role for MMS4 in the processing of recombination intermediates during meiosis in Saccharomyces cerevisiae. *Genetics* 2001;**159**:1511–25.
452. de los Santos T, Hunter N, Lee C, Larkin B, Loidl J, Hollingsworth NM. The Mus81/Mms4 endonuclease acts independently of double-Holliday junction resolution to promote a distinct subset of crossovers during meiosis in budding yeast. *Genetics* 2003;**164**:81–94.
453. Dendouga N, Gao H, Moechars D, Janicot M, Vialard J, McGowan CH. Disruption of murine Mus81 increases genomic instability and DNA damage sensitivity but does not promote tumorigenesis. *Mol Cell Biol* 2005;**25**:7569–79.
454. McPherson JP, Lemmers B, Chahwan R, et al. Involvement of mammalian Mus81 in genome integrity and tumor suppression. *Science* 2004;**304**:1822–6.
455. Liu Y, Masson JY, Shah R, O'Regan P, West SC. RAD51C is required for Holliday junction processing in mammalian cells. *Science* 2004;**303**:243–6.
456. Sharan SK, Kuznetsov SG. Resolving RAD51C function in late stages of homologous recombination. *Cell Div* 2007;**2**:15.
457. Ip SC, Rass U, Blanco MG, Flynn HR, Skehel JM, West SC. Identification of Holliday junction resolvases from humans and yeast. *Nature* 2008;**456**:357–61.
458. Fekairi S, Scaglione S, Chahwan C, et al. Human SLX4 is a Holliday junction resolvase subunit that binds multiple DNA repair/recombination endonucleases. *Cell* 2009;**138**:78–89.
459. Svendsen JM, Smogorzewska A, Sowa ME, et al. Mammalian BTBD12/SLX4 assembles a Holliday junction resolvase and is required for DNA repair. *Cell* 2009;**138**:63–77.
460. Munoz IM, Hain K, Declais AC, et al. Coordination of structure-specific nucleases by human SLX4/BTBD12 is required for DNA repair. *Mol Cell* 2009;**35**:116–27.
461. Andersen SL, Bergstralh DT, Kohl KP, LaRocque JR, Moore CB, Sekelsky J. Drosophila MUS312 and the vertebrate ortholog BTBD12 interact with DNA structure-specific endonucleases in DNA repair and recombination. *Mol Cell* 2009;**35**:128–35.
462. Fricke WM, Brill SJ. Slx1-Slx4 is a second structure-specific endonuclease functionally redundant with Sgs1-Top3. *Genes Dev* 2003;**17**:1768–78.
463. Reymer A, Frykholm K, Morimatsu K, Takahashi M, Norden B. Structure of human Rad51 protein filament from molecular modeling and site-specific linear dichroism spectroscopy. *Proc Natl Acad Sci U S A* 2009;**106**:13248–53.
464. Matsuoka S, Ballif BA, Smogorzewska A, et al. ATM and ATR substrate analysis reveals extensive protein networks responsive to DNA damage. *Science* 2007;**316**:1160–6.
465. Harfe BD, Jinks-Robertson S. DNA mismatch repair and genetic instability. *Annu Rev Genet* 2000;**34**:359–99.
466. Kolodner RD, Marsischky GT. Eukaryotic DNA mismatch repair. *Curr Opin Genet Dev* 1999;**9**:89–96.
467. Kolodner R. Biochemistry and genetics of eukaryotic mismatch repair. *Genes Dev* 1996;**10**:1433–42.
468. Datta A, Adjiri A, New L, Crouse GF, Jinks Robertson S. Mitotic crossovers between diverged sequences are regulated by mismatch repair proteins in Saccaromyces cerevisiae. *Mol Cell Biol* 1996;**16**:1085–93.
469. Datta A, Hendrix M, Lipsitch M, Jinks-Robertson S. Dual roles for DNA sequence identity and the mismatch repair system in the regulation of mitotic crossing-over in yeast. *Proc Natl Acad Sci U S A* 1997;**94**:9757–62.

470. Nicholson A, Hendrix M, Jinks-Robertson S, Crouse GF. Regulation of mitotic homeologous recombination in yeast. Functions of mismatch repair and nucleotide excision repair genes. *Genetics* 2000;**154**:133–46.
471. Welz-Voegele C, Jinks-Robertson S. Sequence divergence impedes crossover more than noncrossover events during mitotic gap repair in yeast. *Genetics* 2008;**179**:1251–62.
472. Spell RM, Jinks-Robertson S. Examination of the roles of Sgs1 and Srs2 helicases in the enforcement of recombination fidelity in Saccharomyces cerevisiae. *Genetics* 2004;**168**:1855–65.
473. Myung K, Datta A, Chen C, Kolodner RD. SGS1, the Saccharomyces cerevisiae homologue of BLM and WRN, suppresses genome instability and homeologous recombination. *Nat Genet* 2001;**27**:113–6.
474. Larocque JR, Jasin M. Mechanisms of recombination between diverged sequences in wild-type and BLM-deficient mouse and human cells. *Mol Cell Biol* 2010;**30**:1887–97.
475. Saydam N, Kanagaraj R, Dietschy T, et al. Physical and functional interactions between Werner syndrome helicase and mismatch-repair initiation factors. *Nucleic Acids Res* 2007;**35**:5706–16.

Overview for the Histone Codes for DNA Repair

Elizabeth A. Williamson,[*] Justin W. Wray,[†] Pranshu Bansal,[†] and Robert Hromas[*]

[*]*Department of Medicine, University of Florida, Gainesville, Florida, USA*

[†]*Department of Medicine, University of New Mexico, Albuquerque, New Mexico, USA*

DNA damage occurs continuously as a result of various factors—intracellular metabolism, replication, and exposure to genotoxic agents, such as ionizing radiation and chemotherapy. If left unrepaired, this damage could result in changes or mutations within the cell genomic material. There are a number of different pathways that the cell can utilize to repair these DNA breaks. However, it is of utmost interest to know how the DNA damage is signaled to the various DNA pathways. As DNA damage occurs within the chromatin, we postulate that modifications of histones are important for signaling the position of DNA damage, recruiting the DNA repair proteins to the site of damage, and creating an open structure such that the repair proteins can access the site of damage. We discuss the modifications that occur on the histones and the manner in which they relate to the type of damage that has occurred as well as the DNA repair pathways that are activated.

Cells are exposed to many agents that can cause DNA damage and in order to maintain genomic stability, this damage must be repaired. The signaling pathways are well described from ataxia telangiectasia mutated/ATM and Rad3 related (ATM/ATR) through Chk1 and Chk2, which signal the cell in which damage has occurred and needs to be repaired, but the signals that occur at the site of damage are only now becoming better understood. Histones are nuclear proteins that package and organize DNA into nucleosomes. There are five types of histones: H1, H2A, H2B, H3, and H4. Histone H1 is involved in the

Progress in Molecular Biology and Translational Science, Vol. 110
DOI: 10.1016/B978-0-12-387665-2.00008-0

higher order structure of chromatin whereas the other four histone proteins associate with the DNA to form nucleosomes. Each nucleosome is composed of 146 bp of DNA and eight histone molecules, two copies each of histone H2A, H2B, H3, and H4. The DNA is wrapped around the histones. The N terminus of each histone contains a number of lysine (K) residues. These residues are positively charged and these positively charged residues can then interact with the negatively charged phosphates in DNA. When this positive charge is neutralized, say by acetylation, then the binding affinity between the histones and the DNA is reduced. This modification, acetylation, is important in the regulation of gene transcription.

Little is known about the role of histone modifications in relation to DNA repair. There are a number of potential modifications that histones could undergo, such as acetylation, phosphorylation, and ubiquitylation. We review the most recent studies, which suggest that DNA repair is influenced and affected by histone modifications. Each of the five DNA repair pathways is discussed.

I. Histone Modifications of Homologous Recombination Repair

Homologous recombination repair (HRR) is one of two primary pathways in the repair of DNA double-strand breaks (DSBs). Mechanistically, HRR utilizes a homologous template, such as a sister chromatid or homologous chromosome, to repair broken DNA. This process is generally considered error-free repair, though it may lead to loss of heterozygosity and further chromosomal instability. HRR functions primarily during the latter portions of the cell cycle including S and G2, likely because of the presence of homologous sister chromatids. A complete review of HRR is presented elsewhere in this text.

HRR plays important roles in multiple oncogenic processes. Its most infamous role was described over a decade ago in breast cancer. The breast cancer genes (BRCA) 1 and 2 have since been thoroughly implicated in HRR and account for a significant proportion of familial inherited breast cancers. Different portions of the HRR pathway have further been shown to be mutated in a variety of nonfamilial, somatic, sporadic breast cancers and this is the basis for the current Phase I, II, and III clinical trials with several molecules that inhibit a parallel pathway including, specifically, PARP-1.[1] HRR has been shown to be mutated in a variety of other cancers including leukemia, ovarian, pancreatic, colon, and uterine cancers. This pathway has also been shown to be involved in several genetically inherited diseases including ataxia telangiectasia, and Werner's, Bloom's, and Cockayne syndromes. All told, this process has a number of implications in a number of different pathological processes.[1]

HRR is a complex process that must have significant access to DNA and involves the disruption of chromatin structure, at least temporarily. The first step in HRR is the processing of DNA ends to produce 3′ DNA ends for RAD51 binding. Then, RAD51 induces the search for homology that must disrupt base pairing and, by proxy, the chromatin structure. Next, the synthesis portion of HRR requires significant access to DNA and will not be possible without chromatin access, similar to that seen during replication. Finally, after the entire process has occurred, there must be a process responsible for the reorganization of the chromatin structure to resolve the enzymatic sequence of events. It can be predicted that each of these steps has associated pathways to regulate the intricate nature of modulating chromatin. In the current literature, there are three basic types of regulation of chromatin described, including histone variants, histone modifications, and ATP-dependent chromatin remodeling. In this section, we describe these events in their basic temporal activation during HRR.

Histone structure has been well described elsewhere, but in short, there are the canonical replication inserted histones: H2A, H2B, H3, and H4. Histone H2A has two variants that have been associated with DNA repair: H2AX and H2AZ. The classical event in DNA DSB repair is phosphorylation of histone H2AX, a variant of histone 2A, on Serine 139 (termed γ-H2AX) in mammals. This event has been shown to be mediated by Mec1 and Tel 1 in *Saccharomyces cerevisiae* or ATM, ATR, and DNA PKcs in mammalian cells. γ-H2AX phosphorylation results from the production of DSBs by extrinsic or intrinsic factors such as ionizing radiation (IR), chemotherapeutic drugs, or spontaneous cell damage as a result of replication fork arrest or oxidative damage, and is involved in the recognition of DNA DSBs. In mammalian cells, H2AX represents 2–25% of the H2A variants in the overall chromatin depending on cell type. Yeast cells that express unmodifiable H2AX are hypersensitive to treatment with drugs such as miracle mineral supplement (MMS) or phleomycin, and mammalian stem cells lacking γ-H2AX are sensitive to IR and prone to genomic instability.[2]

The exact function of γ-H2AX is still not completely understood, though several important roles have been well described. γ-H2AX is not necessary for the initial steps in repair, but is involved in localizing and concentrating repair factors at the site of the DSB including chromatin modifiers such as the histone acetyl transferases (HAT) NuA4, which is implicated in the relaxation of chromatin surrounding a break. Further, γ-H2AX is an important binding site for MDC1, an adapter protein that serves as a landing bay of sorts for a plethora of proteins including BRCA1, the MRN (Mre11/Rad50/Nbs1) complex, and

53BP1.[3] Interestingly, an AA substitution for glutamic acid, mimicking phosphorylation, does not induce localization of repair factors indicative that γ-H2AX is only one step in a more complicated process.[4]

No text on γ-H2AX in the context of chromatin would be complete without a discussion on γ-H2AX's role in the relaxation of heterochromatin. It is still up for debate whether the phosphorylation of H2AX induces chromatin relaxation, or whether relaxed chromatin is more likely to be marked by phosphorylation. The first model is that the phosphorylation event following damage recruits proteins that relax heterochromatin in preparation for repair. This is supported by the general finding that γ-H2AX foci are rarely seen in heterochromatin, indicating that the structure is relaxed rapidly after phosphorylation has occurred, as has been reported around a DSB. Second, it could be that heterochromatin is structurally less likely to be marked. As heterochromatin is more compacted, it yields less access to kinases, less phosphorylation, and further, is less likely to be damaged by oxidative damage produced during IR or by chemotherapy.[5]

In either of the current models, γ-H2AX has become a marker for the presence and/or location of DSBs within the cell. It is a commonly used antibody for determining the presence of DSBs using multiple laboratory techniques including ChIP, Western blotting, and immunofluorescence. It has been established that there are near 1 to 1 ratios (0.7–0.9 depending on cell type) of DNA DSBs to γ-H2AX nuclear foci in mammalian cells, which is especially useful in localizing other factors to sites of DNA DSBs. Of note, γ-H2AX is phosphorylated in most DNA DSBs including those produced during DNA replication and those frank DSBs that will be repaired via nonhomologous end-joining (NHEJ), which is discussed separately in this chapter.

The other H2A variant, H2AZ, is less well defined in DNA repair. In transcription, H2AZ is inserted in most RNA polymerase II promoters to allow easier eviction. Its role in repair is thought to be similar to this. It has been shown in yeast to be inserted by SWR1 around DSBs depending on sumoylation at K126 and K133 in the C-terminal tail, and this relocalization is important for further steps in the HRR process including RAD51 function.[6] This process likely enhances chromatin eviction in preparation for HRR and has been shown to occur during the initial steps in the process to influence access to chromatin.

The initial steps in the DSB pathway involve Tip60 (Tat-interacting protein, or NuA4 in yeast), MRN, and ATM, which have been more recently elucidated to shed light on the events leading to γ-H2AX production.[7] The current model begins with undamaged, normal chromatin compacted and stabilized by the heterochromatin protein 1 (HP1) and KRAB-associated protein (KAP-1). These proteins sit on heterochromatin and maintain the condensed state.

HP1 is an adapter molecule that is primarily responsible for maintaining heterochromatin. KAP-1 was shown to actively regulate the decondensation of chromatin surrounding DSBs when deactivated, likely by phosphorylation. When a DSB occurs, the DNA is structurally opened and the MRN complex recognizes this occurrence by an unknown mechanism. MRN binds the free DNA ends. Then this action unmasks trimethylated H3 at K9 (H3K9me3) in the vicinity of the DSB. In undamaged cells, Tip60 and ATM form a stable complex and Tip60 is capable of recognizing H3K9me3 via its chromodomain, which allows its recruitment to the DSB, along with ATM. Tip60 then acetylates ATM, which activates the preliminary kinase functions, and allows autophosphorylation and the production of γ-H2AX, which begins a cascade that results in megabases of γ-H2AX production around the DSB. In the same time frame as ATM is activated, Tip60 begins to hyperacetylate H3 and H4 at multiple residues, resulting in further chromatin relaxation[8] (Fig. 1).

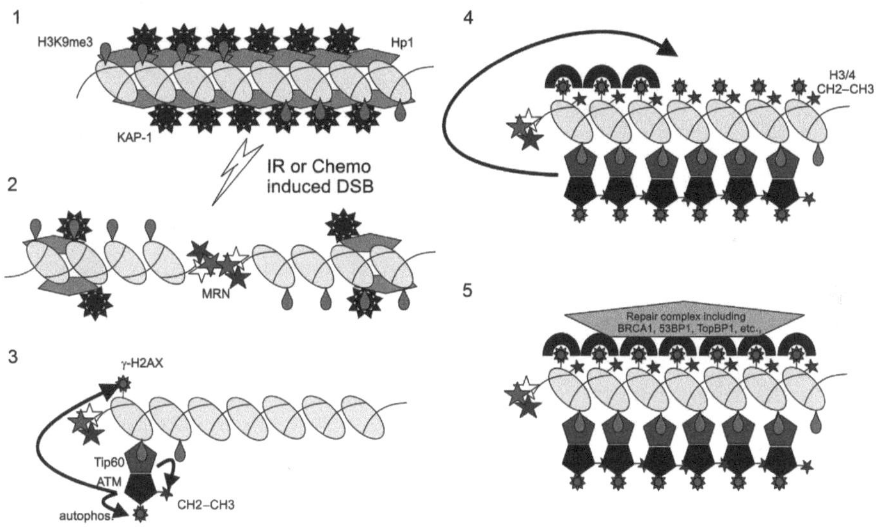

Fig. 1. (1) DNA bound to chromatin tightly compacted and maintained by Hp1 and KAP-1. (2) DSB occurs and MRN is recruited to DSB displacing Hp1 and KAP-1, revealing H3K9me3. (3) Tip60 and ATM are recruited to H3K9me3. Tip60 acetylates and activates ATM. ATM induces γ-H2AX and autophosphorylates. (4) MDC1 is recruited to γ-H2AX, ATM is activated and extends γ-H2AX, Tip60 acetylates H3/4. (5) MDC1 expands and binds to γ-H2AX which recruits a plethora of repair factors in conjunction with acetylated and methylated histones. Multiple steps are then ready to proceed, including HRR end resection, chromatin remodeling and the rest of the HRR mechanism. (See Color Insert.)

At this point in the activation cascade, γ-H2AX has been activated and repair factors are beginning to be recruited. This recruitment is facilitated by a further decondensation of chromatin, which results in a number of chromatin modifications becoming exposed. Dot1 trimethylates H3 at K79 and is important in the recruitment of 53BP1, which activates checkpoint factors simultaneously to HRR.[9,10] This modification is considered constitutive, but becomes unmasked during chromatin remodeling and functions as a binding site for 53BP1. Another methylation event that is constitutive, but important in HRR is H4K20me. This modification is also unmasked during DNA repair and results in the recruitment of multiple factors by their Tudor domains that recognize H4K20me, which appears to be species specific.[11]

There are a number of acetylation events that occur on H3 and H4 that have been associated with the early events in HRR. These include H3 lysines (K9, 14, 18, 23, and 27) and H4 lysines (K5, 8, 12, and 16). Multiple HATs have been implicated in this process including GCN5, NuA4, and HAT1.[12,13] These modifications increase near the DSB and are thought to be involved in loosening the chromatin structure for proper repair. Interestingly, when the four acetylated K residues on H4 are all mutated, DSB repair is abolished overall, but their individual functions are not well established. H3 acetylation has been most associated with HRR.[14] GCN5, an H3 HAT, has been well established in the eviction of chromosomes during transcription and has more recently been implicated in HRR, leading to the INO80-mediated chromosome eviction.[15,16] Specifically, H3 K14 and K23 appear to be the most critical events and are stimulated by phosphorylation of H3 S10.[17] Mutation in these residues also confers sensitivity to MMS, a chemotherapy which produces lesions primarily repaired by HRR. Further, the competing pathway, NHEJ, protein DNA PKcs has been reported to phosphorylate GCN5 in human cells and inactivate its HAT domain,[18] indicating that GCN5 plays a pro-HRR role in DSB repair. GCN5 also interacts with and coregulates BRCA1 depending on its HAT activity, which further establishes it in the HRR pathway.[19]

Protein ubiquitylation is an important process in cell biology rivaling the significance of phosphorylation events. Ubiquitin requires a chain of events to occur including several steps mediated by classes of proteins labeled E1, E2, and E3 ligases. In short, ubiquitin is prepared by E1, ligated by E2, and specificity is mediated by E3. There are more than 500 E3 ubiquitinating enzymes and multiple mono- and polyubiquitin chain products. With the complexity that has been described for this pathway, it has inevitably found a role in chromatin modification that affects HRR. Not only is H2AX modified by phosphorylation, it is also modified by ubiquitin dependent on the RNF8 (E3) and UBC13 (E2) proteins, and this modification is present in colocalization with γ-H2AX.[20]This modification occurs downstream of γ-H2AX and is dependent on the binding of MDC1 to γ-H2AX and the activity of p400, an SWI/SNF

ATPase chromatin-remodeling enzyme.[7] MDC1 then recruits RNF8, which ubiquitylates H2AX. In the best described model, this signals a transition from early to later recruitment of repair factors. The BRCA1 complex was previously reported to be dependent on γ-H2AX it is now known to fall later in the pathway and be dependent upon ubiquitylation of H2AX as well. Once BRCA1 is recruited, it is further responsible for the maintenance of H2AX ubiquitylation by means of its own E3 ligase function.[21] This pathway then remains intact to continue to recruit and maintain the damage response.[22,23]

As mentioned earlier, not only are histone variants and modifications important in the process of HRR, but histone remodeling also plays a central role in allowing access to DNA. There are multiple enzymes that play a role in histone remodeling in HRR. These include INO80, SWI/SNF, and SWR1. Each of these proteins is recruited to the break in a DSB-dependent manner and function to allow HRR to occur.

The ATPase motifs of the chromatin-remodeling proteins are in the SWI/SNF family. Within this family are several subdivisions including the INO80 family, which is present from yeast to mammals and includes INO80 and SWR1 in *S. cerevisiae* and INO80, Snf2-related CBP activator protein, and p400 in mammals. This family of proteins has been most described in transcription, but also plays a significant role in DSB repair and HRR.[24]

Both INO80 and SWR1 contain an Arp4 subunit that directly interacts with γ-H2AX and is required for recruitment to the DSB.[25,26] Both INO80 and SWR1 are required for HRR. SWR1 functions in concert with the HAT NuA4 (Tip60 in mammals) to introduce H2AZ into the chromatin around the break,[27] and when H2AZ is deleted, less single-stranded DNA is produced, though the kinetics of HRR are not significantly altered.[15] This process allows chromatin eviction to occur more readily and promotes proper processing of DNA to yield access for HRR.

In the last few years, the role of INO80 and chromatin eviction has become convoluted. It is clear that INO80 plays a major role in chromatin dynamics around a DSB because when the Arp8 subunit of INO80 is deleted, significantly less ssDNA is observed at the ends of the DSB.[15] Several studies have indicated that INO80 is responsible for histone eviction surrounding the DSB during DNA processing.[16,28] This is further corroborated by data indicating that the RSC complex also plays a role in chromatin remodeling surrounding the DSB.[29] On the other hand, some data indicate that H2B is not lost during DNA processing, even at the point at which RAD51 is bound to ssDNA.[30] An interesting point is that transcriptional histone eviction leads to five- to tenfold decreases in ChIP signal whereas in DSB repair, only two- to fourfold decreases are seen.[31] Taking all of this into account, it is clear that chromatin is significantly remodeled around a DSB due to INO80, but the amount of true histone eviction that occurs is still cloudy.

Once HRR has been activated, there is a defined set of enzymatic reactions that yields the final product. Within the process of homology search including strand invasion, RAD51 is capable of the reaction *in vitro* without any histone remodelers even in the presence of histones[32] and no histone modifications have been associated. *In vitro*, the yeast SWI/SNF chromatin-remodeling complex enhances the process specifically in heterochromatin.[29] This indicates that although RAD51 is capable of performing the essential steps in the HRR mechanism, there are steps in certain situations that require further chromatin processing to be efficient in the nuclear environment. Otherwise, the chromatin regulation associated with basic processes of HRR is still unclear to date.

Finally, the resolution of HRR requires that the chromatin structure be restored. This process is still being elucidated. The most well-described portion of resolution is the dephosphorylation of γ-H2AX and this is consistent in both HRR and NHEJ. This process is mediated by protein phosphatase 2A (PP2A) in mammalian cells and phosphatase Pph3 in yeast or PP4C in mammalians.[33] Inactivation of these phosphatases leads to sustained γ-H2AX and inefficient repair. It is clear that these factors are responsible for the resolution of γ-H2AX, but the activation of this process remains undescribed.

The removal of the acetylation events on H3 and H4 described earlier requires multiple histone deacetylases (HDACs), which are activated in the latter portion of the repair process and facilitate the condensation of chromatin back into its predamage configuration. The specific pathways are reviewed by Huertas *et al.*[11]

The chromatin methylation events H3K79me and H4K20me are a result of the opening of the chromatin structure. It is reasonable to assume that this process is reversed by reformation of the original chromatin structure and these methylations being reburied within the tertiary structure. It is also possible that there are other factors that facilitate the binding and reformation of HP1 and KAP-1 structures associated with H3K9me3, but these have yet to be described.

In the resolution of HRR, chromatin must also be replaced in its original configuration. This is accomplished by the process of chromatin assembly, though it is somewhat controversial because, as already described, it is not completely clear whether the chromatin is truly evicted or not. In the case where chromatin is evicted, it must be reestablished. Biochemically, two factors have been implicated in this process: chromatin assembly factor 1 (CAF-1) and antisilencing factor 1 (Asf1). CAF-1 is present at DSBs and is likely responsible for the insertion of H3.[34] Further, CAF-1 and Asf1 interact and it is proposed that Asf1 is recruited to sites of DNA damage in this fashion. Asf1 also stimulates acetylation of free H3 on K56 via the HAT Rtt109 and this modified H3 inserted into the DNA flanking the repaired DSB is the signal for completion of repair.[35]

Overall, HRR is a complicated process that requires significant alterations to the chromatin structure to be efficient. In review, this includes insertion of variants of H2; modifications to all of the core histones and their variants including the production of phosphorylation, acetylation, and ubiquitylation, and the recognition of methylation; and the structural remodeling of chromatin in both the initiation and completion of HRR.

II. Histone Modifications of NHEJ

NHEJ is an error-prone pathway for repairing DNA DSBs. As with homologous recombination (HR), the first modification that occurs after DNA damage is the phosphorylation on histone H2Ax on Ser139, resulting in γ-H2Ax. This modified histone recruits the Ku70/80 complex to the DNA ends where it binds to the ends and protects the DNA ends from being degraded as well as acting as a mark of the DSB. The MRN complex is then brought into the DSB. Together these two complexes then recruit the DNA ligase IV/XRCC4 complex to religate the ends and repair the break. There are two main barriers for efficient and effective DNA repair by NHEJ: rapid recruitment of DNA repair proteins to the site of the DSB; and ease of access to this damage. Chromatin packed with nucleosomes forms a barrier to these DNA repair proteins so certain structural issues have to be overcome to allow access for the DNA repair proteins.

Many studies have been carried out on the function of γ-H2Ax and its role for marking a DSB and recruiting DNA repair proteins but there are other histone modifications that have been identified that may be involved in the NHEJ repair pathway. These modifications include methylation (me), acetylation (ac), phosphorylation (p), and ubiquitination (ub) of histones H2A, H2B, H3, and H4. These histone modifications have been postulated to have two functions: they can either mark the site of a DNA break to help recruit the DNA repair proteins or they can be involved in changing and opening the chromatin structure so that the DNA repair proteins can gain access to the DSB. As will be discussed, histone modifications do have one or both functions in NHEJ.

A recent study identified dimethylation of lysine 36 on histone H3 (H3K36me2) as being a mark of DNA damage and subsequent recruitment of NHEJ proteins.[36] *In vitro* studies have identified Metnase, a DNA repair protein with a histone methylase SET domain, as being the methylase for this histone H3 residue.[37] Using a model system that allows the generation of a single DSB in a cell to study NHEJ, it was shown that Metnase could indeed dimethylate H3K36 *in vivo* at/around a DSB, and that the presence of this histone modification both recruited and stabilized DNA repair proteins at the

DSB. Moreover, the levels of H3K36me2 were proportional to the efficiency of the repair. Therefore, this suggests that H3K36me2 is in a category with phosphorylated H2Ax and ubiquitylated H2A in that they are all DNA damage-induced modifications that lead to recruitment of DNA repair proteins and can then improve the efficiency of DNA repair.[38–42]

Many histone modifications have been identified as being related to or marking sites of active transcription. One such modification is H3K4me3. This modification is carried out by Set1p methyltransferase and is found in newly created DSBs.[43] The recruitment was mediated by the RSC chromatin-remodeling complex, and monoubiquitylation of histone H2B (H2BK123ub) was also required for the methylation of H3K4. When H3K4me3 was not present, there was a decrease in NHEJ activity, suggesting that H3K4me3 is modulating chromatin at the site of DNA damage.

In addition to methylation and phosphorylation, acetylation of histones at sites of DNA damage can also assist in the repair process. Acetylation by HATs is an important and critical chromatin modification in DNA repair. Histones H3 and H4 are acetylated on their N-terminal lysines (K5, K8, K12, K16) after DNA damage and these modifications are essential for NHEJ.[44] In addition, NuA4-Tip60 HAT acetylates histone H4 at DSBs and improves the repair of the DSBs.[13,25,44] Chromatin structure and nucleosomes create a barrier to DNA repair proteins and acetylation of the histones facilitates the relaxation of chromatin and nucleosome repositioning. The SWI/SNF2 superfamily of remodeling complexes (INO80, SWR1, SWI/SNF, and RSC) is also involved in this relaxation and repositioning. INO80, SWR1, and RSC complexes are also necessary for Ku70/80 to be recruited to the DSB.[26,45] The RSC complex localizes to the DSB and also interacts with and recruits Mre11; then ATPase remodeling occurs which opens the chromatin and allows NHEJ proteins to gain access to the DSB.[46] CBP and p300 acetylate histones H3 (K18) and H4 (K5, K8, K12, K16) after recruitment to the DSB, and then cooperate with the SWI/SNF complex to recruit Ku70/80.[47] These studies together suggest that histone acetylation at a DSB is important for the recruitment of NHEJ proteins mainly via chromatin relaxation and nucleosome repositioning while at the same time being important for the recruitment of the early-response NHEJ proteins.

In many studies, a single modification or type of modification appears to be important for enhancing NHEJ repair of DSBs. However, some studies suggest that there may be a combination of modifications in a specific order on a single histone protein. In yeast, acetylation of histone H4 on K16 (H4K16ac) is important for NHEJ.[48] This modification is removed by the Sin3p/Rdp3p HDAC complex. Histone H4 is also phosphorylated on serine residue 1 in response to DNA damage and this is important for NHEJ and inhibits the acetylation of H4 by NuA4.[49,50] These studies indicate that there are a series of coordinated histone modifications that occur for DNA repair. H4K16ac by

NuA4 relaxes the chromatin structure on the region surrounding the DSB; then this is removed by Sin3p/Rdp3p, which is then followed by the H4Ser1 phosphorylation that inhibits the reacetylation of H4K16 by NuA4.[51]

γ-H2Ax is phosphorylated after DNA damage. It can be phosphorylated by ATM and by DNA-dependent protein kinase (DNA PKcs), a protein important in NHEJ. It has been shown that DNA PKcs can phosphorylate H2Ax within the nucleosome environment. Studies are conflicting as to whether this phosphorylation is influenced by histone acetylation.[52,53] Phosphorylated H2Ax does not affect nucleosome conformation but it does affect nucleosome stability as well as impairing histone H1 binding.[53]

The ubiquitylation of histones can also regulate DNA damage responses. The pattern appears to be primarily monoubiquitylation rather than polyubiquitylation. It appears that ubiquitylation of specific histone residues is required prior to methylation on other histone residues.[43,54] H2BK123ub is also required for H3K36 and H3K79 methylation.

There are multiple histone modifications that can occur to aid in this DNA repair pathway. From the studies, it appears that there is cross talk between different modifications, and even a specific order of modifications that is followed. Much more work needs to be done to further define these modifications in mammalian systems and to understand the way they relate to enhancement or inhibition of the NHEJ DNA repair pathway.

III. Histone Modifications of Nucleotide Excision Repair

The nucleotide excision repair (NER) pathway acts to remove DNA double-helix distorting lesions, including cyclobutane pyrimidine dimmers and 6-4 photoproducts, produced by ultraviolet (UV) light, as well as several kinds of bulky adducts induced by chemical agents such as cisplatin and 4-nitroquinoline oxide, which interfere with base pairing and block DNA duplication and transcription. NER is divided into two subpathways: global genomic NER (GG-NER) and transcription-coupled NER (TC-NER). GG-NER acts mainly on damage in nontranscribed regions of DNA, whereas TC-NER acts on damage in actively transcribed DNA. The first step of the NER reaction, the recognition of the DNA lesion, differs considerably between the GG-NER and TC-NER subpathways. In GG-NER, the xeroderma pigmentosa group C (XPC)–hHR23B complex (human homolog of yeast RAD 23) is responsible for the crucial damage sensing and repair recruitment step. This step also regulates the rate at which NER is carried out, while in TC-NER, the first signal for repair activity seems to be the blockage of transcription elongation by RNA polymerase II in front of DNA lesions.[55] The NER machinery is thought to be recruited to the stalled RNA polymerase II by the TC-NER-specific "coupling" factors CSA and CSB.[56] The subsequent steps in GG-NER and TC-NER are similar to each other and to those in NER in prokaryotes. XPB and XPD, which are subunits of transcription factor TFIIH, have helicase activity and unwind the DNA at the sites of damage. XPG protein has a structure-specific endonuclease activity, which makes an incision 3′ to the damaged DNA. Subsequently, the XPF–ERCC1 complex makes the 5′ incision during the NER. The dual incision leads to the removal of an ssDNA with a single-strand gap of 25–30 nucleotides. The resulting gap in DNA is filled by DNA ploymerase δ or ε by copying the undamaged strand. Proliferating cell nuclear antigen (PCNA) assists the DNA polymerase in the reaction, and replication protein A (RPA) protects the other DNA strand from degradation during NER. Finally, DNA ligase seals the nicks to finish NER.

Histone acetylation is an important event in NER. Several investigators have shown that treatment of human cells with sodium butyrate (an inhibitor of HDACs), stimulates the initial rate of NER *in vivo* and appears to correlate with an increase in the highest acetylated form of histone H4.[57–59] These authors have also shown that DNA repair synthesis occurring early after UV irradiation in mammalian cells is significantly enhanced in hyperacetylated mononucleosomes and seems not to result from increased UV damage in hyperacetylated chromatin.[57–59]

A study has shown that the TATA-box-binding protein-free TAF-containing complex (TFTC), a Gcn5-containing HAT complex and a transcriptional coactivator, is involved in the NER pathway.[60] TFTC has also been previously reported to acetylate histone H3 both *in vitro* and *in vivo*.[61] These authors demonstrated that TFTC shares strong homology with a subunit of UV-damaged DNA-binding factor DDB1, which is recruited to UV-induced DNA lesions *in vivo*.[60] TFTC binds preferentially to naked UV-damaged DNA and to nucleosomes assembled on UV-damaged DNA *in vitro*. Moreover, TFTC preferentially acetylates nucleosomes assembled on UV-damaged DNA templates *in vitro*.[60] A more recent study also demonstrated that UV irradiation leads to a dramatic increase in H3 acetylation *in vivo*.[62] As TFTC acts as coactivator in transcription, it might be associated mainly with TC-NER.

There have also been studies showing that other transcription factors take part in the acetylation process of histones as well, thereby leading to NER. It was recently discovered that the E2F1 transcription factor accumulates at sites of UV-induced DNA damage and directly stimulates NER through a nontranscriptional mechanism by associating with the GCN5 acetyltransferase in response to UV radiation and thus recruiting GCN5 to sites of damage. UV radiation induces the acetylation of histone H3 lysine 9 (H3K9) and this requires both GCN5 and E2F1.[63] Moreover, as previously observed for E2F1, knockdown of GCN5 results in impaired recruitment of NER factors to sites of damage and inefficient DNA repair. These findings demonstrate a direct role for GCN5 and E2F1 in NER involving H3K9 acetylation and increased accessibility to the NER machinery.[63]

Histone methylation occurs on lysine and arginine side chains but does not alter the charge on histone protein as done by acetylation and phosphorylation. Not much research has been done to find the role of histone methylation in NER in humans, but one study has shown that lysine methylation in histones is required for efficient NER in *S. cerevisiae*[64]

Although phosphorylation is important for some histones and DNA repair pathways, there are still no studies which clearly prove some role of histone phosphorylation on NER. One study showed that histone H2A phosphorylation controls Crb2 (a cell cycle checkpoint protein) recruitment at DNA breaks, maintains checkpoint arrest, and influences DNA repair in fission yeast.[65] Another study on human primary fibroblasts proposed that histone H2AX phosphorylation occurs after UV-induced NER starts operating.[66]

As described for other DNA repair pathways, monoubiquitylation of histones happens in NER. UV-induced monoubiquitylation of H2A is dependent on functional NER and occurs after incision of the damaged strand.[67] However, although this modification has been shown to occur at UV, it still has to be seen how this modification affects the process of NER.

IV. Histone Modifications of Base Excision Repair

Base excision repair (BER) is the primary DNA repair pathway that corrects base lesions that arise due to oxidation, alkylation, deamination, and depurinatiation/depyrimidination damage. BER facilitates the repair of damaged DNA via two general pathways: short and long patch. The short-patch BER pathway leads to a repair tract of a single nucleotide. Alternatively, the long-patch BER pathway produces a repair tract of at least two nucleotides. The BER pathway is initiated by one of many DNA glycosylases, which recognize and catalyze the removal of damaged bases. The completion of the BER pathway is accomplished by the coordinated action of at least three additional enzymes. These downstream enzymes carry out strand incision, gap filling, and ligation.

There are only a few studies published about covalent histone modification in BER pathways. Though several repair factors involved in BER pathways, such as Fen1, DNA-polymerase-β, and TDG, are *in vitro* and *in vivo* substrates for phosphorylation and acetylation,[68–72] phosphorylation, methylation, or acetylation of histones in BER pathways has not been reported so far.

However, for BER, there are many studies that indicate that histones are covalently modified by mono(ADP)-ribose in response to DNA damage. It has been shown that when cells were exposed to damage by ·OH radicals or methylating/alkylating agents, total covalent mono(ADP ribosyl)ation of histones increased by factors of 5–12, while the levels of histone H1-linked mono (ADP-ribosyl) groups were elevated even more than 30-fold.[73–75] Initial reports suggested that histone H1 is covalently mono(ADP ribosyl)ated at E2, E14, E116, or R34, and histone H2B at E2.[76–79] In addition, some other studies have shown that mono(ADP ribosyl)ation occurs also at glutamic acid residues of H2A and at arginine residues of H3 and H4.[80] Interestingly, butyrate exposure elevated basal levels of histone mono(ADP ribosyl)ation on H4 but reduced subsequent mono(ADP ribosyl)ation of histones initiated by DNA damage,[81] suggesting an antagonistic cross talk of histone mono(ADP ribosyl)ation and histone acetylation in BER pathways. Still, no nuclear mono(ADP ribosyl) transferase responsible for these modifications has been identified so far in higher eukaryotic cells. Because of this missing link, many authors, especially in the poly(ADP ribose)polymerase field, are quoting these modifications as poly (ADP ribosyl)ation. Interestingly, a recent study showed evidence that SIR2-like proteins (a family of β-nicotinamide adenine dinucleotide (NAD)-dependent HDACs) may function as mono(ADP-ribosyl) transferases in these processes.[82] SIR2-like proteins have been shown to take part in a wide range of cellular events, including chromosome silencing, chromosome segregation, DNA repair, DNA recombination, and the determination of life span.[83] TbSIR2RP1, a SIR2-related protein from the protozoan parasite *Trypanosoma brucei*, has been shown to catalyze mono(ADP ribosyl)ation of histones, particularly H2A and H2B, *in vitro*.[82] Under- or overexpression of TbSIR2RP1 decreased or increased cellular resistance to oxidizing DNA damage, respectively.[82] Remarkably, treatment of trypanosomal nuclei with a DNA-alkylating agent resulted in a significant increase in the level of histone mono(ADP ribosyl)ation, in particular H2A and H2B, and a concomitant increase in chromatin sensitivity to micrococcal nuclease.[82] Both of these responses correlated with the level of TbSIR2RP1 expression. Moreover, these studies supported previous evidence for a link between deacetylation and mono (ADP ribosyl)ation.

V. DNA Mismatch Repair and Histone Modifications

DNA mismatch repair (MMR) is a highly conserved DNA repair system, which is essential to all organisms. It has an important role in maintaining genomic instability. MMR targets mismatches that arise during replication and homologous recombination (HR), as well as repairing mismatches that occur in DNA following treatment with alkylating agents.[83,84] In terms of disease, a loss of the MMR pathway is associated with an increase in genomic instability and has been linked to hereditary nonpolyposis colorectal cancer.[85,86]

MMR has been well characterized in bacteria, yeast, and mammalian cells. In mammalian cells, the MMR pathway involves two heterodimers called MutS and MutL. MutS heterodimers contain MSH2 complexed with either MSH6 or MSH3. MutL heterodimers contain MLH1 complexed with either PMS2 or MLH3. In addition to these two complexes, a number of other proteins are required including RFC, EXO1, RPA, and PCNA. MutS heterodimers are responsible for "sensing" or recognizing the mismatch followed by recruitment of the MutL heterodimer to the chromatin.[87] The two heterodimers then bind ATP and undergo a conformational change, resulting in a protein that can more freely diffuse along the DNA in either direction. This suggests that the MutS–MutL complex may be functioning as a "sliding clamp" on the DNA. The MutS–MutL complex diffuses across the region containing the DNA mismatch and meets either RPA at the 5′ terminus or PCNA at the 3′ terminus. After this, it loads and activates EXO1 to degrade the faulty strand until the mismatch is removed. Finally, DNA polymerase δ fills the gap and DNA ligase I completes the process.

MMR is strongly associated with replication, a process that induces significant chromatin remodeling and has a number of associated histone modification events.[88–90] However, it is not yet fully defined as to whether MMR actually requires any specific chromatin remodeling or any histone modifications. Most biochemical studies have been carried out on naked DNA substrates so the effect of chromatin on this process is not well studied.[91] However, studies using the yeast genome have demonstrated that MMR has different efficiencies in different regions, which is suggestive of chromosome structure being an influence on MMR.[92]

DNA modifications resulting from treatment with alkylating agents are known to activate MMR, but there are subtle differences depending on the dosage. High doses lead to a G2 cell cycle arrest after one round of replication and PCNA is no longer present in the chromatin-bound MutS–MutL complex, but with low doses, the same effects are seen only after the second round of replication.[87] This latter effect might be a method of maintaining the MMR complexes on the chromatin longer such that other proteins can be recruited and activated.

Although aberrant DNA methylation has been shown to be responsible for the silencing of MMR genes, studies on hypoxia and stem cells have suggested that epigenetic chromatin inactivation is responsible for silencing the MMR genes MLH1 and MSH6.[93] Chromatin immunoprecipitation identified hypoacetylated/hypermethylated histone H3 lysine 9 (H3K9me3) as being present, which was associated with impaired SP1 binding to the promoters of these genes. Treatment with HDAC inhibitors increased histone H3K9 acetylation, which in turn increased SP1 activity on the promoters, and hence increased expression of MLH1 and MSH6. Therefore, a deregulated MMR pathway may result in genomic instability in stem cells, thus leading to their malignant transformation into cancer stem cells.

The MMR pathway operates within the nucleosomal environment but the impact of nucleosomal organization on MMR or vice versa is not understood. Nucleosomes are highly stable protein–DNA complexes that sterically block other DNA-binding proteins from accessing the chromatin. This could be a hindrance to the diffusion of the MutS–MutL heterodimers. A recent study has shown that the MMR complexes can cause disassembly or disruption of the nucleosome.[94] Acetylation of histone H3 lysine 56 within the nucleosomes both reduces their affinity for DNA and enhances their disruption by the diffusing MMR complexes. This disruption is passive as it requires only binding but not hydrolysis of ATP. Therefore, the MMR complex carries out two functions: it specifically recognizes mismatched DNA; and it creates a nucleosome-free environment to allow other DNA repair proteins access when the MMR sliding clamp performs passive DNA lesion-dependent chromatin remodeling.

While histone modifications have not been fully elucidated for the efficacy or activities of MMR, recent studies do suggest that chromatin/histone modifications are important in the MMR response to DNA damage.

As seen from this overview, there are multiple modifications that can occur on all histones, including phosphorylations, acetylations, methylations, and ubiquitylations. These modifications are common to all the DNA repair pathways. However, ribosylation appears to occur only for the DNA BER pathway. For the base excision and the mismatch repair pathways, there are few studies showing a link between histone modifications and the function of those pathways, but recent studies suggest that there is some role for histone modifications and their activity, so more studies should be carried out. Histone modifications appear to have many functions. They can mark the site of damage and recruit specific DNA repair proteins to the damage. They can also change the conformation of the histone such that it results in the opening of the chromatin around the DSB to allow better access for the DNA repair proteins. There can also be cross talk between some histone modifications to regulate the repair process, from opening the chromatin to allow repair to occur to closing the chromatin again once the DSB is fully repaired. There are also modifications that occur that activate different repair pathways. The repair pathway that becomes activated may therefore be a combination of the histone modification and the type of damage that has happened—DSB versus an interstrand cross-link versus a mismatched base. Further studies should be able to determine which histone modifications are important and how cross talk between them affects the function of these DNA repair pathways.

References

1. Evers B, Helleday T, Jonkers TJ. Targeting homologous recombination repair defects in cancer. *Trends Pharmacol Sci* 2010;**31**:372–80.

2. Moore JD, Krebs JE. Histone modifications and DNA double-strand break repair. *Biochem Cell Biol* 2004;**82**:446–52.
3. Xie A, Hartlerode A, Stucki M, Odate S, Puget N, Kwok A, et al. Distinct roles of chromatin-associated proteins MDC1 and 53BP1 in mammalian double-strand break repair. *Mol Cell* 2007;**28**:1045–57.
4. Yuan J, Adamski R, Chen J. Focus on histone variant H2AX: to be or not to be. *FEBS Letts* 2010;**584**:3717–24.
5. Cann KL, Dellaire G. Heterochromatin and the DNA damage response: the need to relax. *Biochem Cell Biol* 2011;**89**:45–60.
6. Kalocsay M, Hiller NJ, Jentsch S. Chromosome-wide Rad51 spreading and SUMO-H2A.Z-dependent chromosome fixation in response to a persistent DNA double-strand break. *Mol Cell* 2009;**33**:335–43.
7. Xu Y, Sun Y, Jiang X, Ayrapetov MK, Moskwa P, Yang S, et al. The p400 ATPase regulates nucleosome stability and chromatin ubiquitination during DNA repair. *J Cell Biol* 2010;**191**: 31–43.
8. Sun Y, Jiang X, Price BD. Tip60: connecting chromatin to DNA damage signaling. *Cell Cycle* 2010;**9**:930–6.
9. Giannattasio M, Lazzaro F, Plevani P, Muzi-Falconi M. The DNA damage checkpoint response requires histone H2B ubiquitination by Rad6-Bre1 and H3 methylation by Dot1. *J Biol Chem* 2005;**280**:9879–86.
10. Wysocki R, Javaheri A, Allard S, Sha F, Cote J, Kron SJ. Role of Dot1-dependent histone H3 methylation in G1 and S phase DNA damage checkpoint functions of Rad9. *Mol Cell Biol* 2005;**25**:8430–43.
11. Huertas D, Sendra R, Munoz P. Chromatin dynamics coupled to DNA repair. *Epigenetics* 2009;**4**:31–42.
12. Tamburini BA, Tyler JK. Localized histone acetylation and deacetylation triggered by the homologous recombination pathway of double-strand DNA repair. *Mol Cell Biol* 2005;**25**: 4903–13.
13. Murr R, Loizou JI, Yang YG, Cuenin C, Li H, Wang ZQ, et al. Histone acetylation by Trrap-Tip60 modulates loading of repair proteins and repair of DNA double-strand breaks. *Nat Cell Biol* 2006;**8**:91–9.
14. Qin S, Parthun MR. Histone H3 and the histone acetyltransferase Hat1p contribute to DNA double-strand break repair. *Mol Cell Biol* 2002;**22**:8353–65.
15. van Attikum H, Fritsch O, Hohn B, Gasser SM. Recruitment of the INO80 complex by H2A phosphorylation links ATP-dependent chromatin remodeling with DNA double-strand break repair. *Cell* 2004;**119**:777–88.
16. Tsukuda T, Fleming AB, Nickoloff JA, Osley MA. Chromatin remodelling at a DNA double-strand break site in Saccharomyces cerevisiae. *Nature* 2005;**438**:379–83.
17. Cheung P, Tanner KG, Cheung WL, Sassone-Corsi P, Denu JM, Allis CD. Synergistic coupling of histone H3 phosphorylation and acetylation in response to epidermal growth factor stimulation. *Mol Cell* 2000;**5**:905–15.
18. Barlev NA, Poltoratsky V, Owen-Hughes T, Ying C, Liu L, Workman JL, et al. Repression of GCN5 histone acetyltransferase activity via bromodomain-mediated binding and phosphorylation by the Ku-DNA-dependent protein kinase complex. *Mol Cell Biol* 1998;**18**:1349–58.
19. Oishi H, Kitagawa H, Wada O, Takezawa S, Tora L, Kouzu-Fujita M, et al. An hGCN5/TRRAP histone acetyltransferase complex co-activates BRCA1 transactivation function through histone modification. *J Biol Chem* 2006;**281**:20–6.
20. Mailand N, Bekker-Jensen S, Faustrup H, Melander F, Bartek J, Lukas C, et al. RNF8 ubiquitylates histones at DNA double-strand breaks and promotes assembly of repair proteins. *Cell* 2007;**131**:887–900.

21. Morris JR, Solomon E. BRCA1: BARD1 induces the formation of conjugated ubiquitin structures, dependent on K6 of ubiquitin, in cells during DNA replication and repair. *Hum Mol Genet* 2004;**13**:807–17.
22. Bennett EJ, Harper JW. DNA damage: ubiquitin marks the spot. *Nat Struct Mol Biol* 2008;**15**: 20–2.
23. Hofmann K. Ubiquitin-binding domains and their role in the DNA damage response. *DNA Repair* 2009;**8**:544–56.
24. Morrison AJ, Shen X. Chromatin remodelling beyond transcription: the INO80 and SWR1 complexes. *Nat Rev Mol Cell Biol* 2009;**10**:373–84.
25. Downs JA, Allard S, Jobin-Robitaille O, Javaheri A, Auger A, Bouchard N, et al. Binding of chromatin-modifying activities to phosphorylated histone H2A at DNA damage sites. *Mol Cell* 2004;**16**:979–90.
26. van Attikum H, Fritsch O, Gasser SM. Distinct roles for SWR1 and INO80 chromatin remodeling complexes at chromosomal double-strand breaks. *EMBO J* 2007;**26**:4113–25.
27. Kusch T, Florens L, Macdonald WH, Swanson SK, Glaser RL, Yates 3rd JR, et al. Acetylation by Tip60 is required for selective histone variant exchange at DNA lesions. *Science* 2004;**306**: 2084–7.
28. Tsukuda T, Lo YC, Krishna S, Sterk R, Osley MA, Nickoloff JA. INO80-dependent chromatin remodeling regulates early and late stages of mitotic homologous recombination. *DNA Repair* 2009;**8**:360–9.
29. Sinha M, Watanabe S, Johnson A, Moazed D, Peterson CL. Recombinational repair within heterochromatin requires ATP-dependent chromatin remodeling. *Cell* 2009;**138**:1109–21.
30. Shroff R, Arbel-Eden A, Pilch D, Ira G, Bonner WM, Petrini JH, et al. Distribution and dynamics of chromatin modification induced by a defined DNA double-strand break. *Curr Biol* 2004;**14**:1703–11.
31. Sinha M, Peterson CL. Chromatin dynamics during repair of chromosomal DNA double-strand breaks. *Epigenomics* 2009;**1**:371–85.
32. Sinha M, Peterson CL. A Rad51 presynaptic filament is sufficient to capture nucleosomal homology during recombinational repair of a DNA double-strand break. *Mol Cell* 2008;**30**:803–10.
33. Chowdhury D, Keogh MC, Ishii H, Peterson CL, Buratowski S, Lieberman J. gamma-H2AX dephosphorylation by protein phosphatase 2A facilitates DNA double-strand break repair. *Mol Cell* 2005;**20**:801–9.
34. Nabatiyan A, Szuts D, Krude T. Induction of CAF-1 expression in response to DNA strand breaks in quiescent human cells. *Mol Cell Biol* 2006;**26**:1839–49.
35. Chen CC, Carson JJ, Feser J, Tamburini B, Zabaronick S, Linger J, et al. Acetylated lysine 56 on histone H3 drives chromatin assembly after repair and signals for the completion of repair. *Cell* 2008;**134**:231–43.
36. Fnu S, Williamson EA, De Haro LP, Brenneman M, Wray J, Shaheen M, et al. Methylation of histone H3 lysine 36 enhances DNA repair by nonhomologous end-joining. *Proc Natl Acad Sci U S A* 2011;**108**:540–5.
37. Lee SH, Oshige M, Durant ST, Rasila KK, Williamson EA, Ramsey H, et al. The SET domain protein Metnase mediates foreign DNA integration and links integration to nonhomologous end-joining repair. *Proc Natl Acad Sci U S A* 2005;**102**:18075–80.
38. Escargueil AE, Soares DG, Salvador M, Larsen AK, Henriques JA. What histone code for DNA repair? *Mutat Res* 2008;**658**:259–70.
39. Sedelnikova OA, Pilch DR, Redon C, Bonner WM. Histone H2AX in DNA damage and repair. *Cancer Biol Ther* 2003;**2**:233–5.

40. Huen MS, Grant R, Manke I, Minn K, Yu X, Yaffe MB, et al. RNF8 transduces the DNA damage signal via histone ubiquitylation and checkpoint protein assembly. *Cell* 2007;**131**: 901–14.
41. Sobhian B, Shao G, Lilli DR, Culhane AC, Moreau LA, Xie B, et al. RAP80 target BRCA1 to specific ubiquitin structures at DNA damage sites. *Science* 2007;**316**:1198–202.
42. Kim H, Chen J, Yu X. Ubiquitin-binding protein RAP80 mediates BRCA1-dependent DNA damage response. *Science* 2007;**316**:1202–5.
43. Faucher D, Wellinger RJ. Methylated H3K4, a transcription-associated histone modification, is involved in the DNA damage response. *PLoS Genet* 2010;**6**:e1001082.
44. Bird AW, Yu DY, Pray-Grant MG, Qiu Q, Harmon KE, Megee PC, et al. Acetylation of histone H4 by Esa1 is required for DNA double-strand break repair. *Nature* 2002;**419**:411–5.
45. Shim EY, Ma JL, Oum JH, Yanez Y, Lee SE. The yeast chromatin remodeler RSC complex facilitates end joining repair of DNA double-strand breaks. *Mol Cell Biol* 2005;**25**:3934–44.
46. Shim EY, Hong SJ, Oum JH, Yanez Y, Zhang Y, Lee SE. RSC mobilizes nucleosomes to improve accessibility of repair machinery to the damaged chromatin. *Mol Cell Biol* 2007;**27**:1602–13.
47. Ogiwara H, Ui A, Otsuka A, Satoh H, Yokomi I, Nakajima S, et al. Histone acetylation by CBP and p300 at double-strand break sites facilitates SWI/SNF chromatin remodeling and the recruitment of non-homologous end joining factors. *Oncogene* 2011;**30**:2135–46.
48. Jazayeri A, McAinsh AD, Jackson SP. Saccharomyces cerevisiae Sin3p facilitates double-strand break repair. *Proc Natl Acad Sci U S A* 2004;**101**:1644–9.
49. Cheung WL, Turner FB, Krishnamoorthy T, Wolner B, Ahn SH, Foley M, et al. Phosphorylation of histone H4 serine 1 during DNA damage requires casein kinase II in *S. cerevisiae*. *Curr Biol* 2005;**15**:656–60.
50. Utley RT, Lacoste N, Jobin-Robitaille O, Allard S, Cote J. Regulation of NuA4 histone acetyltransferase activity in transcription and DNA repair by phosphorylation of histone H4. *Mol Cell Biol* 2005;**25**:8179–90.
51. Ataian Y, Krebs JE. Five repair pathways in one context: chromatin modification during DNA repair. *Biochem Cell Biol* 2006;**84**:490–504.
52. Park EJ, Chan DW, Park JH, Oettinger MA, Kwon J. DNA-PK is activated by nucleosomes and phosphorylated H2AX within the nucelosomes in an acetylation-dependent manner. *Nucleic Acids Res* 2003;**31**:6819–27.
53. Li A, Yu Y, Lee SC, Ishibashi T, Lees-Miller SP, Ausio J. Phosphorylation of hisotne H2A.X by DNA-dependent protein kinase is not affected by core histone acetylation but it alters nucleosome stability and histone H1 binding. *J Biol Chem* 2010;**285**:17778–88.
54. Yan Q, Dutt S, Xu R, Graves K, Juszczynski P, Manis JP, et al. BBAP monoubiquitylates histone H4 at lysine 91 and selectively modulates the DNA damage response. *Mol Cell* 2009;**36**:110–20.
55. Sugasawa K, Ng JM, Masutani C, Iwai S, van der Spek PJ, Eker AP, et al. Xeroderma pigmentosum group C protein complex is the initiator of global genome nucleotide excision repair. *Mol Cell* 1998;**2**:223–32.
56. van Gool AJ, Citterio E, Rademakers S, van Os R, Vermeulen W, Constantinou A, et al. The Cockayne syndrome B protein, involved in transcription-coupled DNA repair, resides in an RNA polymerase II-containing complex. *EMBO J* 1997;**16**:5955–65.
57. Dresler SL. Stimulation of deoxyribonucleic acid excision repair in human fibroblasts pretreated with sodium butyrate. *Biochemistry* 1985;**24**:6861–9.
58. Ramanathan B, Smerdon MJ. Enhanced DNA repair synthesis in hyperacetylated nucleosomes. *J Biol Chem* 1989;**264**:11026–34.
59. Smerdon MJ, Lan SY, Calza RE, Reeves R. Sodium butyrate stimulates DNA repair in UV-irradiated normal and xeroderma pigmentosum human fibroblasts. *J Biol Chem* 1982;**257**:13441–7.

60. Brand M, Moggs JG, Oulad-Abdelghani M, Lejeune F, Dilworth FJ, Stevenin J, et al. UV-damaged DNA-binding protein in the TFTC complex links DNA damage recognition to nucleosome acetylation. *EMBO J* 2001;**20**:3187–96.
61. Brand M, Yamamoto K, Staub A, Tora L. Identification of TATA-binding protein-free TAFII-containing complex subunits suggests a role in nucleosome acetylation and signal transduction. *J Biol Chem* 1999;**274**:18285–9.
62. Allison SJ, Milner J. Loss of p53 has site-specific effects on histone H3 modification, including serine 10 phosphorylation important for maintenance of ploidy. *Cancer Res* 2003;**63**:6674–9.
63. Guo R, Chen J, Mitchell DL, Johnson DG. GCN5 and E2F1 stimulate nucleotide excision repair by promoting H3K9 acetylation at sites of damage. *Nucleic Acids Res* 2010;**2010**:1–8.
64. Chaudhuri S, Wyrick JJ, Smerdon MJ. Histone H3 Lys70 methylation is required for efficient nucleotide excision repair in a silenced locus of *Saccharomyces cerevisiae*. *Nucleic Acids Res* 2009;**37**:1690–700.
65. Nakamuran TM, Du LL, Redon C, Russel P. Histone H2A phosphorylation controls Crb2 recruitment at DNA breaks, maintains checkpoint arrest, and influences DNA repair in fission yeast. *Mol Cell Biol* 2004;**24**:6215–30.
66. Matsumoto M, Yaginuma K, Igarashi A, Imura M, Hasegawa M, Iwabuchi K, et al. Perturbed gap-filling synthesis in nucleotide excision repair causes histone H2AX phosphorylation in human quiescent cells. *J Cell Sci* 2007;**120**:1104–12.
67. Bergink S, Salomons FA, Hoogstraten D, Groothuis TA, de Waard H, Wu J, et al. DNA damage triggers nucleotide excision repair-dependent monoubiquitylation of histone H2A. *Genes Dev* 2006;**20**:1343–52.
68. Bhakat KK, Hazra TK, Mitra S. Acetylation of the human DNA glycosylase NEIL2 and inhibition of its activity. *Nucleic Acids Res* 2004;**32**:3033–9.
69. Hasan S, Stucki M, Hassa PO, Imhof R, Gehrig P, Hunziker P, et al. Regulation of human flap endonuclease-1 activity by acetylation through the transcriptional coactivator p300. *Mol Cell* 2001;**7**:1221–31.
70. Hasan S, El-Andaloussi N, Hardeland U, Hassa PO, Burki C, Imhof R, et al. Acetylation regulates the DNA end-trimming activity of DNA polymerase beta. *Mol Cell* 2002;**10**:1213–22.
71. Henneke G, Koundrioukoff S, Hubscher U. Phosphorylation of human Fen1 by cyclin-dependent kinase modulates its role in replication fork regulation. *Oncogene* 2003;**22**:4301–13.
72. Tini M, Benecke A, Um SJ, Torchia J, Evans RM, Chambon P. Association of CBP/p300 acetylase and thymine DNA glycosylase links DNA repair and transcription. *Mol Cell* 2002;**9**:265–77.
73. Adamietz P, Rudolph A. ADP-ribosylation of nuclear proteins in vivo. Identification of histone H2B as a major acceptor for mono- and poly(ADP-ribose) in dimethyl sulfate-treated hepatoma AH 7974 cells. *J Biol Chem* 1984;**259**:6841–6.
74. Kreimeyer A, Wielckens K, Adamietz P, Hilz H. DNA repair-associated ADP-ribosylation in vivo. Modification of histone H1 differs from that of the principal acceptor proteins. *J Biol Chem* 1984;**259**:890–6.
75. Wielckens K, Bredehorst R, Adamietz P, Hilz H. Mono ADP-ribosylation and poly ADP-ribosylation of proteins in normal and malignant tissues. *Adv Enzyme Regul* 1982;**20**:23–37.
76. Burzio LO, Riquelme PT, Koide SS. ADP ribosylation of rat liver nucleosomal core histones. *J Biol Chem* 1979;**254**:3029–37.
77. Ogata N, Ueda K, Hayaishi O. ADP-ribosylation of histone H2B. Identification of glutamic acid residue 2 as the modification site. *J Biol Chem* 1980;**255**:7610–5.
78. Ogata N, Ueda K, Kagamiyama H, Hayaishi O. ADP-ribosylation of histone H1. Identification of glutamic acid residues 2, 14, and the COOH-terminal lysine residue as modification sites. *J Biol Chem* 1980;**255**:7616–20.

79. Ushiroyama T, Tanigawa Y, Tsuchiya M, Matsuura R, Ueki M, Sugimoto O, et al. Amino acid sequence of histone H1 at the ADP-ribose-accepting site and ADP-ribose X histone-H1 adduct as an inhibitor of cyclic-AMP-dependent phosphorylation. *Eur J Biochem* 1985;**151**:173–7.
80. Golderer G, Grobner P. ADP-ribosylation of core histones and their acetylated subspecies. *Biochem J* 1991;**277**:607–10.
81. Bohm L, Schneeweiss FA, Sharan RN, Feinendegen LE. Influence of histone acetylation on the modification of cytoplasmic and nuclear proteins by ADP-ribosylation in response to free radicals. *Biochim Biophys Acta* 1997;**1334**:149–54.
82. Garcia-Salcedo JA, Gijon P, Nolan DP, Tebabi P, Pays E. A chromosomal SIR2 homologue with both histone NADdependent ADP-ribosyltransferase and deacetylase activities is involved in DNA repair in *Trypanosoma brucei*. *EMBO J* 2003;**22**:5851–62.
83. Marra G, Schar P. Recognition of DNA alterations by the mismatch repair system. *Biochem J* 1999;**338**:1–13.
84. Hsieh P. Molecular mechanisms of DNA mismatch repair. *Mutat Res* 2001;**486**:71–87.
85. Kunkel TA, Erie DA. DNA mismatch repair. *Annu Rev Biochem* 2005;**74**:681–710.
86. Jirincy J. The multifaceted mismatch-repair system. *Nat Rev Mol Cell Biol* 2006;**7**:335–46.
87. Mastrocola AS, Heinen CD. Nuclear reorganization of DNA mismatch repair proteins in response to DNA damage. *DNA Repair* 2010;**9**:120–33.
88. Flores-Rozas H, Clark D, Kolodner RD. Proliferating cell nuclear antigen and Msh2p-Msh6p interact to form an active misrepair recognition complex. *Nat Genet* 2000;**26**:375–8.
89. Kleczkowska HE, Marra G, Lettieri T, Jirincy J. hMSH3 and hMSH6 interact with PCNA and colocalize with it to replication foci. *Genes Dev* 2001;**15**:724–36.
90. Groth A, Rocha W, Verreault A, Almouzni G. Chromatin challenges during DNA replication and repair. *Cell* 2007;**128**:721–33.
91. Zhang Y, Yuan F, Presnell SR, Tian K, Gao Y, Tomkinson AE, et al. Reconstitution of 5′-directed human mismatched repair in a purified system. *Cell* 2005;**122**:693–705.
92. Hawk JD, Stefanovic L, Boyer JC, Petes TD, Farber RA. Variation in efficiency of DNA mismatch repair at different sites in the yeast genome. *Proc Natl Acad Sci U S A* 2005;**102**:8639–43.
93. Rodriguez-Jimenez FJ, Moreno-Manzano V, Lucas-Dominguez R, Sanchez-Puelles JM. Hypoxia causes downregulation of mismatch repair system and genomic instability in stem cells. *Stem Cells* 2008;**26**:2052–62.
94. Javaid S, Manohar M, Punja N, Mooney A, Ottesen JJ, Poirer MG. Nucleosome remodeling by hMSH2-hMSH6. *Mol Cell* 2011;**36**:1086–94.

The RSC and INO80 Chromatin-Remodeling Complexes in DNA Double-Strand Break Repair

Anna L. Chambers and
Jessica A. Downs

MRC Genome Damage and Stability Centre, University of Sussex, Falmer, Brighton, United Kingdom

In eukaryotes, DNA is packaged into chromatin and is therefore relatively inaccessible to DNA repair enzymes. In order to perform efficient DNA repair, ATP-dependent chromatin-remodeling enzymes are required to alter the chromatin structure near the site of damage to facilitate processing and allow access to repair enzymes. Two of the best-studied remodeling complexes involved in repair are RSC (Remodels the Structure of Chromatin) and INO80 from *Saccharomyces cerevisiae*, which are both conserved in higher eukaryotes. RSC is very rapidly recruited to breaks and mobilizes nucleosomes to promote phosphorylation of H2A S129 and resection. INO80 enrichment at a break occurs later and is dependent on phospho-S129 H2A. INO80 activity at the break site also facilitates resection. Consequently, both homologous recombination and nonhomologous end-joining are defective in *rsc* mutants, while subsets of these repair pathways are affected in *ino80* mutants.

Progress in Molecular Biology
and Translational Science, Vol. 110
DOI: 10.1016/B978-0-12-387665-2.00009-2

1877-1173/12 $35.00

I. Introduction

DNA double-strand breaks (DSBs) are particularly toxic lesions to cells. Endogenous DSBs, generated through stalling or collapse of replication forks when they encounter single-strand DNA lesions such as abasic sites and single-strand breaks, have been calculated to occur at a frequency of ~50 per cell cycle in human cells and ~1 per 4–5 cells per cell cycle for *Saccharomyces cerevisiae*.[1] DSBs can also be formed by exposure to genotoxic agents such as ionizing radiation (IR). Efficient repair of DSBs is essential in order to maintain genome stability, as a single unrepaired break can result in lethality. There are two major pathways for the repair of DSBs, namely nonhomologous end-joining (NHEJ) and homologous recombination (HR). NHEJ involves processing and religation of broken ends, and is an inherently error-prone process (reviewed in Ref. 2) while HR uses homologous sequences as a template for accurate repair (reviewed in Ref. 3).

In eukaryotes, repair of DSBs and other lesions must occur in the context of chromatin. The basic unit of chromatin is the nucleosome, comprising approximately 146 bp DNA wrapped nearly two times around a histone octamer, formed from a tetramer consisting of $(H3–H4)_2$ and two H2A–H2B dimers. In the presence of linker histones and other scaffold proteins, this "beads-on-a-string" chromatin is able to be packaged into higher-order chromatin structures such as the 30 nm fiber and chromosome loops. Therefore, in order to efficiently repair DNA damage within chromatin, remodeling is necessary to allow repair enzymes to access the lesion, followed by restoration of the original chromatin structure upon completion of repair, termed the "access–repair–restore" model.[4] In the past decade, it has become apparent that chromatin functions not only as a means of packaging DNA, but that it also plays an active role in signaling during repair, modulating both recruitment and activity of repair and checkpoint proteins.

Access to chromatin can be regulated in two ways: by covalent posttranslational modification of histone proteins, including phosphorylation, methylation, acetylation, and ubiquitylation, and by the action of adenosine triphosphate (ATP)-dependent remodeling enzymes. ATP-dependent remodelers are large multi-subunit complexes that couple ATP hydrolysis to movement of histones or nucleosomes. A number of different remodeling activities can be performed by these complexes, including exchange or incorporation of core histones or histone variants, eviction of histones or nucleosomes, and repositioning or sliding of nucleosomes.[5]

Each complex contains a catalytic ATPase subunit that is a member of the Snf2 family of helicases and translocases. Based on conserved motifs within the catalytic subunit, remodeling complexes can be further divided into 24 subfamilies,[6,7] including SWI/SNF2 types (characterized by bromodomains that can bind acetylated lysines), INO80 types (with a split ATPase domain),[8] ISWI types (characterized by SANT and SLIDE domains), and CHD types (characterized by chromodomains that can bind methylated lysines). Complex-specific auxiliary subunits confer different properties to the remodeling complexes, in terms of both regulation of catalytic activity and recruitment through different chromatin-binding domains. Evidence exists for the involvement of a number of chromatin-remodeling complexes in the DNA damage response, such as RSC, INO80, SWR1, and SWI/SNF in budding yeast and PBAF, BAF, ACF, and TIP60 in mammalian cells (reviewed in Ref. 9).

The role of chromatin-remodeling complexes in double-strand break repair (DSBR) has been most extensively studied using the HO system in *S. cerevisiae*.[10] Mating-type switching in budding yeast involves introduction of a DSB at the mating-type locus by HO endonuclease, followed by intrachromosomal recombination with one of two silent donor cassettes, *HML*α or *HMR*a. In the HO system used to study the molecular details of DSBR, the donor cassettes have been deleted and therefore induction of HO introduces a persistent DSB at a single, specific location (*MAT* locus) within the genome that cannot be repaired by HR. Although repair of this break can only occur through NHEJ, the key proteins involved in HR are still recruited to the break. This allows the temporal accumulation of repair and checkpoint proteins, histone modifications, and changes in nucleosome positioning to be followed by chromatin immunoprecipitation (ChIP) and nuclease accessibility experiments.

One of the earliest events that can be detected following DSB induction is phosphorylation of the C terminus of histone H2A on S129.[11] In higher eukaryotes, this phosphorylation motif occurs not on the core H2A, but on the histone variant H2AX[12] that constitutes approximately 10% of the total H2A.[13] Yeast containing H2A that is not able to be phosphorylated have a defect in NHEJ and are mildly sensitive to DNA-damaging agents,[11] while mouse cells lacking H2AX are sensitive to IR and the animals are predisposed to cancer in the absence of p53.[14–17] H2A phosphorylation is carried out by the damage-response phosphatidyl-3-OH-kinase-like kinases Tel1 and Mec1 in *S. cerevisiae*[11] (homologous to ataxia telangiectasia-mutated (ATM) and ATM- and Rad3-related (ATR), respectively, in mammals) that additionally phosphorylate many other targets on SQ/TQ motifs in response to damage.[18–20] H2A phosphorylation spreads over a large region of ~50 kb in budding yeast[21] or megabases in higher eukaryotes,[22] with the levels of phosphorylation greatest at 3–5 kb from the break site, but low in the 1 kb immediately adjacent to the break.[23]

Tel1, in conjunction with the nuclease MRX (Mre11-Rad50-Xrs2), is primarily recruited to unprocessed DSBs. In contrast, Mec1 and its partner protein Ddc1 are recruited to tracts of human replication protein A (RPA)-bound single-stranded DNA. Single-stranded DNA is generated via processing of DSBs through exonucleolytic resection by several nucleases including MRX and Exo1 to initiate HR,[24] but is also present at other types of lesion such as stalled replication forks. Activation of Mec1 and Tel1 results in amplification of the DNA damage signal cascade and leads to recruitment and retention of many repair and checkpoint proteins near the site of the lesion. One of the consequences is arrest of the cell cycle by activation of checkpoints, permitting repair of DNA damage before cell division or replication. In *S. cerevisiae*, the major checkpoint activated in response to DNA damage is at the G2/M boundary and involves the upregulation of the ribonucleotide reductase (*RNR*) genes and phosphorylation of Rad53 in an Mec1-dependent manner. Any of the steps of the DNA damage response, that is, damage recognition, resection, H2A phosphorylation, checkpoint activation, or binding and retention of downstream effectors, could conceivably be affected by chromatin structure and hence requires the action of chromatin-remodeling complexes.

This review focuses on the roles of two ATP-dependent chromatin remodelers, namely RSC (PBAF or SWI/SNF-B in mammalian cells) and INO80 in DSBR.

II. RSC

A. Subunit Composition and Structure

The RSC complex is the most abundant ATP-dependent chromatin-remodeling complex in budding yeast (~1000–2000 molecules per cell)[25] and is composed of 16 subunits: Sth1 (the catalytic subunit),[26,27] Sfh1,[28] Rsc3,[29,30] Rsc30,[29] Rsc4,[31,32] Rsc58,[33] Rsc6,[34] Rsc7(Npl6),[35] Rsc8(Swh3),[36] Rsc9, Arp7 (Rsc11),[37] Arp9 (Rsc12),[37] Rsc14 (Ldb7), Htl1,[38–40] Rtt102, and either Rsc1 or the highly similar Rsc2 subunit[41] (see Table I). The catalytic subunit, Sth1, is essential for viability and its ATPase domain is closely related to that of the Swi2/Snf2 subunit of the SWI/SNF complex. Three other RSC subunits are also found in the related SWI/SNF complex, namely Rtt102, Arp7, and Arp9 (see Table I). Another three subunits are homologous to subunits of SWI/SNF, namely Sfh1, Rsc8, and Rsc6 (see Table I), suggesting that these form the core of the complex with Sth1, while the remaining subunits are specific to RSC.

Two isoforms of the RSC complex exist, containing either the Rsc1 subunit or the highly similar Rsc2 subunit[41] (46% identity and 64% similarity). Simultaneous deletion of both *rsc1* and *rsc2* is lethal, whereas single deletion of either is viable, suggesting that there is some functional redundancy between the two RSC isoforms. Rsc2 is approximately 10-fold more abundant than Rsc1, but both contain the same domain organization, that is, two bromodomains (likely to be acetyl–lysine-binding domains) separated by a weak nonspecific DNA-binding motif, termed an AT-hook, followed by a Bromo-Adjacent Homology (BAH) domain (important for nucleosome binding).[41,123] Several other

TABLE I
SUBUNIT COMPOSITION OF RSC AND SWI/SNF COMPLEXES

S. cerevisiae		Human	
RSC	**SWI/SNF**	**hSwi/Snf-B (PBAF)**	**hSwi/Snf-A (BAF)**
Sth1	**Snf2**	**Brg1**	**Brm or Brg1**
Arp7	Arp7	β-Actin	β-Actin
Arp9	Arp9	BAF53	BAF53
Rtt102	Rtt102		
Rsc6	Swp73	BAF60a	BAF60a
Sfh1	Snf5	Snf5	Snf5
Rsc8	Swi3	BAF170 and BAF155	BAF170 and BAF155
Rsc3			
Rsc30			
Rsc1 or Rsc2		Polybromo/BAF180	
Rsc4			
Rsc7			
Rsc9			
Ldb7			
Htl1			
Rsc58			
		BAF57	BAF57
	Swi1		BAF250
	Snf6		
	Snf1		
	Swp29/Tfg3/Anc1/Taf30		
	Swp82		

Catalytic subunits are shown in bold. Yeast RSC and SWI/SNF complexes share Arp7, Arp9, and Rtt102 subunits and four more subunits share homology. The human PBAF complex is homologous to RSC, while BAF is homologous to SWI/SNF. PBAF and BAF share seven subunits and are defined by the presence of either BAF180 or BAF250.

subunits of RSC are essential and many contain further chromatin-binding and regulatory domains (see Table II). The lethality that results from deletion of some RSC subunits demonstrates the importance of RSC function in cells, most likely in transcription of some essential genes.

Cryo-electron microscopy structural studies on the RSC complex revealed a large central cavity of sufficient size to accommodate a nucleosome. Incubation of the complex with nucleosomes led to an increased area of density within the central cavity, suggesting that this constitutes a nucleosome-binding pocket.[42–44]

Images of both the Rsc2–RSC complex isoform[42] and a mixture of Rsc1- and Rsc2-containing complexes[43,45] showed predominantly two different conformations, that is, an "open" and a "closed" conformation that varied mainly in the position of the lower arm. Incubation with an acetylated H3 N-terminal peptide stabilized the "closed" conformation of this arm.[45]

TABLE II

Functional Domains of RSC Subunits

Subunit	Domains and properties	References
Sth1	Essential, catalytic subunit containing ATPase domain, has, and post-HSA domains involved in Arp binding and regulation of ATPase activity, bromodomain (acetyl–lysine binding)	26,27
Arp7	Binds as Arp7–Arp9 heterodimer to HSA domain of Sth1	37
Arp9	Binds as Arp7–Arp9 heterodimer to HSA domain of Sth1	37
Rtt102		
Rsc6	Essential, binds Rsc8	34
Sfh1	Essential, phosphorylated in G1, binds directly to Sth1	28
Rsc8	Essential, contains weak DNA-binding SWIRM domain	36
Rsc3	Essential, sequence-specific DNA binding, heterodimer with highly similar Rsc30, zinc cluster, QIREY domain, coiled coils	29,30
Rsc30	Essential, heterodimer with highly similar Rsc30, zinc cluster, QIREY domain, coiled coils	29
Rsc1 or Rsc2	2 Bromodomains, AT-hook, BAH domain, C-terminal region required for assembly into complex	41
Rsc4	2 Bromodomains, interacts with RNA polymerases	31,32
Rsc7	Fungal-specific, binds Rsc8	35
Rsc9		
Rsc14		
Htl1	Substoichiometric subunit, binds Rsc8	38–40
Rsc58	Essential	33

Additionally, a third class of complex was observed in the mixture of both Rsc1 and Rsc2 isoforms, where part of the bottom arm was missing or had reduced density. Given that this conformation accounted for ~16% of the total population of complexes and that Rsc1 is ~10% as abundant as Rsc2, it is interesting to speculate that maybe this conformation is unique to the Rsc1 isoform, indicating differences either in the structures of the Rsc1 and Rsc2 subunits or in affinities of other subunits for the two isoforms. Addition of nucleosomes did not broadly affect the structure of the RSC complexes, but analysis of the density in the central cavity suggested that alterations had occurred to the structure of the nucleosome upon its binding to RSC, including loss of H2A–H2B.[43] Consistent with RSC-dependent changes in nucleosome structure is the finding that binding to RSC rendered nucleosomal DNA sensitive to exonuclease III and DNase I.[46,47]

B. Biochemical Activity

The biochemical activities of the RSC complex have been examined *in vitro* and have been shown to include nucleosome remodeling, repositioning, disassembly, and histone octamer transfer. The ATPase activity of the RSC complex was stimulated by DNA at least 25 bp in length, but was not further stimulated by the addition of nucleosomes. A turnover rate of approximately 7.5 molecules ATP per second was established under optimal conditions.[25,48,49] The Sth1 subunit alone was found to have ~2.5 times lower ATPase activity than the intact Rsc2–RSC complex.[49] ATP hydrolysis by RSC/Sth1 is coupled to 3′–5′ translocase activity. This was demonstrated by DNA length-dependent stimulation of ATPase activity, stimulation by DNA minicircles, and by triplex displacement activity.[49,50]

When DNA and nucleosome binding were tested, the RSC complex bound both with comparable affinity, but the presence of nucleosomes and ATP caused a shift to a slower migrating form (called an “activated” complex, in which nucleosomal DNA was more susceptible to DNase I).[47] The change in nuclease cleavage pattern suggested that the RSC complex was capable of altering nucleosome structure. Another chromatin-remodeling assay follows the cleavage of a restriction enzyme site that is inaccessible on the surface of a nucleosome. Incubation with RSC and ATP induced remodeling, allowing the site to be cut.[25,47,49] Other documented activities of the RSC complex include the ability to reposition nucleosomes, transfer histone octamers onto naked DNA, and dismantle nucleosomes in the presence of a histone chaperone.[51–53]

Sth1 alone was five- to sixfold less active than the RSC complex in nucleosome mobilization assays, showing the importance of other RSC subunits for maximal RSC complex activity.[49] One of the functions of the auxiliary subunits may be to increase the affinity for nucleosomes. Nucleosomes containing tetra-acetylated H3 were remodeled ~16-fold faster than unmodified nucleosomes due to preferential binding of the acetylated nucleosomes.[53] In addition, RSC has been shown to have increased affinity for nucleosomes acetylated by NuA4,[53] suggesting a role for the bromodomain containing subunits (Rsc1, Rsc2, Rsc4, and Sth1) in recruitment to chromatin and maximum remodeling activity. The majority of these *in vitro* assays have been performed using either a mixed population of Rsc1 and Rsc2-containing complexes or only the Rsc2-containing isoform; one outstanding question is whether the two isoforms possess identical biochemical activity.

The molecular mechanism of nucleosome remodeling has been examined using single-molecule experiments. RSC was shown to form relaxed supercoiled loops of around 400–700 bp in an ATP-dependent manner on DNA tethered and stretched at low forces with a magnetic trap.[54] These loops could be visualized using atomic force microscopy. The formation of loops also occurred on nucleosomal templates, resulting in loops of between 20 and 1200 bp (average of ~100 bp), but these loops could form at higher tensions.[55] Slippage of loops was observed on both templates, but the translocation rate was greater on naked DNA than on nucleosome-bound DNA (>500 bp per s compared with 12 bp per s). Together, these data suggest that nucleosome remodeling may occur by formation of a bulge of DNA on the surface of the nucleosome that can be extended by translocation to form a larger loop, which upon dissipation can result in reverse translocation, a jump in the position of the nucleosome, or nucleosome sliding.

C. *In Vivo* Functions

RSC is important for transcriptional control of many genes in *S. cerevisiae* and co-immunoprecipitated with all three RNA polymerases.[29,31,56] It can act to both activate and repress transcription of a group of genes that are distinct from those regulated by SWI/SNF.[26,29,57] ChIP on chip analysis of the genome-wide localization of the RSC complex identified RSC binding at ~700 promoters (~11% of genes), and no difference was seen in the Rsc1 and Rsc2 profiles.[56] Furthermore, around 12% of RNA polymerase II-transcribed genes were up- or downregulated at least twofold in an *rsc4* mutant strain.[31] RSC was bound at a number of tRNA promoters transcribed by RNA polymerase III and was enriched at many genes involved in mitochondrial function, nitrogen and carbon metabolism, and histone promoters.

Transcripts of genes affecting cell wall integrity, cell cycle control, and spindle pole body formation were misregulated in *rsc* mutants, despite not having enriched levels of RSC at their promoters by ChIP–chip, and transcription of some genes with RSC bound at their promoters is unaffected in *rsc* mutants.[29] This demonstrates that the correlation between promoter binding and transcriptional regulation is not a simple one and that different RSC subunits have different effects on transcription. More recently, a putative Rsc3 sequence-specific binding site has been identified approximately 100 bp upstream of the transcription start site of 708 genes (169 of which are essential) and Rsc3 has been connected with nucleosome exclusion from the promoters of these genes.[30] *In vitro*, RSC was also able to facilitate passage of RNA polymerase II through acetylated nucleosomes.[58] To date, none of the known DNA repair genes have been identified as being transcriptionally controlled by RSC.

In addition to transcription, the RSC complex has been shown to be important for proper kinetochore function, adaptation to the spindle assembly checkpoint, plasmid and chromosome maintenance, correct localization of the nuclear pore complex, meiotic sporulation, and sister chromatid cohesion (discussed later).[59–65] Furthermore, progression through the cell cycle is affected in *rsc* mutants. Cell cycle arrest at the G2/M checkpoint occurs in strains containing either a deletion, a ts allele, or a degron allele of genes encoding a number of RSC subunits (Sth1, Sfh1, Rsc3, Rsc4, Rsc6, Rsc8, Rsc9, Rsc58),[26,28,29,66–68] while loss of Sth1 or Htl1 function promotes polyploidy.[26,66,68,69]

The RSC complex was first implicated in the DNA damage response upon the finding that deletion of several of the nonessential subunits of RSC (*rsc1*, *rsc2*, *rsc7*, *rsc30*, *htl1*) or temperature-sensitive mutations in Sth1 confer sensitivity to a range of DNA-damaging agents including methyl methane-suphonate (MMS), phleomycin, bleomycin, hydroxyurea (HU), UV, and IR.[35,41,70–72]

D. The Role of RSC in DSBR

1. Recruitment to DSBs

Direct involvement of RSC in the repair of DSBs was proposed when it was found that Sth1 was recruited to a persistent site-specific break. Using the HO system described in the introduction, ChIP experiments showed enrichment of Sth1 relative to an uninduced sample on both sides of the DSB as early as 10 min after HO induction and, as such, this places RSC recruitment very early in the response to DSBs.[72,73] Sth1 at the *MAT* locus had returned to background levels 2 h after break formation was switched off.[73] Rsc8 and Rsc1 subunits were also recruited to breaks with comparable kinetics, suggesting that the whole RSC complex is recruited.[73,74] However, a subunit of the SWI/SNF complex, Snf5, was detected at the break only after 40 min, that is, later than RSC, revealing distinct temporal profiles of association for different remodelers.[72] Enrichment of the Rsc2 isoform of the RSC complex has not been shown directly but Sth1 recruitment is defective in an *rsc2* strain, although presumably the much less abundant Rsc1-isoform is still recruited under these conditions.[72] On the other hand, the Rsc30 subunit is not essential for Sth1 recruitment.[73] Association of Sth1 or Rsc1 with chromatin is not impaired in strains in which S129 of H2A cannot be phosphorylated, nor is it affected by cell cycle phase.[74,75] Collectively, these data suggest that RSC directly influences DSBR by acting at the break site and that it is required early in repair.

2. Factors Affecting RSC Recruitment

Mre11 and Ku are two of the earliest proteins to bind at a DNA DSB. In the absence of Mre11, Sth1 enrichment at a DSB was not detectable by ChIP while its enrichment was slightly delayed in the absence of yKu70.[73] It has also been reported than Mre11 can interact with Rsc1 and Rsc2.[73] However, as discussed later, RSC remodeling activity was found at a DSB in the absence of Mre11, indicating that there is at least a small amount of RSC present. Notably in that regard, the ChIP data were calculated to determine the enrichment upon break induction, that is, the amount of RSC bound following break induction relative to that bound in the absence of the break. As such, a lack of enrichment does not mean the absence of RSC, rather that there was no further increase in the amount that was there originally. Consistent with association prior to the induction of a DNA DSB, we found that Rsc2 and

Rsc7 are important in establishing normal chromatin structure at *MAT*α in the absence of breaks, which demonstrates that RSC is present and functions at this locus under normal growth conditions.[76]

3. Nucleosome Remodeling at a DSB

Upon break induction, a change in chromatin structure occurs immediately adjacent to the break. Nucleosome positions in chromatin extracted from MNase-treated permeabilized yeast were visualized by indirect end labeling. This assay showed repositioning of six nucleosomes away from the break on the proximal side and a small increase in MNase cleavage immediately distal to the break.[74,76] Consistent with mobilization of nucleosomes in the locality of a break, chromatin isolated from cross-linked nuclei and then subjected to MNase digestion was more efficiently cleaved after DSB induction when analyzed by qPCR.[74] Increased efficiency of restriction enzyme cleavage at sites close to the break is also consistent with enhanced accessibility of this region after break induction.[74] The change in chromatin structure occurred within 30 min, was Mre-11-independent, and occurred in G1, demonstrating that it is an event distinct from subsequent MRX-dependent histone eviction. Nucleosome mobilization on break formation is not unique to the *MAT* locus; it was also shown to occur at an HO break introduced at the *URA3* and *LEU2* loci.[76]

DSB-dependent nucleosome repositioning was found to be dependent on the RSC complex but not on other chromatin-remodeling complexes such as SWI/SNF, INO80, or Rad54.[76] Chromatin in a strain where Sth1 was repressed by the addition of doxycycline did not undergo a dramatic change in MNase sensitivity following break induction.[74] An *rsc2* strain also failed to show increased MNase sensitivity to the same extent as the wild type (WT) following break induction. In contrast, indirect end-labeling analysis established that nucleosome repositioning still occurred in an *rsc2* strain (and *rsc7* and *rsc30*) (although as mentioned earlier, the chromatin structure before HO cleavage was altered in *rsc2* and *rsc7* compared to the WT). However, in this assay, a defect in remodeling in an *rsc1* strain was detected.[76] The apparent discrepancy in the *RSC1* versus *RSC2* dependency of the chromatin remodeling could be explained by differences in the assays used. While the qPCR-based assay analyzing MNase susceptibility may be more sensitive to changes in occupancy within a population or "breathing" of nucleosomes, the indirect labeling assay may be more sensitive to positional changes, and consequently both isoforms of the RSC complex may be involved in the remodeling event. It should also be remembered that by definition *rsc1* and *rsc2* strains retain some RSC function. The relative contribution of the two isoforms could be dependent on cell cycle, chromatin context, or genomic location.

4. H2A S129 Phosphorylation

As mentioned in the introduction, one of the earliest events at a DSB is the phosphorylation of S129 of H2A by Mec1/Tel1. Recruitment of RSC and its remodeling of the surrounding chromatin occur on a similar timescale to H2A phosphorylation, raising the possibility that the two events are connected. Rsc1 enrichment to a DSB is unaffected in a mutant in which H2A S129 cannot be phosphorylated, but H2A S129 phosphorylation is defective in *rsc* mutants, placing RSC upstream of phosphorylation in the DNA damage response.[74–76] Residual H2A phosphorylation remained in *rsc* mutants, both by ChIP at a DSB and by Western blot following MMS treatment, but the efficiency of phosphorylation was diminished. Additionally, enrichment of Mec1 and Tel1 were decreased approximately twofold in *rsc2* strains, consistent with a defect in H2A phosphorylation.[75] A recent report has also implicated RSC in DSB-dependent methylation of H3K4 by Set1,[77] suggesting that remodeling of the chromatin flanking a DNA DSB is necessary for downstream chromatin modifications.

5. Resection

Another of the early events in DSBR is the binding of Mre11 and Ku to the ends of the break. By ChIP, association of Mre11 and Ku70 with the break site was reduced when Sth1 expression was repressed, suggesting that RSC-dependent changes in chromatin structure facilitate Mre11 and Ku binding.[74] One of the predicted consequences of defective Mre11 recruitment would be reduced efficiency of MRX-dependent resection. Resection was slightly compromised in *rsc* mutants as measured by quantitative amplification of single-stranded DNA (QAOS) or cleavage of restriction sites in double-stranded DNA (dsDNA).[68,76] In agreement with a resection defect, RPA enrichment was reduced near the break in an *rsc2* strain and a strain in which Sth1 was repressed, while Rad51 recruitment was slightly delayed.[74,75]

Overall, these data suggest a model for the action of RSC at a DSB (see Fig. 1). Following DSB formation, a small amount of Mre11 or Ku rapidly binds to the ends of the break, which facilitates recruitment of RSC either directly or indirectly, which in turn remodels the chromatin in the region of the break. The remodeled chromatin is more accessible and permissive for the accumulation of more Mre11 and Ku, acting in a positive feedback loop to recruit further RSC. The presence of Mre11 stimulates resection and consequently recruitment of Mec1- to RPA-coated single-stranded DNA and phosphorylation of H2A. This amplification cascade means that although resection and H2A phosphorylation still occur in the absence of RSC, they occur more efficiently in its presence.

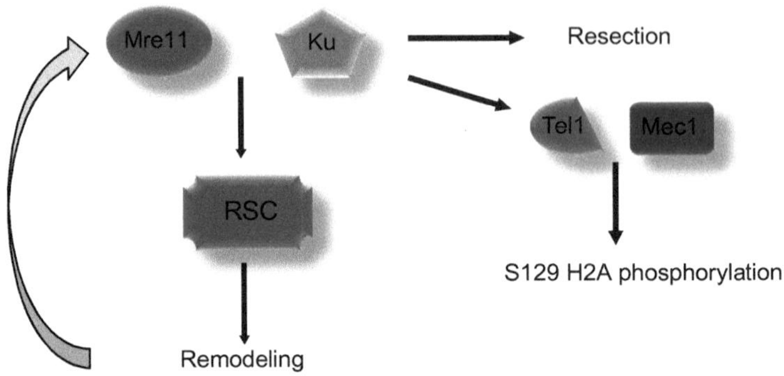

FIG. 1. The initial binding of some Mre11 and Ku to a double-strand break facilitates remodel structure of chromatin (RSC) recruitment and remodeling of chromatin immediately adjacent to the break. The change in chromatin structure promotes binding of more Mre11 and Ku and subsequently more RSC in a positive feedback loop. The result is efficient resection and H2A S129 phosphorylation.

6. Checkpoint Activation

The defect in ssDNA formation and H2A phosphorylation in *rsc* mutants might be expected to lead to reduced checkpoint activity. In support of this, Liang et al. found reduced Rad9 recruitment to the break in G1 and a small reduction in Rad53 phosphorylation in response to MMS in *rsc2* cells. However, when cell cycle arrest following DNA damage was examined directly, we found that both G1 and G2 checkpoints were activated in *rsc1* and *rsc2* deletion strains.[41] Moreover, strains containing degron alleles or deletions of the essential subunits of RSC terminally arrest at the G2/M checkpoint, demonstrating that RSC activity is not essential for cell cycle arrest. Perhaps checkpoint activation is modestly delayed in *rsc* mutants, but it appears that the remaining ssDNA is sufficient to ultimately trigger cell cycle arrest.

7. Effect on Repair: NHEJ

As may be expected, given its role at a DSB, RSC has been shown to be involved in the NHEJ pathway. A genetic screen for NHEJ proteins isolated *rsc8* and *rsc30* mutants as novel NHEJ factors.[73] Survival through error-prone NHEJ of a single DSB at two different loci was decreased in *rsc30*, *rsc1*, and *rsc2* deletion strains,[41,73] and in the case of the *rsc30* deletion strain, altered sequences were found at junctions. Precise NHEJ was also impaired in this strain. Another means of assessing end-joining activity is the plasmid repair assay, in which transformation with a linearized plasmid confers survival on selective media only if it is successfully repaired by NHEJ. Survival is expressed relative to that when an undigested plasmid is used in order to normalize for transformation efficiency. Consistent with a defect in NHEJ, decreased plasmid repair activity was observed in *rsc30* and *htl1* deletion strains.[40,73,78] Contrary to these studies, however, are data from Chai et al. where plasmid repair activity appeared to be substantially greater than WT in *rsc1* and *rsc2* deletion strains and in strains with mutant alleles of *sth1*.[72] This contradiction could be explained by the fact that the efficiency of transformation of uncut and linearized plasmid varies with cell density and that end-joining activity varies with cell cycle phase and cell density.[79] Overall the data, together with the decreased Ku recruitment to a DSB in the absence of Sth1, indicate that the RSC complex is important for enabling efficient end-joining of DSBs.

8. Effect on Repair: HR

It has been reported that *rsc1*, *rsc2*, and *htl1* mutants are defective in HR/SSA (single-stranded annealing) in plasmid gap repair assays.[72,78] During mating-type switching in budding yeast, a DSB introduced at the *MAT* locus is repaired by HR using one of the silent donor cassettes as a template. Although strand invasion and extension of the invading strand were unaffected in an *rsc2* strain, there appeared to be a defect in a postsynaptic step of repair. Using several assays, we find that recombination is affected in *rsc2* deletion strains[41] and, given the links between cohesion and recombination and the defect in cohesion in *rsc* mutants, discussed later, it is tempting to speculate that the function of RSC in recombination is connected to its effect on cohesion.

9. Cohesion

In G1, the cohesin complex (Mcd1/Scc1, Scc3, Smc1, and Smc3) is loaded onto chromosomes by Scc2–Scc4. During replication, cohesion of the two sister chromatids is established in a mechanism involving the Eco1 acetyltransferase, and facilitated by the alternative replication factor C complex containing Ctf18, Ctf8, and Dcc1, which is important for proper chromosome segregation. Cohesion remains until chromosome segregation, when it is removed by proteolytic digestion and phosphorylation of Mcd1. In budding yeast, cohesin is enriched at centromeres and at sites between convergently transcribed genes on chromosome arms (reviewed in Ref. 80). Additional cohesin is also recruited to DSBs, and a large ~100 kb domain of cohesin forms around a break in an H2A S129 phosphorylation-dependent process.[81]

Genetic synthetic interactions were found between *rsc2* and several genes involved in sister chromatid cohesion, namely *scc1-73*, *smc3-4*, *ctf8*, *ctf13*, and *scc2-4* (*eco1*).[64] To investigate whether RSC contributes to sister chromatid cohesion, separation of sister chromatids was monitored using GFP-Tet bound to Tet operator repeats on a chromosome arm. Increased separation of sisters in G2 was seen in *rsc1* and *rsc2* deletion strains and at the restrictive temperature in a ts-*sth1* strain.[64,65] Premature separation of sister chromatids occurred specifically on chromosome arms but not at a centromere.[65] As expected, for a strain with defective sister chromatid cohesion, *rsc2* deletion strains lost chromosomes with greater frequency than for a WT strain.[64,65] An interaction between the cohesin complex and the RSC complex was found by co-immunoprecipitation, but conflicting data exist on whether the role of RSC in sister chromatid cohesion is in loading of cohesin or in establishing cohesion.[64,65] The contribution of RSC to damage-induced cohesion has also been investigated, and enrichment of Mcd1 at an ~20 kb domain surrounding a DSB was reduced in an *rsc2* deletion strain.[75] It is yet to be determined if this reduction is sufficient to lose cohesion of the sister chromatids at the break site, which could result in either decreased or inappropriate recombination.

10. PBAF: The Human Homolog of RSC

As described earlier, RSC is a member of the SWI/SNF subfamily of chromatin-remodeling enzymes. In yeast, RSC is very similar to, and even shares, some subunits with SWI/SNF (Table I). In humans, there are also two complexes in the SWI/SNF subfamily, termed BAF and PBAF (or SWI/SNF-A and SWI/SNF-B). Notably, however, the overlap between the two complexes is far more substantial in humans with only a few defining subunits (Table I). The BAF180 subunit of PBAF contains six bromodomains followed by two BAH domains and appears to be a fusion of the budding yeast Rsc1, Rsc2, and Rsc4 subunits, suggesting that PBAF is the homologous complex to RSC. Frequently, studies are performed using cell lines with mutations or knockdowns of shared subunits, which provide insights into the role of "SWI/SNF," as it is not possible to make any distinction between the two complexes. To add to this complexity, the presence of BRG1 and hBRM in BAF are mutually exclusive and are not likely to be functionally redundant (for review see Ref. 82). Therefore, loss of BRG1 affects PBAF and a subset of BAF complexes, while loss of hBRM affects solely a subset of BAF complexes. This overlap makes it difficult to determine whether PBAF has unique functions in mediating DNA DSB repair in mammalian cells, but what is clear from the studies described later is that when subunits common to both complexes are impaired, DNA DSB responses are affected.

After treatment with IR, expression of an ATPase-defective dominant negative version of BRG1 resulted in defective H2AX phosphorylation, even though ATM activation and foci formation appeared to be normal. This was also apparent when cells with both BRG1 and hBRM knocked down were treated with IR.[83] While not directly in response to DNA DSBs, another study also found defective H2AX phosphorylation in cells depleted of either hSNF5 or BRG1 in response to UV irradiation in an ATM-dependent manner.[84] These reports are consistent with the data from yeast, suggesting a conserved function in promoting H2A(X) phosphorylation at sites of damage. However, in another report, H2AX phosphorylation in response to IR is normal in hSNF5-deficient cells,[85] and we do not see loss of IR-induced H2AX phosphorylation in cells with BAF180 and BRG1 simultaneously knocked down (E. Riballo, JAD Downs, and P.A. Jeggo, unpublished data). In addition, knockdown of BRIT1, which is important for recruitment of SWI/SNF to chromatin (described later), results in increased (rather than decreased) H2AX phosphorylation after IR.[86] It is possible that a greater degree of redundancy exists in mammalian cells and that the contribution of SWI/SNF activity to H2AX phosphorylation is only apparent under certain conditions.

The SWI/SNF complex is recruited to sites of DNA DSBs,[83,86,87] suggesting that as in yeast, it plays a direct role at the DNA lesions. The mechanism by which it is recruited is not entirely clear, but several proteins and protein modifications have been implicated. BRIT1 has three BRCT domains and interacts with SWI/SNF via the BAF170 and BAF155 subunits. When BRIT1 was depleted, the association of BRG1 and hBRM with chromatin, both globally and at a DNA DSB, was impaired,[86] suggesting that it mediates recruitment. Histone acetylation has also been implicated in mediated recruitment of SWI/SNF to chromatin at sites of DNA damage.[87,88] The CBP/p300 histone acetyltransferase was found to acetylate H3 and H4 at DNA DSBs. Ablation of this HAT resulted in impaired recruitment of hBRM to DNA DSBs.[87] Somewhat consistent with this, Lee et al. found evidence for a positive feedback loop, in which IR-induced H3 acetylation promotes BRG1 binding to chromatin, which in turn promotes H2AX phosphorylation, and this stimulates H3 acetylation.[88] This report implicated GCN5 as the responsible HAT, although the contribution of GCN5 and CBP/p300 does not have to be mutually exclusive. It is also possible that BRG1 and hBRM have nonoverlapping roles with different modes of recruitment to DNA DSBs.

What is the impact of SWI/SNF activity on DNA DSB responses? Cells expressing the ATPase inactive dominant negative version of BRG1 show defective survival after DNA damage, increased DNA damage-induced apoptosis, and slower repair of DNA DSBs.[83,89] Similarly, downregulation of both BRG1 and hBRM or inactivation of Snf5 leads to slower repair of DNA DSBs and reduced survival after damage.[83,90] Consistent with a specific role in mediating DSB repair, accumulation of Ku at sites of DNA damage is reduced and NHEJ activity is impaired in cells with hBRM knocked down.[87] Although not a subunit of either core complex, BRIT1 deficiency also showed a defect in DSB repair after treatment with IR and lower levels of HR activity using a reporter system.[86] However, in another study, SNF5-deficient cells did not show hypersensitivity to a variety of DNA-damaging agents (cisplatin, etoposide, and UV) and the cells showed no evidence of chromosomal aberrations as might be expected if DNA DSB repair was defective.[85] This might indicate that the contribution of SWI/SNF to DNA DSB repair is cell type or context dependent.

It is worth noting that in addition to a direct contribution to DNA DSB repair activity at chromatin flanking the break, SWI/SNF has indirect effects as well. Both SWI/SNF complexes are involved in regulating transcription of a large number of genes, and some of these are important for DNA damage responses (e.g., Refs. 91,92). It would be very valuable to identify separation of function mutants of SWI/SNF subunits in order to study the effects on DNA damage responses that are specifically due to remodeling activity at DNA DSBs *in vivo*. Moreover, it will be of great interest to determine what, if any, specialization exists between PBAF and BAF in their DNA repair and genome stability functions.

III. INO80

A. Subunit Composition

The INO80 complex was first purified from *S. cerevisiae* by immunoprecipitation of a tagged version of the catalytic subunit, Ino80, which had been identified as important for transcription of inositol biosynthesis genes.[93,94] Gel filtration analysis identified Ino80 as part of a high-molecular-weight complex and using mass spectroscopy it was found to contain 15 subunits: Ino80, Rvb1, Rvb2, Arp4, Arp5, Arp8, actin, Taf14, Nhp10, and the *I*no-*E*ighty-associated *S*ubunits Ies1–Ies6 (see Table III).[94] Seven of these subunits are conserved in the human INO80 complex, which also contains some metazoan-specific subunits (see Table III).[95] The catalytic subunit, Ino80, shares significant similarity with the Snf2/Swi2 family of ATPase-dependent chromatin remodelers and mutation of a conserved lysine required for ATPase activity failed to complement *ino80* mutant phenotypes.[93,94] The characteristic features of the catalytic subunits of the INO80 subfamily of remodelers are the presence of an insertion within the ATPase domain and the presence of two conserved motifs, the TELY motif near the N terminus and the GTIE motif near the C terminus.

Unique to the INO80 subfamily of chromatin-remodeling complexes (which includes Swr1) is the presence of two other proteins with ATPase activity, Rvb1 and Rvb2. These two subunits are homologous to the RuvB bacterial helicase, a hexameric AAA+ family protein, and appear to be associated with the INO80 complex with a stoichiometry of 6:1.[94] Temperature degron strains of *ino80* and *rvb2* show significant overlap in the genes that are misregulated.[96] The Nhp10 subunit is an HMGB-like protein and therefore could be capable of binding to structured DNA or nucleosomes, and while the Ies1–6 subunits have not been well characterized, the fact that Ies2 and Ies6 are conserved in humans suggests that their function is important for INO80 complex activity. In common with other remodelers, INO80 also contains Arps

TABLE III

SUBUNIT COMPOSITION OF INO80

Yeast INO80	Human INO80
Ino80	**hINO80**
Rvb1	RUVBL1/Tip49a
Rvb2	RUVBL2/Tip49b
Arp4	BAF53a
Arp5	hArp5
Arp8	hArp8
Ies2	hIes2/PAPA-1
Ies6	hIes6/C18orf37
Act1	
Taf14/Anc1	
Ies1	
Ies3	
Ies4	
Ies5	
Nhp10	
	FLJ20309/INO80D
	NFRKB
	MCRS1
	Amida
	YY1
	CCDC95/FLJ90652/INO80E
	Uch37

Catalytic subunits are shown in bold. INO80 complexes from yeast and humans possess a core of conserved subunits and additional yeast- or metazoan-specific subunits. The metazoan-specific subunits associate with the N terminus of hINO80, while the conserved core subunits bind to central and C-terminal regions of the protein.

(actin-related proteins). While actin and Arp4 are also found in the NuA4 histone acetyltransferase complex, Arp5 and Arp8 are unique to INO80. The presence of the Rvb proteins, but not their catalytic activity, is required for the association of Arp5 with the INO80 complex.[96] Arp4, actin, Arp8 (and Taf14) appear to be associated with an N-terminal region of Ino80.[97] Deletion of *INO80* is lethal in some strain backgrounds and is very sick in others; therefore, *arp5* and *arp8* deletion strains have frequently been used to phenocopy *ino80* deletion (as discussed later).

B. Biochemical Activities

Like other Snf2-remodeling complexes, the ATPase activity of INO80 is DNA-stimulated, but unlike RSC, INO80 is also capable of 3′–5′ strand displacement in traditional helicase assays.[94] INO80 bound DNA in electrophoretic mobility shift assays (EMSAs) and facilitated transcription of a reconstituted chromatin template. Cleavage of this template by a restriction enzyme was increased on addition of INO80, demonstrating chromatin-remodeling activity.[94,97] INO80 nucleosome-sliding activity was confirmed by further studies using nucleosomes assembled *in vitro* onto a fragment of the *INO1* promoter. There are three preferred nucleosome positions on this template, which can be resolved on native gels, and INO80 was shown to mobilize these nucleosomes in an ATP-dependent manner.[97] More recently, the sliding activity of INO80 has been characterized in more detail and was found to act to space nucleosomes in an array requiring a minimum of >30 bp (and optimally 70 bp) of linker DNA for nucleosome movement.[98]

As SWR1, which is in the same subfamily as INO80, possesses histone exchange activity, that is, replaces H2A with the variant H2AZ, it had been proposed that INO80 may also possess exchange activity. This was recently demonstrated *in vitro*, where incorporation of H2A in place of H2AZ (i.e., the reverse activity to SWR1) was shown to be catalyzed by INO80 but not other remodelers.[99] Moreover, loss of INO80 activity resulted in altered H2AZ incorporation *in vivo*.[99] To date, nucleosome eviction by INO80 has not been found *in vitro* under conditions in which other remodelers catalyze nucleosome loss.[98,99]

As mentioned previously, *arp5* and *arp8* deletion strains are commonly used in place of *ino80* deletions. Analysis of INO80 complexes purified from *arp5* or *arp8* deletion strains showed the complexes to remain reasonably intact with only the loss of Arp5 from the *arp5* deletion and the loss of Arp4 and actin in addition to Arp8 from the *arp8* deletion strain.[97] Biochemical characterization of these complexes found loss of chromatin-remodeling activity and reduced nucleosome-stimulated ATPase activity to approximately 10–20% of WT levels.[97] DNA-binding activity of the complexes was also reduced to 50% and 3% of WT for *arp5* and *arp8*, respectively. Unexpectedly, the INO80 complex purified from an *nhp10* deletion strain (which also lacks Ies3) was still able to reposition nucleosomes, despite having DNA-binding activity of only 24% of the WT complex.[97]

C. *In Vivo* Functions

The INO80 complex was initially isolated based on its function in regulating transcription of some inositol synthesis genes. It has since been found to regulate transcription of genes including stress-responsive genes, such as *PHO5*, *GAL1*, *CYC1*, and *ICL1*.[93,96,100] Analysis of gene expression in an *ino80* deletion strain identified 1156 genes with at least a 1.5-fold change in expression compared to WT, while an *arp8* deletion strain had 395 genes with altered expression, indicating that *arp8* deletion strains still retain partial INO80 function.[101] No known DNA repair or checkpoint genes were downregulated in either of the deletion strains, and MMS-dependent transcriptional induction of checkpoint, repair, and INO80 subunit genes was largely unaffected in the mutant strains.[101]

Outside the scope of this review, INO80 has been shown to bind to origins of replication and be important for the recovery of stalled replication forks.[102–104] It has also been implicated in regulation of telomeres[105] and in NER of some loci following UV exposure.[106]

The first connection between INO80 and DNA repair was made when an *ino80* strain was found to be sensitive to a number of DNA-damaging agents. Sensitivity was greatest to HU and MMS but also occurred following treatment with UV and IR.[94] Deletion of the nonessential subunits *arp5* or *arp8* resulted in sensitivity to MMS and HU, and hypomorphic *arp4* alleles displayed sensitivity to MMS.[97,101,107,108] Given the lack of transcriptional misregulation of repair and checkpoint genes in INO80 complex mutants, this prompted investigation of whether INO80 has a direct role in DNA repair.

D. The Role of INO80 in DSBR

1. Recruitment to DSBs

As for RSC, a direct role of INO80 in DSBR was indicated by its recruitment to DSBs. Using the HO system to generate a persistent DSB, accumulation of INO80 at a break was monitored over time by ChIP. Some recruitment of the Ino80 catalytic subunits Arp5 and Arp8 could be detected 30 min after break induction, but enrichment continued to increase at least up to 4 h.[101,108,109] In addition, Rvb1, present in both INO80 and SWR1 complexes, was enriched adjacent to the break 45 min after break induction and Arp4, present in both INO80 and NuA4 complexes, was enriched at the break after 2 h.[108] Further analysis of the recruitment of INO80 revealed spreading of the complex over a region of approximately 10 kb on either side of the break after 4 h, but was greatest in the immediate vicinity of the break.[107,109] Accumulation of INO80 was also detected after 2 h at an HO-induced DSB elsewhere in the genome, although the profile of enrichment differed from that found at the *MAT* locus in that it was relatively uniform across the 5 kb adjacent to the break.[107]

2. Factors Affecting INO80 Recruitment

Unlike *rsc* mutants, phosphorylation of H2A S129 appears to be broadly unaffected in *ino80* complex mutants. At 2 h following HO break formation, when INO80 is significantly enriched near the break, H2A phosphorylation was similar to WT levels as measured by ChIP in *ino80*, *arp5*, *arp8*, and *nhp10* deletion strains, and levels detected in Western blots after a 2-h treatment with MMS were unaffected in *ino80*, *arp8*, and *nhp10* strains.[101,107,109,110] However, a strain containing a deletion of ~900 bp at the N terminus of *ino80* possessed roughly a sixfold reduction in the phospho-S129 H2A ChIP signal near the break 2 h after induction, indicating that this mutant does not behave identically to the *ino80* deletion.[111] Two hours is a relatively long time after break induction, and there could be subtle delays in accumulation of phosphorylated H2A S129 in some *ino80* mutants. However, this could also be partly due to delays in DSB formation that have been observed in some mutant strains.[107]

Purification of the INO80 complex from cells treated with MMS resulted in co-immunoprecipitation of phosphorylated H2A.[109] This interaction was unaffected when Arp5 or Arp8, Arp4, and actin were absent from the complex, but was reduced when *nhp10* was deleted (although affinity for bulk histones was slightly reduced too). The INO80 complex from the *nhp10* deletion strain also lacked Ies3 and notably the recruitment of this mutant complex to a DSB was defective *in vivo*.[109] Furthermore, enrichment of Ino80 was found to be reduced in a strain in which H2A could not be phosphorylated[101] and in an *mec1/tel1/sml1* deletion strain.[109] In addition, myc-tagged Ino80 (and Arp4) was pulled down from cell extracts by a phosphorylated S129 H2A peptide.[108] Together, these data demonstrate that Ino80 is recruited to chromatin containing phospho-S129 H2A adjacent to a DSB and both Nhp10/Ies3 and Arp4 have been implicated in contributing to this interaction.

3. Chromatin Remodeling at a DSB

Like RSC, INO80 recruitment to a DSB has been linked to changes in chromatin close to the break. However, these remodeling events occur much later than the nucleosome mobilization catalyzed by the RSC complex. The first report connecting INO80 to chromatin remodeling at a break observed a decrease in the ChIP signal of H2B and H3 near the break relative to that before the break was formed.[110] The authors interpreted this as eviction of histones and found that eviction was reduced, but not eliminated in *arp8* and *mre11* deletion strains, while it was largely unaffected in a strain in which H2A could not be phosphorylated. In their study, resection was found to be unaffected in an *arp8* strain and this led them to conclude that INO80 remodeling promotes loss of histones from the region containing phosphorylated H2A in both an MRX-dependent manner and a resection-independent manner. However, it has since been reported that the loss of histones observed at a break is tightly coupled to resection.[112] When the H3 ChIP signal was normalized to input DNA at each time point, that is, accounting for loss of DNA to amplify due to resection, it was discovered that there was no change in the histone to DNA ratio.[112] This is in agreement with what had previously been found for H2B.[21] These more recent studies would suggest that INO80 is not required to evict histones from chromatin flanking a DNA DSB (but the histones are instead lost as a consequence of resection), which is consistent with the fact that INO80 is not able to catalyze histone eviction *in vitro*. However, a study from van Attikum et al. did normalize to input DNA and still found loss of H3 after 2–4 h close to the break at the *MAT* locus, and at another locus (*PDR10*), the H3 signal first increased and then fell to below the level before break induction.[107] The loss of H3 at the *MAT* locus was slightly delayed in an *nhp10* deletion strain, but this mutant had no effect on the chromatin changes at the *PDR10* locus. It is difficult to reconcile the results from these different studies, but one possibility is that changes in chromatin structure due to remodeling or binding of other proteins may mask the epitopes recognized by the antibodies used in the ChIP assays.

In light of recent biochemical studies of the INO80 complex that found no detectable nucleosome eviction activity, the role of INO80 at a DSB may instead be to increase accessibility by either sliding nucleosomes or via histone exchange.[99] Shortly after DNA DSB formation, SWR1-dependent deposition of H2AZ was detected at a break.[113] However, this enrichment of H2AZ was found to be transient and 2 h after break induction, the signal had almost returned to basal levels.[107,113] Given its ability to replace H2AZ with H2A and its localization at breaks, INO80 could be responsible for the removal of H2AZ at DNA DSBs and this conceivably could be important for regulating chromatin accessibility during repair.

4. Resection

Single-stranded DNA accumulates at a DSB with similar kinetics to INO80 enrichment and so the effect of INO80 activity on resection was examined. After DSB induction, less single-stranded DNA was detected in *arp8* and *nhp10* deletion strains than in the WT by a qualitative QAOS assay.[101,107] A decrease in single-stranded DNA was also detected in *arp8*, *nhp10*, and *ino80* deletion strains by Southern blot and restriction enzyme digest assays,[114] while in ChIP assays, slightly more input DNA was amplified from an *arp8* deletion strain than from the WT.[112] Taken as a whole, this suggests that there is a small but detectable defect in resection in *ino80* mutants. In agreement with this, 1 h after break formation, Mre11 recruitment to a break was reduced in an *arp8* deletion. Similarly, Mec1, which binds to RPA-coated single-stranded DNA displayed reduced association with the break in an *arp8* strain.[107] This defect may not be sufficiently large to be detected in all assays or strain backgrounds, as no problems with resection were seen in an *arp8* strain or a strain lacking 900 bp from the start of *ino80* in restriction enzyme assays, although a delay in strand invasion was found in the *ino80* strain.[110,111]

Overall, *ino80* mutants appear to have a slight defect in single-stranded DNA formation, suggesting that INO80 activity at chromatin next to a break facilitates efficient resection. This could, at least in part, explain the potential effect of INO80 on histone eviction, as INO80 could indirectly promote histone eviction by allowing more efficient nuclease activity. This could be achieved either by histone sliding or by removal of H2AZ as described earlier.

Resection may result in loss of histones, but equally a decrease in histone ChIP signal would occur if the histones remained associated with single-stranded DNA. An apparent anomaly exists in that INO80 does not significantly accumulate at a break until 1–2 h after break formation, yet Mre11 recruitment, which is defective in *ino80* mutants, occurs very quickly. This could be explained by a feedback loop similar to that proposed for the RSC complex, where some rapid Mre11 and Tel1 binding to ends can trigger H2A phosphorylation, enabling INO80 recruitment and its remodeling activity. Remodeling could promote resection to allow further recruitment of Mre11 and Mec1. Therefore, while initial Mre11 binding and H2A phosphorylation would be - INO80-independent, subsequent retention and accumulation would be INO80-dependent. As H2A phosphorylation and INO80 have distinct localization profiles, it may also be the case that initial recruitment of INO80 to the break site is not via phosphorylated H2A directly, but via another mechanism and that phosphorylated H2A is then required for INO80 retention. This hypothetical initial recruitment signal may not be present immediately after break formation, hence the delay in INO80 recruitment.

5. Checkpoint Activation

Cells containing *ino80* mutations are proficient in checkpoint activation in response to DNA-damaging agents, but are compromised in response to a DSB. Cell cycle arrest, *RNR* induction, and Rad53 activation were normal in response to HU in *ino80* and *arp5* deletion strains (although downregulation of Rad53 phosphorylation was delayed).[103,109] In response to MMS, Rad53 was activated, if not hyperactivated, in *ino80*, *arp5*, and *arp8* deletion strains and MMS-dependent induction of *RNR* genes occurred.[101] In contrast, when *arp8* was deleted, Rad53 activation was impaired following HO cleavage to create a DSB.[107] This would suggest that resection is required in order to efficiently activate the checkpoint in response to a DSB, but not in response to treatment with HU or MMS, which activates the intra-S checkpoint. However, despite the delay in Rad53 activation in response to a DSB, the checkpoint is eventually activated sufficiently to arrest the cell cycle at G2/M.[111] INO80 also appears to play a role in adaptation to checkpoint arrest, which permits continued progression through the cell cycle, despite the remaining presence of DNA damage.[111] Finally, Mec1/Tel1 phosphorylation of the Ies4 subunit of INO80 was implicated in modulating a checkpoint in response to replication stress in a redundant manner to Tof1.[114]

6. Effect on Repair: NHEJ

Given its recruitment and activity at a DSB, it might be expected that INO80 plays a role in DNA repair. Moreover, mutants in the HR pathway show synthetic sickness and additive sensitivity with *ino80* mutants.[109,110] Breaks formed at the *MAT* locus in a strain with the *HML* and *HMR* donor cassettes deleted can only be repaired by NHEJ. After a pulse of HO expression and then growth on glucose plates to switch off expression, error-free NHEJ allows survival. In this assay, a strain with the first 900 bp of *ino80* deleted, *arp8* and *nhp10* deletion strains did not have a defect in error-free NHEJ.[107,110,111] An *arp8* deletion strain also behaved like the WT in a plasmid repair assay, another assay for error-free NHEJ.[110] Conversely, in the presence of constant HO expression, by plating on galactose plates (or plates without doxycycline using a tet-HO expression plasmid), survival is dependent on error-prone NHEJ to prevent cleavage of the HO site again. In this assay, *arp5* and *arp8* deletion strains and the *ino80* mutant lacking the first 900 bp of the gene displayed a mild defect in error-prone NHEJ.[101,107,111] This could indicate a decreased ability to process DNA ends in *ino80* mutants.

7. Effect on Repair: HR

As INO80 appears to facilitate resection at a DSB, it may be expected to affect repair by HR. To initiate HR, single-stranded DNA generated by resection is first coated with RPA, which is then exchanged for Rad51 that is capable of homology searching and strand invasion catalyzed by Rad52. Analysis of recruitment of RPA to the break site in an *arp8* deletion strain with the mating-type donor cassettes deleted was found to be comparable to the WT, but a slight delay in Rad51 loading was observed.[110] However, an *arp8* strain with the donor cassettes present, which is therefore proficient for mating-type switching, did not have reduced Rad51 at the *MAT* locus, but possessed a decreased amount of Rad51 at the *HMRa* locus.[115] This *arp8* strain and one with the first 900 bp of *ino80* deleted displayed a delay in strand invasion at *HMLα*.[111,115] This was attributed to diminished nucleosome repositioning at the *HMRa* locus in the *arp8* deletion strain. Despite these differences with the WT, *ino80* mutants do not have a dramatic defect in mating-type switching, demonstrating that efficiency of HR is not affected or compensated.[107,111,115]

Mating-type switching is a very specific type of HR, involving gene conversion without a crossover. INO80 has been found to have a role in other types of recombination. In haploid *arp8* cells, spontaneous sister chromatid recombination occurred at the same frequency as in the WT, but while the rate of recombination of the WT strain increased following treatment with MMS, the rate in the *arp8* strain remained the same.[116] Similarly, the rate of spontaneous recombination between alleles on different chromosomes is the same for WT and *arp8* cells, but the rate of MMS-induced HR is reduced in an *arp8* strain.[116] A modest decrease in allelic recombination was detected in *arp8* diploids, but the end product of recombination was more markedly affected by *arp8* deletion, most notably in the high frequency of discontinuous gene conversion tracts.[115] Thus, it appears that HR is not globally impaired in *ino80* mutants, but that specific types of recombination may be affected.

E. Human INO80

As mentioned above, there are eight subunits that are conserved between the human and yeast INO80 complexes, including the Ino80 ATPase itself (see Table III). In addition, the human INO80 complex possesses seven metazoan-specific subunits (FLJ20309/INO80E, NJRKB, MCRS1, Amida, YY1, CCDC95/FLJ90652/INO80E, and Uch37). Interestingly, these assemble onto the N terminus of hIno80 and are dispensable for chromatin-remodeling activity of the complex *in vitro*.[95] The YY1 subunit of hINO80 is a zinc finger-containing Polycomb group transcription factor termed Yin Yang-1 (YY1). While YY1 has no clear counterpart in lower eukaryotes, the fission yeast INO80 complex contains another zinc finger-containing transcription factor termed Iec1,[117] suggesting that the function of YY1 may be conserved outside of metazoans.

Loss of YY1 results in gross chromosomal aberrations, including polyploidy, and MEFs lacking YY1 are hypersensitive to DNA damage. This prompted Shi and colleagues to investigate whether INO80 also functions in maintenance of genome stability.[118] Using a reporter-based system, they found that cells lacking either INO80 or YY1 showed impaired HR activity. Intriguingly, the authors found that YY1 is capable of binding to Holliday junction structures *in vitro*, and that these structures are a preferred binding substrate when compared with dsDNA. These data raise the possibility that hINO80 is targeted to recombination intermediates via YY1 and functions to promote HR.

Consistent with a direct role in mediating DNA DSB repair, hINO80 was found to be recruited to sites of DNA damage by immunofluorescence.[119] Moreover, the authors find that recruitment is impaired when the ARP8 subunit is knocked down, while *arp8* mutation in yeast has no effect on recruitment,[101,109] suggesting that the mechanism of recruitment to DNA DSBs is not conserved. More recently, INO80 was found to go to DNA breaks by ChIP, and this is important for 53BP1 foci formation.[124]

As expected for a complex that promotes recombination, cells lacking WT levels of hINO80 are hypersensitive to IR and show delayed repair.[120] However, in this study, the authors also found that the Rad54B and XRCC3 repair genes were transcriptionally downregulated in the hINO80 knockdown cells. By reintroducing transcription of these genes, the authors found that the DSB repair defect in hINO80-deficient cells was largely rescued. Notably, loss of YY1 had no apparent effect on DNA repair genes by microarray analysis.[118] It seems reasonable that hINO80 may work both directly and indirectly to promote repair of DNA DSBs in mammalian cells, and perhaps the relative contribution of each mechanism varies by cell type and growth conditions.

IV. Perspectives

The importance of the RSC and INO80 chromatin-remodeling complexes in the DNA damage response has been clearly demonstrated but a number of questions remain. These two remodelers are not the only ones involved in DNA repair. Other remodelers shown to contribute to efficient repair and checkpoint responses in budding yeast include SWI/SNF (localizes to a DSB and remodels the donor template[72]) and SWR1 (important for error-free NHEJ[107]). In higher eukaryotes, the list is even longer and, in addition to INO80 and BAF/PBAF (SWI/SNF), includes ACF1, TIP60, and WICH. The restoration of chromatin structure following completion of repair is also likely to require ATP-dependent chromatin-remodeling complexes, but to date, this has not been studied in any detail. It seems probable that the full extent of involvement of chromatin remodeling and the list of complexes involved has not yet been uncovered.

Regulated sequential recruitment of multiple chromatin-remodeling complexes to a DSB occurs, raising the question of why there is so much complexity if their function is purely to open chromatin to allow access to repair proteins. It is already apparent that changes in chromatin structure catalyzed by remodeling complexes act not only to allow access to repair proteins but also to facilitate processing and binding of signaling and checkpoint proteins. Different remodeling events may be required for different steps of the repair process. It remains to be explored how the INO80 remodeling event is coordinated with the rapid changes in chromatin structure catalyzed by RSC that also promote resection or how these events integrate with the functions of the other remodelers recruited to the break. Perhaps INO80 remodeling at a break first requires the action of RSC, or INO80 may only be required at persistent breaks where repair has failed to occur.

Alternatively, it could be that not all breaks are processed and remodeled in the same way. One can envisage breaks formed in heterochromatin or regions with higher-order chromatin structure having a greater dependency on chromatin-remodeling complexes in order to open their chromatin than those in euchromatin or in highly transcribed regions. Similarly, the relative importance of different complexes may vary with the cell cycle phase either through changes in expression or through posttranslational modifications and could even help to channel repair down a particular pathway, for example, HR versus NHEJ. Different lesions such as UV-induced 6-4 photoproducts and cyclopyrimidine dimmers (CPDs), alkylated bases, abasic sites, and interstrand cross-links are also likely to require chromatin remodeling for their repair and could have differing requirements for specific remodeling complexes. It seems to be the case that chromatin remodelers have both redundant and nonoverlapping roles in DNA repair.

Chromatin-remodeling complexes contain many subunits with chromatin-binding domains such as chromodomains, bromodomains, and BAH domains. In concert with posttranslational modifications of histones that occur as part of the DNA damage response, this provides a means to target chromatin-remodeling complexes to the site of damage in a location- and timing-specific manner. Binding of these targeting domains to chromatin modifications is generally of low affinity but may contribute to retention of remodelers, if not their initial recruitment. Equally, the chromatin-remodeling complexes themselves could be posttranslationally modified and this has the potential to alter their activity; indeed, phosphorylation of the Ies4 subunit of INO80 by Mec1/Tel1 has been identified and Rsc2 has been found to be sumoylated.[121] To date, the specific functions of many of the individual subunits of remodeling complexes have yet to be elucidated. It may be these subsidiary subunits that provide specificity and regulation.

One of the challenges for the future is to integrate the growing numbers of DNA damage chromatin modifications and the multiple chromatin-remodeling complexes involved in the DNA damage response to generate a mechanistic picture of their function in DNA repair. As chromatin-remodeling complexes are involved in many other processes in the cell including transcription and replication, the interplay of these cellular activities with the DNA damage response must be taken into consideration. There is a high degree of conservation between yeast and human remodelers and so many of the findings in *S. cerevisiae* are likely to be applicable to human cells. While the mechanism is not entirely known, it is clear that genes encoding subunits of chromatin-remodeling complexes are frequently mutated or misregulated in cancer (e.g., Ref. 122). As a result, this family of enzymes may provide diagnostic and therapeutic targets for cancer and genome instability disorders.

Acknowledgments

This work was supported by Cancer Research UK (CEAC7905/A8265). We apologize to the authors whose work we have not cited due to limitations in the size and scope of the review.

References

1. Vilenchik MM, Knudson AG. Endogenous DNA double-strand breaks: production, fidelity of repair, and induction of cancer. *Proc Natl Acad Sci USA* 2003;**100**:12871–6.

2. Daley JM, Palmbos PL, Wu D, Wilson TE. Nonhomologous end joining in yeast. *Annu Rev Genet* 2005;**39**:431–51.
3. Paques F, Haber JE. Multiple pathways of recombination induced by double-strand breaks in *Saccharomyces cerevisiae*. *Microbiol Mol Biol Rev* 1999;**63**:349–404.
4. Smerdon MJ. DNA repair and the role of chromatin structure. *Curr Opin Cell Biol* 1991;**3**:422–8.
5. Clapier CR, Cairns BR. The biology of chromatin remodeling complexes. *Annu Rev Biochem* 2009;**78**:273–304.
6. van Vugt JJ, Ranes M, Campsteijn C, Logie C. The ins and outs of ATP-dependent chromatin remodeling in budding yeast: biophysical and proteomic perspectives. *Biochim Biophys Acta* 2007;**1769**:153–71.
7. Flaus A, Martin DM, Barton GJ, Owen-Hughes T. Identification of multiple distinct Snf2 subfamilies with conserved structural motifs. *Nucleic Acids Res* 2006;**34**:2887–905.
8. Bakshi R, Prakash T, Dash D, Brahmachari V. In silico characterization of the INO80 subfamily of SWI2/SNF2 chromatin remodeling proteins. *Biochem Biophys Res Commun* 2004;**320**:197–204.
9. Huertas D, Sendra R, Munoz P. Chromatin dynamics coupled to DNA repair. *Epigenetics* 2009;**4**:31–42.
10. Haber JE. Lucky breaks: analysis of recombination in *Saccaromyces*. *Mutat Res* 2000;**451**: 53–69.
11. Downs JA, Lowndes NF, Jackson SP. A role for Saccharomyces cerevisiae histone H2A in DNA repair. *Nature* 2000;**408**:1001–4.
12. Rogakou EP, Pilch DR, Orr AH, Ivanova VS, Bonner WM. DNA double-stranded breaks induce histone H2AX phosphorylation on serine 139. *J Biol Chem* 1998;**273**:5858–68.
13. Redon C, Pilch DR, Rogakou E, Sedelnikova OA, Newrock K, Bonner WM. Histone H2A variants H2AX and H2AZ. *Curr Opin Genet Dev* 2002;**12**:162–9.
14. Bassing CH, Chua KF, Sekiguchi J, Suh H, Whitlow SR, Fleming JC, et al. Increased ionizing radiation sensitivity and genomic instability in the absence of histone H2AX. *Proc Natl Acad Sci USA* 2002;**99**:8173–8.
15. Bassing CH, Suh H, Ferguson DO, Chua KF, Manis J, Eckersdorf M, et al. Histone H2AX: a dosage-dependent suppressor of oncogenic tranlocations and tumors. *Cell* 2003;**114**:359–70.
16. Celeste A, Petersen S, Romanienko PJ, Fernandez-Capetillo O, Chen HT, Sedelnikova OA, et al. Genomic instability in mice lacking histone H2AX. *Science* 2002;**296**:922–7.
17. Celeste A, Difilppantonio S, Difilppantonio MJ, Fernandez-Capetillo O, Pilch DR, Sedelnikova OA, et al. H2AX haploinsufficiency modifies genomic stability and tumor susceptibility. *Cell* 2003;**114**:371–83.
18. Matsuoka S, Ballif BA, Smogorzewska A, McDonald 3rd ER, Hurov KE, Luo J, et al. ATM and ATR substrate analysis reveals extensive protein networks responsive to DNA damage. *Science* 2007;**316**:1160–6.
19. Smolka MB, Albuquerque CP, Chen SH, Zhou H. Proteome-wide identification of in vivo targets of DNA damage checkpoint kinases. *Proc Natl Acad Sci USA* 2007;**104**:10364–9.
20. Chen SH, Albuquerque CP, Liang J, Suhandynata RT, Zhou H. A proteome-wide analysis of kinase-substrate network in the DNA damage response. *J Biol Chem* 2010;**285**:12803–12.
21. Shroff R, Arbel-Eden A, Pilch D, Ira G, Bonner WM, Petrini JH, et al. Distribution and dynamics of chromatin modification induced by a defined DNA double-strand break. *Curr Biol* 2004;**14**:1703–11.
22. Rogakou EP, Boon C, Redon C, Bonner WM. Megabase chromatin domains involved in DNA double-strand breaks in vivo. *J Cell Biol* 1999;**146**:905–15.
23. Chen HT, Bhandoola A, Difilippantonio MJ, Zhu J, Brown MJ, Tai X, et al. Response to RAG-mediated V(D)J cleavage by NBS1 and γ-H2AX. *Science* 2000;**290**:1962–4.

24. Shim EY, Chung WH, Nicolette ML, Zhang Y, Davis M, Zhu Z, et al. Saccharomyces cerevisiae Mre11/Rad50/Xrs2 and Ku proteins regulate association of Exo1 and Dna2 with DNA breaks. *EMBO J* 2010;**29**:3370–80.
25. Cairns BR, Lorch Y, Li Y, Zhang M, Lacomis L, Erdjument-Bromage H, et al. RSC, an essential, abundant chromatin-remodeling complex. *Cell* 1996;**87**:1249–60.
26. Du J, Nasir I, Benton BK, Kladde MP, Laurent BC. Sth1p, a *Saccharomyces cerevisiae* Snf2p/Swi2p homolog, is an essential ATPase in RSC and differs from Snf/Swi in its interactions with histones and chromatin-associated proteins. *Genetics* 1998;**150**:987–1005.
27. Szerlong H, Hinata K, Viswanathan R, Erdjument-Bromage H, Tempst P, Cairns BR. The HSA domain binds nuclear actin-related proteins to regulate chromatin-remodeling ATPases. *Nat Struct Mol Biol* 2008;**15**:469–76.
28. Cao Y, Cairns BR, Kornberg RD, Laurent BC. Sfh1p, a component of a novel chromatin-remodeling complex, is required for cell cycle progression. *Mol Cell Biol* 1997;**17**:3323–34.
29. Angus-Hill ML, Schlichter A, Roberts D, Erdjument-Bromage H, Tempst P, Cairns BR. A Rsc3/Rsc30 zinc cluster dimer reveals novel roles for the chromatin remodeler RSC in gene expression and cell cycle control. *Mol Cell* 2001;**7**:741–51.
30. Badis G, Chan ET, van Bakel H, Pena-Castillo L, Tillo D, Tsui K, et al. A library of yeast transcription factor motifs reveals a widespread function for Rsc3 in targeting nucleosome exclusion at promoters. *Mol Cell* 2008;**32**:878–87.
31. Soutourina J, Bordas-Le Floch V, Gendrel G, Flores A, Ducrot C, Dumay-Odelot H, et al. Rsc4 connects the chromatin remodeler RSC to RNA polymerases. *Mol Cell Biol* 2006;**26**:4920–33.
32. Kasten M, Szerlong H, Erdjument-Bromage H, Tempst P, Werner M, Cairns BR. Tandem bromodomains in the chromatin remodeler RSC recognize acetylated histone H3 Lys14. *EMBO J* 2004;**23**:1348–59.
33. Taneda T, Kikuchi A. Genetic analysis of RSC58, which encodes a component of a yeast chromatin remodeling complex, and interacts with the transcription factor Swi6. *Mol Genet Genomics* 2004;**271**:479–89.
34. Treich I, Ho L, Carlson M. Direct interaction between Rsc6 and Rsc8/Swh3, two proteins that are conserved in SWI/SNF-related complexes. *Nucleic Acids Res* 1998;**26**:3739–45.
35. Wilson B, Erdjument-Bromage H, Tempst P, Cairns BR. The RSC chromatin remodeling complex bears an essential fungal-specific protein module with broad functional roles. *Genetics* 2006;**172**:795–809.
36. Da G, Lenkart J, Zhao K, Shiekhattar R, Cairns BR, Marmorstein R. Structure and function of the SWIRM domain, a conserved protein module found in chromatin regulatory complexes. *Proc Natl Acad Sci USA* 2006;**103**:2057–62.
37. Szerlong H, Saha A, Cairns BR. The nuclear actin-related proteins Arp7 and Arp9: a dimeric module that cooperates with architectural proteins for chromatin remodelling. *EMBO J* 2003;**22**:3175–87.
38. Lu YM, Lin YR, Tsai A, Hsao YS, Li CC, Cheng MY. Dissecting the pet18 mutation in Saccharomyces cerevisiae: HTL1 encodes a 7-kDa polypeptide that interacts with components of the RSC complex. *Mol Genet Genomics* 2003;**269**:321–30.
39. Romeo MJ, Angus-Hill ML, Sobering AK, Kamada Y, Cairns BR, Levin DE. *HTL1* encodes a novel factor that interacts with the RSC chromatin remodeling complex in *Saccharomyces cerevisiae*. *Mol Cell Biol* 2002;**22**:8165–74.
40. Florio C, Moscariello M, Ederle S, Fasano R, Lanzuolo C, Pulitzer JF. A study of biochemical and functional interactions of Htl1p, a putative component of the Saccharomyces cerevisiae, Rsc chromatin-remodeling complex. *Gene* 2007;**395**:72–85.
41. Cairns BR, Schlichter A, Erdjument-Bromage H, Tempst P, Kornberg RD, Winston F. Two functionally distinct forms of the RSC nucleosome-remodeling complex, containing essential AT hook, BAH, and bromodomains. *Mol Cell* 1999;**4**:715–23.

42. Leschziner AE, Saha A, Wittmeyer J, Zhang Y, Bustamante C, Cairns BR, et al. Conformational flexibility in the chromatin remodeler RSC observed by electron microscopy and the orthogonal tilt reconstruction method. *Proc Natl Acad Sci USA* 2007;**104**:4913–8.
43. Chaban Y, Ezeokonkwo C, Chung WH, Zhang F, Kornberg RD, Maier-Davis B, et al. Structure of a RSC-nucleosome complex and insights into chromatin remodeling. *Nat Struct Mol Biol* 2008;**15**:1272–7.
44. Asturias FJ, Chung WH, Kornberg RD, Lorch Y. Structural analysis of the RSC chromatin-remodeling complex. *Proc Natl Acad Sci USA* 2002;**99**:13477–80.
45. Skiniotis G, Moazed D, Walz T. Acetylated histone tail peptides induce structural rearrangements in the RSC chromatin remodeling complex. *J Biol Chem* 2007;**282**:20804–8.
46. Lorch Y, Maier-Davis B, Kornberg RD. Mechanism of chromatin remodeling. *Proc Natl Acad Sci USA* 2010;**107**:3458–62.
47. Lorch Y, Cairns BR, Zhang M, Kornberg RD. Activated RSC-nucleosome complex and persistently altered form of the nucleosome. *Cell* 1998;**94**:29–34.
48. Boyer LA, Logie C, Bonte E, Becker PB, Wade PA, Wolffe AP, et al. Functional delineation of three groups of the ATP-dependent family of chromatin remodeling enzymes. *J Biol Chem* 2000;**275**:18864–70.
49. Saha A, Wittmeyer J, Cairns BR. Chromatin remodeling by RSC involves ATP-dependent DNA translocation. *Genes Dev* 2002;**16**:2120–34.
50. Saha A, Wittmeyer J, Cairns BR. Chromatin remodeling through directional DNA translocation from an internal nucleosomal site. *Nat Struct Mol Biol* 2005;**12**:747–55.
51. Lorch Y, Maier-Davis B, Kornberg RD. Chromatin remodeling by nucleosome disassembly in vitro. *Proc Natl Acad Sci USA* 2006;**103**:3090–3.
52. Lorch Y, Zhang M, Kornberg RD. Histone octamer transfer by a chromatin-remodeling complex. *Cell* 1999;**96**:389–92.
53. Ferreira H, Flaus A, Owen-Hughes T. Histone modifications influence the action of Snf2 family remodelling enzymes by different mechanisms. *J Mol Biol* 2007;**374**:563–79.
54. Lia G, Praly E, Ferreira H, Stockdale C, Tse-Dinh YC, Dunlap D, et al. Direct observation of DNA distortion by the RSC complex. *Mol Cell* 2006;**21**:417–25.
55. Zhang Y, Smith CL, Saha A, Grill SW, Mihardja S, Smith SB, et al. DNA translocation and loop formation mechanism of chromatin remodeling by SWI/SNF and RSC. *Mol Cell* 2006;**24**:559–68.
56. Ng HH, Robert F, Young RA, Struhl K. Genome-wide location and regulated recruitment of the RSC nucleosome-remodeling complex. *Genes Dev* 2002;**16**:806–19.
57. Floer M, Wang X, Prabhu V, Berrozpe G, Narayan S, Spagna D, et al. A RSC/nucleosome complex determines chromatin architecture and facilitates activator binding. *Cell* 2010;**141**:407–18.
58. Carey M, Li B, Workman JL. RSC exploits histone acetylation to abrogate the nucleosomal block to RNA polymerase II elongation. *Mol Cell* 2006;**24**:481–7.
59. Desai P, Guha N, Galdieri L, Hadi S, Vancura A. Plc1p is required for proper chromatin structure and activity of the kinetochore in Saccharomyces cerevisiae by facilitating recruitment of the RSC complex. *Mol Genet Genomics* 2009;**281**:511–23.
60. Rossio V, Galati E, Ferrari M, Pellicioli A, Sutani T, Shirahige K, et al. The RSC chromatin-remodeling complex influences mitotic exit and adaptation to the spindle assembly checkpoint by controlling the Cdc14 phosphatase. *J Cell Biol* 2010;**191**:981–97.
61. Yukawa M, Koyama H, Miyahara K, Tsuchiya E. Functional differences between RSC1 and RSC2, components of a for growth essential chromatin-remodeling complex of Saccharomyces cerevisiae, during the sporulation process. *FEMS Yeast Res* 2002;**2**:87–91.
62. Titus LC, Dawson TR, Rexer DJ, Ryan KJ, Wente SR. Members of the RSC chromatin-remodeling complex are required for maintaining proper nuclear envelope structure and pore complex localization. *Mol Biol Cell* 2010;**21**:1072–87.

63. Wong MC, Scott-Drew SR, Hayes MJ, Howard PJ, Murray JA. RSC2, encoding a component of the RSC nucleosome remodeling complex, is essential for 2 microm plasmid maintenance in Saccharomyces cerevisiae. *Mol Cell Biol* 2002;**22**:4218–29.
64. Baetz KK, Krogan NJ, Emili A, Greenblatt J, Hieter P. The ctf13-30/CTF13 genomic haploinsufficiency modifier screen identifies the yeast chromatin remodeling complex RSC, which is required for the establishment of sister chromatid cohesion. *Mol Cell Biol* 2004;**24**:1232–44.
65. Huang J, Hsu J, Laurent BC. The RSC nucleosome-remodeling complex is required for cohesin's association with chromosome arms. *Mol Cell* 2004;**13**:739–50.
66. Campsteijn C, Wijnands-Collin AM, Logie C. Reverse genetic analysis of the yeast RSC chromatin remodeler reveals a role for RSC3 and SNF5 homolog 1 in ploidy maintenance. *PLoS Genet* 2007;**3**:e92.
67. Tsuchiya E, Hosotani T, Miyakawa T. A mutation in NPS1/STH1, an essential gene encoding a component of a novel chromatin-remodeling complex RSC, alters the chromatin structure of Saccharomyces cerevisiae centromeres. *Nucleic Acids Res* 1998;**26**:3286–92.
68. Tsuchiya E, Uno M, Kiguchi A, Masuoka K, Kanemori Y, Okabe S, et al. The Saccharomyces cerevisiae NPS1 gene, a novel CDC gene which encodes a 160 kDa nuclear protein involved in G2 phase control. *EMBO J* 1992;**11**:4017–26.
69. Lanzuolo C, Ederle S, Pollice A, Russo F, Storlazzi A, Pulitzer JF. The HTL1 gene (YCR020W-b) of Saccharomyces cerevisiae is necessary for growth at 37 degrees C, and for the conservation of chromosome stability and fertility. *Yeast* 2001;**18**:1317–30.
70. Bennett CB, Lewis LK, Karthikeyan G, Lobachev KS, Jin YH, Sterling JF, et al. Genes required for ionizing radiation resistance in yeast. *Nat Genet* 2001;**29**:426–34.
71. Koyama H, Itoh M, Miyahara K, Tsuchiya E. Abundance of the RSC nucleosome-remodeling complex is important for the cells to tolerate DNA damage in Saccharomyces cerevisiae. *FEBS Lett* 2002;**531**:215–21.
72. Chai B, Huang J, Cairns BR, Laurent BC. Distinct roles for the RSC and Swi/Snf ATP-dependent chromatin remodelers in DNA double-strand break repair. *Genes Dev* 2005;**19**:1656–61.
73. Shim EY, Ma JL, Oum JH, Yanez Y, Lee SE. The yeast chromatin remodeler RSC complex facilitates end joining repair of DNA double-strand breaks. *Mol Cell Biol* 2005;**25**:3934–44.
74. Shim EY, Hong SJ, Oum JH, Yanez Y, Zhang Y, Lee SE. RSC mobilizes nucleosomes to improve accessibility of repair machinery to the damaged chromatin. *Mol Cell Biol* 2007;**27**:1602–13.
75. Liang B, Qiu J, Ratnakumar K, Laurent BC. RSC Functions as an early double-strand-break sensor in the cell's response to DNA damage. *Curr Biol* 2007;**17**:1432–7.
76. Kent NA, Chambers AL, Downs JA. Dual chromatin remodeling roles for RSC during DNA double strand break induction and repair at the yeast MAT locus. *J Biol Chem* 2007;**282**:27693–701.
77. Faucher D, Wellinger RJ. Methylated H3K4, a transcription-associated histone modification, is involved in the DNA damage response pathway. *PLoS Genet* 2010;**6**: e1001082.
78. Moscariello M, Florio C, Pulitzer JF. Accurate repair of non-cohesive, double strand breaks in Saccharomyces cerevisiae: enhancement by homology-assisted end-joining. *Yeast* 2010;**27**: 837–48.
79. Karathanasis E, Wilson TE. Enhancement of Saccharomyces cerevisiae end-joining efficiency by cell growth stage but not by impairment of recombination. *Genetics* 2002;**161**:1015–27.
80. Sherwood R, Takahashi TS, Jallepalli PV. Sister acts: coordinating DNA replication and cohesion establishment. *Genes Dev* 2010;**24**:2723–31.
81. Unal E, Heidinger-Pauli JM, Kim W, Guacci V, Onn I, Gygi SP, et al. A molecular determinant for the establishment of sister chromatid cohesion. *Science* 2008;**321**:566–9.

82. Mohrmann L, Verrijzer CP. Composition and functional specificity of SWI2/SNF2 class chromatin remodeling complexes. *Biochim Biophys Acta* 2005;**1681**:59–73.
83. Park JH, Park EJ, Lee HS, Kim SJ, Hur SK, Imbalzano AN, et al. Mammalian SWI/SNF complexes facilitate DNA double-strand break repair by promoting gamma-H2AX induction. *EMBO J* 2006;**25**:3986–97.
84. Ray A, Mir SN, Wani G, Zhao Q, Battu A, Zhu Q, et al. hSNF5/INI1, a component of the human SWI/SNF chromatin remodeling complex, promotes nucleotide excision repair by influencing ATM recruitment and downstream H2AX phosphorylation. *Mol Cell Biol* 2009;**29**:6206–19.
85. McKenna ES, Sansam CG, Cho YJ, Greulich H, Evans JA, Thom CS, et al. Loss of the epigenetic tumor suppressor SNF5 leads to cancer without genomic instability. *Mol Cell Biol* 2008;**28**:6223–33.
86. Peng G, Yim EK, Dai H, Jackson AP, Burgt I, Pan MR, et al. BRIT1/MCPH1 links chromatin remodelling to DNA damage response. *Nat Cell Biol* 2009;**11**:865–72.
87. Ogiwara H, Ui A, Otsuka A, Satoh H, Yokomi I, Nakajima S, et al. Histone acetylation by CBP and p300 at double-strand break sites facilitates SWI/SNF chromatin remodeling and the recruitment of non-homologous end joining factors. *Oncogene* 2011;**30**:2135–46.
88. Lee HS, Park JH, Kim SJ, Kwon SJ, Kwon J. A cooperative activation loop among SWI/SNF, gamma-H2AX and H3 acetylation for DNA double-strand break repair. *EMBO J* 2010;**29**:1434–45.
89. Park JH, Park EJ, Hur SK, Kim S, Kwon J. Mammalian SWI/SNF chromatin remodeling complexes are required to prevent apoptosis after DNA damage. *DNA Repair (Amst)* 2009;**8**:29–39.
90. Klochendler-Yeivin A, Picarsky E, Yaniv M. Increased DNA damage sensitivity and apoptosis in cells lacking the Snf5/Ini1 subunit of the SWI/SNF chromatin remodeling complex. *Mol Cell Biol* 2006;**26**:2661–74.
91. Harte MT, O'Brien GJ, Ryan NM, Gorski JJ, Savage KI, Crawford NT, et al. BRD7, a subunit of SWI/SNF complexes, binds directly to BRCA1 and regulates BRCA1-dependent transcription. *Cancer Res* 2010;**70**:2538–47.
92. Burrows AE, Smogorzewska A, Elledge SJ. Polybromo-associated BRG1-associated factor components BRD7 and BAF180 are critical regulators of p53 required for induction of replicative senescence. *Proc Natl Acad Sci USA* 2010;**107**:14280–5.
93. Ebbert R, Birkmann A, Schuller H-J. The product of the *SNF2/SWI2* paralogue *INO80* of *Saccharomyces cerevisiae* required for efficient expression of various yeast structural genes is part of a high-molecular-weight protein complex. *Mol Microbiol* 1999;**32**:741–51.
94. Shen X, Mizuguchi G, Hamich A, Wu C. A chromatin remodeling complex involved in transcription and DNA processing. *Nature* 2000;**406**:541–4.
95. Chen L, Cai Y, Jin J, Florens L, Swanson SK, Washburn MP, et al. Subunit organization of the human INO80 chromatin remodeling complex: an evolutionarily conserved core complex catalyzes ATP-dependent nucleosome remodeling. *J Biol Chem* 2011;**286**:11283–9.
96. Jonsson ZO, Jha S, Wohlschlegel JA, Dutta A. Rvb1p/Rvb2p recruit Arp5p and assemble a functional Ino80 chromatin remodeling complex. *Mol Cell* 2004;**16**:465–77.
97. Shen X, Ranallo R, Choi E, Wu C. Involvement of actin-related proteins in ATP-dependent chromatin remodeling. *Mol Cell* 2003;**12**:147–55.
98. Udugama M, Sabri A, Bartholomew B. The INO80 ATP-dependent chromatin remodeling complex is a nucleosome spacing factor. *Mol Cell Biol* 2011;**31**:662–73.
99. Papamichos-Chronakis M, Watanabe S, Rando OJ, Peterson CL. Global regulation of H2A.Z localization by the INO80 chromatin-remodeling enzyme is essential for genome integrity. *Cell* 2011;**144**:200–13.

100. Klopf E, Paskova L, Sole C, Mas G, Petryshyn A, Posas F, et al. Cooperation between the INO80 complex and histone chaperones determines adaptation of stress gene transcription in the yeast Saccharomyces cerevisiae. *Mol Cell Biol* 2009;**29**:4994–5007.
101. van Attikum H, Fritsch O, Hohn B, Gasser SM. Recruitment of the INO80 complex by H2A phosphorylation links ATP-dependent chromatin remodeling with DNA double-strand break repair. *Cell* 2004;**119**:777–88.
102. Papamichos-Chronakis M, Peterson CL. The Ino80 chromatin-remodeling enzyme regulates replisome function and stability. *Nat Struct Mol Biol* 2008;**15**:338–45.
103. Shimada K, Oma Y, Schleker T, Kugou K, Ohta K, Harata M, et al. Ino80 chromatin remodeling complex promotes recovery of stalled replication forks. *Curr Biol* 2008;**18**:566–75.
104. Vincent JA, Kwong TJ, Tsukiyama T. ATP-dependent chromatin remodeling shapes the DNA replication landscape. *Nat Struct Mol Biol* 2008;**15**:477–84.
105. Yu EY, Steinberg-Neifach O, Dandjinou AT, Kang F, Morrison AJ, Shen X, et al. Regulation of telomere structure and functions by subunits of the INO80 chromatin remodeling complex. *Mol Cell Biol* 2007;**27**:5639–49.
106. Sarkar S, Kiely R, McHugh PJ. The Ino80 chromatin-remodeling complex restores chromatin structure during UV DNA damage repair. *J Cell Biol* 2010;**191**:1061–8.
107. van Attikum H, Fritsch O, Gasser SM. Distinct roles for SWR1 and INO80 chromatin remodeling complexes at chromosomal double-strand breaks. *EMBO J* 2007;**26**:4113–25.
108. Downs JA, Allard S, Jobin-Robitaille O, Javaheri A, Auger A, Bouchard N, et al. Binding of chromatin-modifying activities to phosphorylated histone H2A at DNA damage sites. *Mol Cell* 2004;**16**:979–90.
109. Morrison AJ, Highland J, Krogan NJ, Arbel-Eden A, Greenblatt JF, Haber JE, et al. INO80 and g-H2AX interaction links ATP-dependent chromatin remodeling to DNA damage repair. *Cell* 2004;**119**:767–75.
110. Tsukuda T, Fleming AB, Nickoloff JA, Osley MA. Chromatin remodelling at a DNA double-strand break site in Saccharomyces cerevisiae. *Nature* 2005;**438**:379–83.
111. Papamichos-Chronakis M, Krebs JE, Peterson CL. Interplay between Ino80 and Swr1 chromatin remodeling enzymes regulates cell cycle checkpoint adaptation in response to DNA damage. *Genes Dev* 2006;**20**:2437–49.
112. Chen CC, Carson JJ, Feser J, Tamburini B, Zabaronick S, Linger J, et al. Acetylated lysine 56 on histone H3 drives chromatin assembly after repair and signals for the completion of repair. *Cell* 2008;**134**:231–43.
113. Kalocsay M, Hiller NJ, Jentsch S. Chromosome-wide Rad51 spreading and SUMO-H2A.Z-dependent chromosome fixation in response to a persistent DNA double-strand break. *Mol Cell* 2009;**33**:335–43.
114. Morrison AJ, Kim JA, Person MD, Highland J, Xiao J, Wehr TS, et al. Mec1/Tel1 phosphorylation of the INO80 chromatin remodeling complex influences DNA damage checkpoint responses. *Cell* 2007;**130**:499–511.
115. Tsukuda T, Lo YC, Krishna S, Sterk R, Osley MA, Nickoloff JA. INO80-dependent chromatin remodeling regulates early and late stages of mitotic homologous recombination. *DNA Repair (Amst)* 2009;**8**:360–9.
116. Kawashima S, Ogiwara H, Tada S, Harata M, Wintersberger U, Enomoto T, et al. The INO80 complex is required for damage-induced recombination. *Biochem Biophys Res Commun* 2007;**355**:835–41.
117. Hogan CJ, Aligianni S, Durand-Dubief M, Persson J, Will WR, Webster J, et al. Fission yeast Iec1-ino80-mediated nucleosome eviction regulates nucleotide and phosphate metabolism. *Mol Cell Biol* 2010;**30**:657–74.

118. Wu S, Shi Y, Mulligan P, Gay F, Landry J, Liu H, et al. A YY1-INO80 complex regulates genomic stability through homologous recombination-based repair. *Nat Struct Mol Biol* 2007;**14**:1165–72.
119. Kashiwaba S, Kitahashi K, Watanabe T, Onoda F, Ohtsu M, Murakami Y. The mammalian INO80 complex is recruited to DNA damage sites in an ARP8 dependent manner. *Biochem Biophys Res Commun* 2010;**402**:619–25.
120. Park EJ, Hur SK, Kwon J. Human INO80 chromatin-remodelling complex contributes to DNA double-strand break repair via the expression of Rad54B and XRCC3 genes. *Biochem J* 2010;**431**:179–87.
121. Wykoff DD, O'Shea EK. Identification of sumoylated proteins by systematic immunoprecipitation of the budding yeast proteome. *Mol Cell Proteomics* 2005;**4**:73–83.
122. De Cristofaro MF, Betz BL, Rorie CJ, Reisman DN, Wang W, Weissman BE. Characterization of SWI/SNF protein expression in human breast cancer cell lines and other malignancies. *J Cell Physiol* 2001;**186**:136–45.
123. Chambers AL, Brownlee PM, Durley SC, Beacham T, Kent NA, Downs JA. *PLoS One* 2012;**7**(2):e32016.
124. Gospodinov A, Vaissiere T, Krastev DB, Legube G, Anachkova B, Herceg Z. Mammalian Ino80 mediates double-strand break repair through its role in DNA end strand resection. *Mol Cell Biol* 2011;**31**(23):4735–45.

Mechanistic Links Between ATM and Histone Methylation Codes During DNA Repair

Ye Xu, Chang Xu, and Brendan D. Price

Division of Genome Stability and DNA Repair, Department of Radiation Oncology, Dana-Farber Cancer Institute, Harvard Medical School, Boston, Massachusetts, USA

The ataxia telangiectasia-mutated (ATM) protein kinase is the master regulator of the DNA double-strand break (DSB) repair pathway. The activation of ATM involves its recruitment to the DSB through interaction with the mre11–rad50–nbs1 complex, followed by the acetylation of ATM by the Tip60 acetyltransferase. This acetylation of ATM within its regulatory domain is essential for activating ATM's kinase activity. Further work has now revealed that Tip60 is activated through direct interaction between Tip60's chromodomain and histone H3 trimethylated on lysine 9 (H3K9me3). The loading of Tip60 onto the chromatin at DSBs therefore represents the primary mechanism for activation of Tip60's acetyltransferase activity in response to DNA damage. The ability of H3K9me3 at DSBs to regulate the activity of Tip60 and the subsequent activation of ATM emphasizes the crucial role played by chromatin architecture in regulating DSB repair. Further, histone methylation

Progress in Molecular Biology
and Translational Science, Vol. 110
DOI: 10.1016/B978-0-12-387665-2.00010-9

1877-1173/12 $35.00

and chromatin structure are disrupted in human cancers, implying that altered chromatin structure in tumor cells may impact DSB repair, increasing genomic instability and contributing to the progression of cancer.

I. The DNA Damage Response

Mammalian cells are constantly exposed to genotoxic events which can directly damage the DNA in the cell. Endogenous DNA damage can occur through the production of oxygen radicals during aerobic respiration or through errors created during DNA replication. However, a significant source of DNA damage is exposure to exogenous DNA-damaging agents. For example, aflatoxins, heavy metals, cigarette smoke, radiation (e.g. UV exposure) and many environmental and man-made chemicals can modify or cross-link bases on the DNA double helix. Consequently, cells must continuously monitor the integrity of the DNA and rapidly remove these lesions. The failure to correctly maintain genome integrity can lead to the generation of mutations and deletions, constituting the underlying mechanism of carcinogenesis.

To maintain genome integrity, cells have multiple DNA repair pathways that are designed to detect and remove specific lesions. These include mismatch repair (removal of mismatched bases), base excision repair (removal of simple DNA adducts), nucleotide excision repair (bulky lesions and UV damage), the Fanconi anemia pathway (interstrand cross-links), and DNA double-strand break (DSB) repair (caused by ionizing radiation (IR) or replication fork collapse) (reviewed in Ref. 1). Collectively, these pathways are referred to as the DNA damage response (DDR). Although these diverse DNA repair pathways focus on distinct lesions, there is considerable functional overlap between the individual proteins so that many DDR proteins function in multiple DNA repair pathways. There is, therefore, significant cross talk between DDR proteins during DNA repair.

In this review, we focus on the repair of DNA DSBs. DSBs can occur in cells through several mechanisms. For example, stalling of replication forks due to DNA damage can lead to fork collapse, creating DSBs that must be repaired before replication can continue. In addition, exposure of cells to IR generates large numbers of DNA strand breaks, including DSBs. IR is widely used to treat cancer, and derives its therapeutic effect by generating DSBs within the tumor mass, thereby preferentially killing the rapidly dividing tumor cells. However, tumor responsiveness to IR is variable, and there is considerable urgency to understand the molecular pathway by which cells detect and repair DSBs, with the aim of identifying differences between normal and tumor cells that can be exploited to improve the therapeutic index of radiation therapy. In fact, intensive study of the mechanism of DSB repair has led to a detailed understanding of how cells detect and repair these potentially toxic lesions. Two distinct pathways for DSB repair have been identified: nonhomologous end-joining (NHEJ) and homology-directed repair (HR). These two pathways differ in both the mechanism and fidelity of DNA repair. NHEJ (reviewed in Ref. 2) relies on direct ligation of the DNA ends at the DSB, often after partial processing of the DNA, so that the final religated product may contain small deletions or mutations. In contrast, HR (reviewed in Refs. 3,4) utilizes the adjacent sister chromatid present during S-phase to direct homology-mediated repair of DSBs. In this case, the fidelity of DSB repair by HR is much higher than that by NHEJ due to the former's ability to use the homologous sequence as a template for repair. However, because HR requires the sister chromatid for repair, its use is effectively limited to the replicative phase of the cell cycle. Nondividing cells are therefore dependent on NHEJ for DSB repair.

Recent work has emphasized that DSB repair is a chromatin-based event that can be heavily influenced by the chromatin architecture at the DSB.[5,6] For example, NHEJ requires end-processing of the DNA and HR requires unwinding of the sister chromatid DNA. Both these processes involve significant remodeling of the local chromatin architecture, including eviction of nucleosomes and unwinding of the DNA. In addition, the complexity of chromatin structure plays a key role in the ability of cells to detect DSBs, with compacted regions of the chromatin requiring more complex remodeling events before repair can be initiated.[7,8] Here, we discuss how the structure of chromatin impacts the ability of cells to detect and repair DSBs, with particular emphasis on understanding the critical role that chromatin structure plays in the activation of the ATM kinase, the master regulator of the DSB repair pathway.

II. *Ataxia telangiectasia* and the ATM Kinase

The ATM protein kinase is the product of the *ataxia telangiectasia* (*A-T*) gene.[9] *A-T* is a rare human disease characterized by cerebellar degeneration, leading to progressive ataxia and loss of motor function. Patients exhibit impaired antibody production, are prone to infection, and have characteristic telangiectasias in the eye. In addition to these devastating effects, *A-T* patients who survive into their teens have a highly elevated risk for cancer, in particular a 30- to 50-fold increase in leukemias and lymphomas.[10] Work over the last 15 years since the cloning of the full-length *A-T* gene has now revealed that ATM functions as the master regulator of the cells' response to DSBs.

A. Cellular Defects in *ataxia telangiectasia*

Early studies demonstrated that cells derived from *A-T* patients were extremely sensitive to IR, but exhibited normal responses to UV or other types of DNA damage, indicating a specific defect in DSB repair.[11] Additional work established that *A-T* cells lacked the normal activation of cell cycle checkpoints following IR exposure, and continued to progress through S-phase in the presence of DSBs.[12,13] Work on *A-T* was greatly stimulated by the cloning of the *A-T* gene, which identified ATM as a member of the PI 3-kinase (PIK) family of DNA damage-activated kinases,[14,15] a family which includes the DNA-PKcs and ATR kinases.[16] Sequence analysis of the *A-T* gene demonstrated that the majority of mutations in *A-T* patients were truncating mutations,[17,18] so that *A-T* patients essentially express no ATM protein. The increased cancer risk and sensitivity to IR in *A-T* patients can therefore be attributed to the absence of functional ATM protein in these patients. It was subsequently demonstrated that ATM was a kinase whose activity was increased by DNA damage and that ATM directly phosphorylated the p53 protein.[19] Work from many laboratories has now identified several hundred proteins as potential targets for ATM,[20] including key components of the DSB repair pathway such as p53, nbs1, chk2, brca1, and H2AX (reviewed in Refs. 1,9). Overall, this work demonstrated that ATM's kinase activity and its ability to phosphorylate target proteins were critical for cells to both activate cell cycle checkpoints and coordinate the repair of DSBs.

We now have a clear picture of the events that are triggered through the activation of ATM's kinase by IR. One of the earliest detectable events is phosphorylation of the histone variant H2AX (termed γH2AX[21]) by ATM within minutes of exposure to IR. γH2AX is phosphorylated on chromatin domains surrounding the DSB which can extend up to 500 kb on either side of the break.[22] The mdc1 scaffold protein then binds directly to γH2AX, creating a platform to recruit other DNA repair proteins, including 53BP1, RNF8, and ultimately brca1, to the DSB.[23–27] This results in the concentration of DNA repair proteins onto chromatin domains extending away from the DSB, which can be visualized by immunofluorescence techniques with antibodies against components of these complexes.[25–27] Further, many of these proteins, including mdc1[26] and brca1,[28] are also phosphorylated by ATM, indicating a complex and highly regulated role for ATM's kinase activity in controlling the cells' response to DSBs. Thus, activation of ATM's kinase activity initiates a signal transduction pathway, which promotes recruitment of DNA repair complexes to DSBs, activates cell cycle checkpoints, and creates an environment for promoting DSB repair. ATM therefore plays a pivotal role in the earliest steps involved in the detection and repair of DSBs.

B. Activation of the ATM Kinase by DSBs

A crucial issue is to understand how the introduction of a DSB into the chromatin is converted into a signal which activates the kinase activity of the ATM protein. A key step in elucidating ATM activation was the observation that ATM undergoes a rapid (within 1 min) autophosphorylation on serine 1981 following DNA damage.[29] Further, using an antibody which specifically detects ATM autophosphorylated on serine 1981, it was clearly shown that phosphorylated ATM accumulated at DSBs within the first minute postirradiation.[29] Subsequently, several additional autophosphorylation sites were identified on ATM.[30,31] This led to the proposal that autophosphorylation of ATM was required to convert inactive ATM dimers to active ATM monomers,[29] a mechanism supported by the observation that mutation of ATM autophosphorylation sites blocks ATM activation[29,30] and dimer–monomer transition.[32] This suggested a model in which ATM was recruited to DSBs as an inactive dimer. ATM's kinase activity was then activated, leading to autophosphorylation of ATM at multiple sites, and the conversion of ATM from an inactive dimer to an active monomer.[29] However, when this was tested in mouse models in which one or more of the ATM autophosphorylation sites were mutated, no significant defect in either ATM activation or ATM function at the whole organism level was seen.[33,34] This discrepancy could have several explanations. One possibility is that the murine system used to express the non-phosphorylatable ATM construct in mice has influenced the outcome.[33,34] A second possibility is that there are differences in ATM function and activation between mice and humans. For example, mice in which ATM is deleted replicate both the increased radiosensitivity and increased tumor formation seen in patients with *A-T*.[35,36] However, the $ATM^{-/-}$ mice do not exhibit the pronounced cerebellar degeneration or other significant neurological defects which are the hallmark of the human disease. In addition, studies on human ATM using biochemical approaches have shown that ATM's kinase activity can be activated, *in vitro*, in the absence of detectable autophosphorylation of ATM.[37,38] Thus, although it is clear that ATM is recruited to DSBs and that this leads to activation of ATM's kinase activity, how autophosphorylation and dimer–monomer conversion contribute to forming the fully active ATM kinase remains unclear. The differences between murine and human ATM, and between cell-based and biochemical systems suggest that complex layers of ATM regulation contribute to ATM activation under different cellular conditions in different cell types.

C. ATM and the mre11–rad50–nbs1 Complex

In addition to examining how ATM's kinase activity is upregulated by DNA damage, several groups have clearly demonstrated that the mre11–rad50–nbs1 (MRN) complex is required for activation of ATM's kinase activity.[39–43] MRN is a trimeric DNA-binding complex, which is rapidly recruited to DSBs.[44] The mre11 protein contains both a DNA-binding and a nuclease domain.[45] Rad50 is a SWI/SNF-related ATPase and functions as a structural component to hold the two ends of the DSB together.[44] Finally, nbs1 functions as a regulatory subunit of MRN, which also undergoes regulatory phosphorylation by ATM.[39] Mutations in either mre11 (AT-like disease (ATLD)[46]) or nbs1 (Nijmegen break syndrome (NBS)[47]) have clinical defects similar to *A-T*, including neurological defects, increased cancer incidence, and increased radiosensitivity. Further, ATM interacts with the c-terminal of nbs1,[43,48] and mutation or deletion of either mre11 or nbs1in *ATLD* and *NBS* cells significantly reduces the activation of ATM's kinase activity by DNA damage *in vivo.*[40–43] Taken together, this work indicates that the MRN complex plays a key role in activation of ATM's kinase activity. This idea was further strengthened by biochemical studies, which showed that purified MRN was sufficient to activate ATM's kinase activity *in vitro.*[37,38,49] The gene inactivation and biochemical studies therefore indicate that MRN is upstream of ATM. This led to the idea that MRN was rapidly recruited to DSBs through the ability of the mre11 subunit to bind to the broken ends of the DNA. Subsequently, interaction between ATM and the nbs1 subunit of the MRN complex activated ATM's kinase activity, allowing ATM to phosphorylate its target proteins. Importantly, both mre11 and nbs1 are phosphorylated by ATM,[50] underscoring the complexity of the regulatory interactions in the DDR response, in which upstream regulators such as MRN are also targets for phosphorylation by downstream effectors such as ATM.

D. Other Contributors to ATM Activation

In addition to ATM, there is evidence that other factors contribute to the activation of ATM's kinase activity. For example, inactivation of the MRN complex does not completely abolish ATM activation in cells. Loss of MRN function is associated with a significant delay in ATM activation at low levels of DNA damage[40–43,51]; however, at high levels of damage loss of MRN function has limited impact on the activation of ATM's kinase activity.[52–54] This type of observation indicates that, while MRN is critical for ATM activation at low levels of DNA damage, high levels of DSBs may bypass the need for MRN to activate ATM's kinase activity. This implies that, *in vivo*, factors in addition to MRN are involved in modulating ATM activation at DSBs. For example, the ATM-interacting protein ATMIN[55,56] is important for ATM activation through a pathway that is independent of the nbs1 protein. Other studies have shown that chromatin structure adjacent to the site of DNA damage can have a profound impact on both ATM activation and DNA repair itself. For example, ATM can be activated by agents that directly alter chromatin condensation.[29] Molecular-imaging studies indicate that tethering of DNA repair proteins to the chromatin was sufficient to activate ATM and the DDR without the need to create DSBs.[57] Further, deletion of the histone H1 linker protein, which leads to global decompaction, significantly amplified the DDR and increased DNA repair.[58] Similarly, deletion of either the chromatin structural protein HMGN1[59] or the Tip60 acetyltransferase, which acetylates chromatin and ATM,[60] impairs ATM activation. Together, this work indicates that the unique architecture of the chromatin adjacent to DSBs can have a profound impact on the efficiency of ATM activation.

III. The Tip60 Acetyltransferase

The previous section outlined the evidence for a link between ATM activation and chromatin structure. Here, we review the evidence that a chromatin-modifying enzyme, the Tip60 acetyltransferase, plays a central role in linking the detection of DSBs to the activation of the ATM kinase.

A. Tip60 Is a Ubiquitously Expressed Acetyltransferase

The Tip60 (KAT5) acetyltransferase plays an essential role in a wide range of signaling pathways, including transcriptional regulation, steroid receptor function, chromatin remodeling, histone acetylation, DNA repair, and maintenance of stem cell function.[61–63] Tip60 acetylates the ε-amino groups of lysine residues on both histone and nonhistone proteins,[63,64] including histones H2A and H4,[65–68] the androgen receptor,[69] p53,[70,71] enzymes involved in glucose metabolism,[72] the ATM kinase,[32,60,51,73] and many others.[61,72] Protein acetylation by Tip60 can directly impact protein function, including regulating p53's apoptotic function[70,71] or activating ATM's kinase activity.[60] Tip60, like most acetyltransferases, has low sequence specificity for protein acetylation, and can promiscuously acetylate many proteins *in vitro*. However, Tip60 interacts specifically with several protein complexes, and it is these protein-binding partners that target Tip60 to specific substrates to promote their acetylation.[61,63] Further, Tip60 is a highly connected "hub" protein which interacts with multiple proteins in complex regulatory pathways.[74] Consequently, inactivation of Tip60 is embryonic lethal in mice,[75] underscoring the central role of Tip60 in regulating cellular pathways required for both development and maintenance of cell viability.

B. Tip60 and Cancer

Although Tip60 functions as an acetyltransferase in diverse signaling systems, here we focus on its critical role in DNA repair. Tip60 is required to maintain genomic integrity and regulate DSB repair in both mammalian and yeast cells.[60,67,76] Loss of Tip60 function leads to increased sensitivity to

IR[32,60,67,51,77] and the accumulation of multiple chromosomal aberrations.[51] Analysis of several tumor types indicates that Tip60 expression levels are altered in prostate, breast, and colorectal cancer.[78–81] In particular, Tip60 functions as a haplo-insufficient tumor suppressor in breast cancer, with reduced levels of nuclear Tip60 frequently identified in breast tumor samples.[82] The observation that Tip60 protein levels are reduced in tumors is consistent with the idea that Tip60 is an essential hub gene, regulating many distinct pathways in the cell. Reduction in the levels of Tip60 protein, rather than mutational inactivation or deletion, would allow tumor cells to limit Tip60 activity directed toward the DDR while maintaining essential transcriptional and regulatory functions. This, in turn, would allow tumors to maintain the elevated levels of genomic instability associated with tumor progression. Although the mechanism by which Tip60 levels are reduced in tumors is not known, methylation of the Tip60 promoter or expression of microRNAs targeting Tip60 expression is likely to dominate in tumor cells. Continued studies in this area are therefore likely to yield important information on the role of Tip60 in tumorigenesis.

Thus, a key function of Tip60 is to protect cells from genomic instability by regulating the DDR and suppressing potentially transforming events that lead to cancer. We now know that Tip60 controls genomic stability through its ability to regulate two key components of the DDR—chromatin remodeling at DSBs by the NuA4–Tip60 complex and acetylation and activation of the ATM kinase. NuA4 is a multi-subunit chromatin-remodeling complex containing p400, a large motor ATPase, and Tip60, an acetyltransferase activity.[83] Several subunits of NuA4, including p400 and Tip60, are recruited to DSBs in both yeast[84] and mammalian cells.[5,77,85] The recruitment of NuA4 to the chromatin at DSBs leads to increased acetylation of histones H2AX and H4[66,77,84,86,87] and a reduction in the stability of nucleosomes at DSBs.[5,77,85] This allows the NuA4–Tip60 complex to create open, mobile chromatin domains that facilitate the recruitment and loading of DDR proteins (including brca1 and 53BP1) onto the chromatin at DSBs (reviewed in Refs. 5,77,85). Chromatin acetylation by the NuA4–Tip60 complex is required for DSB repair; however, this review focuses on the ability of Tip60 to directly acetylate and activate the ATM kinase.

C. Tip60 and ATM Acetylation

The ability of Tip60 to directly acetylate the ATM kinase represents a new type of mechanism for kinase regulation. Previous work had shown that Tip60's acetyltransferase activity was essential for the DDR.[32,60,67,51,77] Recently, it was shown that Tip60's acetyltransferase activity was increased in response to IR and that this led to direct acetylation of ATM by Tip60,[60] a finding confirmed by several other groups.[59,88,89] Further, inactivation of Tip60 inhibited both ATM acetylation and activation of ATM's kinase activity.[6,60] Subsequently, a single lysine residue, at amino acid 3016 near the c-terminal kinase domain, was identified as the site on ATM that was acetylated by Tip60.[32,60,90] Mutation of this lysine blocked the activation of ATM's kinase activity by DNA damage, indicating that Tip60-dependent acetylation of ATM is a key step in the activation of ATM's kinase activity. The ATM acetylation site is highly conserved among higher eukaryotes,[32] whereas the autophosphorylation sites are not,[29,33,34] suggesting that acetylation is an evolutionarily conserved event in ATM regulation.

The positioning of the Tip60 acetylation site adjacent to the kinase domain has provided powerful insights into how acetylation promotes activation of the ATM kinase. ATM is a member of the PIK family, which includes the ATM, DNA-PKcs, and Atr and Tor protein kinases, each of which shares a common c-terminal structure.[91,92] PIK proteins have an unusual c-terminal structure, in which the kinase domain is sandwiched between two additional domains, termed the FAT and FATC domains. The FAT–kinase–FATC structure is only found in PIK proteins, and the FAT and FATC domains are not found independent of each other in other protein families. The Tip60 acetylation site at lysine 3016 of ATM is located in the region of the protein which links the kinase domain to the FATC domain. Further, Tip60 binds tightly to the highly conserved FATC domain at the extreme C-terminal of ATM.[60,90] The close proximity between the Tip60-binding site (FATC domain: amino acids 3029–3056) and the acetylation site (lysine 3016) indicates tight coupling between these two events.[32,60,90] This led to the proposal that acetylation of ATM between the kinase domain and FATC domain may activate ATM's kinase activity by altering the conformation of the ATM protein. Studies on other PIK proteins provide support for this idea. The region between the kinase domain and the FATC domain of PIK proteins has been named the PIK regulatory domain (PRD; reviewed in Refs. 16,93,94). This region is weakly conserved between PIK family members, such as ATM, Atr, and DNA-PKcs; however, the PRD plays a critical role in regulating the kinase activity of individual PIK protein family members. For example, binding of TopBP1 (a regulator of Atr function) to the PRD of Atr activates Atr's kinase activity.[93,94] The PRD of DNA-PKcs is required for activation of its kinase activity by DNA damage,[95,96] and the PRD of mTor is essential for its *in vivo* function.[97] Further, Akt phosphorylates the PRD of Tor1, although how this phosphorylation impacts Tor1 function is not yet known.[98] The PRD of PIK proteins therefore plays a key role in regulating the kinase activity of PIK proteins. For ATM, the acetylation of lysine 3016 in the PRD of ATM may directly regulate the kinase activity of the adjacent kinase domain. This may occur through an alteration in the conformation of the kinase–PRD–FATC domain of ATM, potentially increasing substrate access to the kinase domain, regulating ATM's intrinsic kinase activity, or altering the interaction between ATM and its substrate proteins. Continued studies on understanding how acetylation of ATM's PRD regulates ATM's kinase activity, including structural studies, will provide insight into the regulation of this unique kinase.

D. Activation of Tip60 by DSBs

This leads to the key question for understanding how Tip60 regulates ATM—how is the acetyltransferase activity of Tip60 regulated in response to DSBs? One potential mechanism is through posttranslational modification of Tip60. Tip60 is known to undergo both phosphorylation[99] and autoacetylation[100] in response to DNA damage. However, studies in our laboratory (B.D. Price, unpublished observations) indicate that changes in the posttranslational modification of Tip60 are not involved in the upregulation of Tip60's acetyltransferase activity. However, recent work uncovered a key role for the chromodomain of Tip60 in regulating Tip60's acetyltransferase activity.[51]

Chromodomains are unique protein domains containing conserved hydrophobic amino acids that specifically interact with methylated lysine residues.[101] Lysine methylation is a common posttranslational modification of histones in which the ε-amino group of lysine is either mono-, di- or trimethylated,[102,103] with each methylation state encoding specific functional information. For example, methylation of histone H3 on lysines 9 or 27 and histone H4 on lysine 20 is associated with the inactive heterochromatin,[103–105] whereas methylation of histone H3 on lysines 4 and 36 is associated with transcriptionally active genes.[106–108] Similar to other signal transduction pathways, histone methylation is a dynamic process. A large family of lysine methyltransferases (KMTs), which can create mono-, di-, or trimethylated lysine, has been described.[109,110] Similarly, several families of lysine demethylases, which remove one or more methyl groups from methylated lysine residues, have been described.[111,112] Rapid changes in histone methylation have been implicated in diverse cellular processes, including transcription, DNA replication, and DNA repair. Histone methylation, therefore, represents a dynamic signaling system that creates specific methylation marks for recruitment of chromatin-modifying complexes to the chromatin.

Because Tip60 is recruited to DSBs,[60] Sun *et al.* proposed that this would promote interaction between Tip60's chromodomain and methylated lysine residues on the chromatin.[51] There is now strong evidence that Tip60's chromodomain interacts with a specific histone modification—histone H3 trimethylated on lysine 9 (H3K9me3; Ref. 51). Both *in vitro* and *in vivo* analyses indicate that binding of H3K9me3 to Tip60 functions as an allosteric regulator, increasing the catalytic activity of Tip60. Mutations in the highly conserved hydrophobic amino acids of Tip60's chromodomain block both binding of H3K9me3 to Tip60 and activation of Tip60's acetyltransferase activity by DNA damage. Corresponding experiments *in vivo* demonstrated that inactivating mutations in Tip60's chromodomain inhibited both the activation of Tip60's acetyltransferase activity by DNA damage and the subsequent acetylation and activation of ATM's kinase activity.[51] Further, when global H3K9me3 levels were reduced, either by overexpression of KDM4D, a H3K9me3 demethylase,[112] or by genetic inactivation of the major H3K9 methyltransferases Suv39h1 and Suv39h2,[113] Tip60 activation following DNA damage was significantly decreased.[51] The recruitment of Tip60 to DSBs therefore leads to productive interactions between Tip60's chromodomain and methylated lysine residues on histones at DSBs, and this interaction activates Tip60's HAT activity. The chromodomain of Tip60 therefore functions as the sensor for activation of Tip60 by DNA damage. Further, this work demonstrated that H3K9me3 plays an essential role in the DDR. This conclusion is supported by the observation that cells with reduced levels of H3K9 methylation have defective activation of ATM kinase activity and increased genomic instability.[51,113] Overall, this indicates that recruitment of ATM and Tip60 to the chromatin at DSBs promotes interaction between Tip60's chromodomain and H3K9me3 on adjacent histones. This leads to activation of Tip60's acetyltransferase activity, acetylation of ATM, and activation of ATM's kinase activity.

E. Interaction Between ATM Tip60 and the MRN Complex

Although these results demonstrate a key role for acetylation of the PRD domain of ATM by Tip60 in activating ATM's kinase activity, it is critical to incorporate the published data on the role of the MRN complex into this model. As discussed earlier, both cell-based[40–43] and biochemical systems[37,38,49] support the hypothesis that the MRN complex makes a crucial contribution to activating ATM's kinase activity. The MRN complex physically interacts with ATM,[43,48] and this interaction is sufficient (in reconstituted systems) for activation of ATM's kinase activity.[40–43] In addition, ATM and Tip60 can be co-immunoprecipitated from cells, and this interaction requires the FATC domain of ATM.[60,90] However, despite extensive *in vitro* biochemical analysis, we have been unable to demonstrate direct binding of Tip60 to ATM (BD Price, unpublished data). This implies the presence of additional proteins connecting Tip60 and the FATC domain of ATM. Several studies have indicated that Tip60 can bind to the MRN complex,[114,115] implying that the MRN complex is a potential candidate for bringing ATM and Tip60 together. This raises the possibility that, because both ATM and Tip60 interact with MRN, they are co-recruited to DSBs through the ability of MRN to bind to DNA ends. This is supported by the observation that inactivation of the MRN complex led to defective recruitment of Tip60 to DSBs, and significant delays in the acetylation and activation of ATM by Tip60.[51] Further, this would suggest that the FATC domain of ATM interacts with MRN, as the FATC domain of ATM was required for co-immunoprecipitation of both ATM and Tip60.[90] Determining the nature of the interactions between MRN, Tip60, and ATM, and understanding how this complex functions to activate both ATM and Tip60 will require careful purification and biochemical analysis of these complexes.

These results suggest the following model to describe how the MRN complex and Tip60 function together to promote the full activation of the ATM kinase in response to DSBs. The initial step is the recruitment of the MRN complex to the DSB, which is mediated by the interaction of the mre11 subunit of MRN with the broken DNA ends. Subsequently, ATM and Tip60 are recruited to the MRN complex, concentrating these factors at the DSB. Currently, it is unclear if ATM and Tip60 are recruited as a single ATM–Tip60 complex, or if Tip60 and ATM bind separately to the MRN complex. In either case, localization of ATM and Tip60 onto the chromatin allows the chromodomain of Tip60 to interact with the nearby histone H3, which is trimethylated on lysine 9 (H3K9me3). This interaction activates Tip60's acetyltransferase activity through allosteric regulation of the acetyltransferase domain, leading to acetylation of lysine 3016 of ATM and activation of ATM's kinase activity. However, the relative contributions of the MRN complex and Tip60 to the production of the fully active ATM kinase is currently unclear. MRN could simply serve to recruit and concentrate the inactive ATM–Tip60 complex at DSBs, and thus stabilize the interaction between Tip60 and H3K9me3 at DSBs. This would lead to activation of Tip60's acetyltransferase activity and acetylation and activation of ATM kinase activity. However, purified MRN can activate ATM in an *in vitro* biochemical system which appears to lack Tip60,[38,49] indicating that MRN plays an active role in upregulating ATM kinase activity. We favor the following model to account for activation of ATM by both Tip60 and MRN. First, the binding of the MRN complex to DNA may alter the conformation of the MRN complex, a process that may involve the ATPase activity of the rad50 subunit.[38] This leads to recruitment of ATM and Tip60 to MRN, either as a single ATM–Tip60 complex or as separate proteins. Interaction between MRN and ATM then provides the initial activation of ATM's kinase activity. Tip60 activation through interaction with adjacent H3K9me3 then leads to acetylation of the PRD domain of ATM. This acetylation then fixes ATM in the active conformation, allowing ATM to maintain kinase activity even after dissociation from MRN. It is also possible that interaction between MRN and ATM alters ATM's conformation to promote direct acetylation by Tip60. In purified systems, although MRN activates ATM, the absence of the normal mechanism for ATM inactivation (such as phosphatases and histone deacetylases[116]) may eliminate the normal *in vivo* requirement for acetylation of ATM by Tip60. In this model, both MRN and Tip60 contribute to the creation of the fully active ATM kinase in response to DSBs in cells. However, it should be noted that ATM activation in response to DNA damage is complex and subject to regulation at multiple levels. The contribution of other factors, including the degree of chromatin compaction at breaks,[58] chromatin structural proteins such as HMGN1,[59] and the recently described ATM regulatory protein ATMIN,[55,56] are all likely to make important contributions. In addition, ATM activation can occur independently of DSB production. For example, direct oxidation of ATM during oxidative stress activates ATM by a mechanism which was independent of both MRN and DSBs.[117] Further, ATM activation is implicated in signaling by retinoic acid[118] and insulin signaling,[119,120] underscoring the diverse range of non-DNA damage mechanisms which can mediate full activation of ATM. Unraveling the complexity of ATM activation will continue to provide insight into the regulation of this pivotal protein kinase.

IV. H3K9me3 and DDR

The critical role for the interaction between H3K9me3 and Tip60 in regulating DSBs raises several intriguing questions about the role of H3K9me3 in the overall DDR. Foremost among these is to determine how the chromodomain of Tip60 can locate and interact with H3K9me3 at DSBs. Two broad mechanisms can be proposed. First, Tip60 may utilize H3K9me3, which is already present on nucleosomes adjacent to the DSB. Alternatively, nucleosomes at the break may undergo *de novo* methylation of H3K9 to create H3K9me3 interaction sites for Tip60. Here, we discuss the evidence for these two alternate mechanisms.

A. Tip60 Activation by Preexisting H3K9me3

If Tip60 utilizes preexisting H3K9me3 for activation of its acetyltransferase activity, this would require that H3K9me3 be evenly distributed across the entire chromatin to ensure correct Tip60 activation. Otherwise, chromatin domains lacking significant levels of H3K9me3 would fail to activate Tip60 and consequently fail to activate ATM's kinase activity. In fact, genome-wide ChIP studies indicate that H3K9me3 is predominantly located in the compacted, gene-poor heterochromatic regions of the chromatin,[104,105,121] where it is associated with members of the HP1 protein family.[122] However, H3K9me3 has also been detected in non-heterochromatic regions,[123] indicating that H3K9me3 is not entirely restricted to heterochromatic regions. However, despite the presence of H3K9me3 in non-heterochromatin domains, there are likely to be large chromatin domains (megabases in size) which lack detectable H3K9me3. In this case, DSBs generated in regions lacking

H3K9me3 may not fully activate the ATM–Tip60 pathway for repair of DSBs. Of course, it may be that ATM activation in some chromatin domains is less dependent on H3K9me3 and Tip60 than others, or that Tip60 may interact with other histone modifications, such as H3K36me3 or H3K9me2 (identified as *in vitro* regulators of Tip60's acetyltransferase activity[51]), which would extend the regions of the chromatin containing potentially activating modifications. However, it has been shown that specific loss of H3K9me3 blocks activation of both Tip60 and ATM[51]—an observation that is at odds with the proposal that ATM-Tip60 can be activated in regions lacking H3K9me3. Additional studies indicate that ATM activation can be detected with as little as 0.1 Gy, and that it was maximal at 0.4 Gy.[29] At these doses, fewer than 10 DSBs would be expected per cell. Further, introducing a single DSB with the I-SceI enzyme can fully activate ATM.[29] The chance that creating these low numbers of DSBs would fortuitously always occur in regions of high H3K9me3 density, allowing for ATM activation, seems unlikely. Overall, the restricted distribution of H3K9me3[104,105,121] and the ability of a single DSB to fully activate ATM[29] indicate that Tip60 activation is unlikely to rely solely on the presence of preexisting H3K9me3 on the chromatin.

B. Tip60 Activation by Inducible H3K9 Methylation

An alternative, and more likely mechanism, is that DSBs promote the *de novo* trimethylation of H3K9 on chromatin domains surrounding DSBs. A key implication of this hypothesis is that DSBs would need to promote the accumulation of H3K9me3 methyltransferases in order to increase H3K9me3 density. As activation of ATM and Tip60 are essentially maximal within minutes after DNA damage,[29,51] this recruitment of H3K9 methyltransferases and methylation of histone H3 on lysine 9 would need to occur extremely rapidly (seconds to minutes). However, published studies indicate that the global levels of H3K9me3 are not significantly altered after DNA damage,[51,124] implying that methylation of H3K9 is not regulated by DNA damage. Detecting changes in H3K9 methylation at DSBs against the normal background of H3K9me3 will be challenging. Current approaches using either Western blot techniques or immunofluorescent staining with antibodies[51,124] likely lack the required sensitivity to detect small changes in histone modifications at DSBs because of the normal background levels of these modifications in cells. More sophisticated techniques, such as recently developed ChIP-based assays,[5,85] will be required to address this issue. In fact, two recent reports have demonstrated that histone methylation is detectably increased at DSBs. In the first study, H3K36 methylation was detectably increased immediately adjacent to the DSB by the Metnase methyltransferase.[125] In addition, Pei *et al.* demonstrated methylation of H4K20 by the MMSET methyltransferase,[126] leading to increased levels of H4K20me2 on the chromatin at DSBs. Both these studies indicate that dynamic regulation of histone methylation is a key function of the DDR, and lend support to the idea that H3K9 methylation may be increased at DSBs. Additional studies will be required to determine if H3K9me3 levels are altered at DSBs and to identify the specific methyltransferases involved in this process. Further, the manner in which potential H3K9 methyltransferases are recruited to DSBs, including the role of key DDR proteins such as H2AX, ATM, and mdc1 in this process, remains to be determined.

C. A Special Role for Heterochromatin in ATM Function?

H3K9me3 is predominantly located in the heterochromatin, where it is binds to HP1 proteins. It is therefore of interest to consider how the binding of H3K9me3 to HP1 may impact the ability of Tip60 to associate with the same modification at DSBs.

Chromatin can be broadly divided into two types of structures. Euchromatin represents open, gene-rich, transcriptionally active regions containing histones, which are highly acetylated and are methylated on lysines 4 and 36 of histone H3.[105,107] In contrast, heterochromatin, which constitutes 15–25% of mammalian chromatin, represents condensed regions with low gene density but high levels of repetitive sequences.[121,127] Heterochromatin also has low levels of histone acetylation but high levels of histone H3 methylated on lysines 9 and 36.[105,121,123,128] Heterochromatin is maintained in this repressed state through binding of specific heterochromatin-binding proteins, including HP1α, HP1β, and HP1γ, which bind directly to H3K9me2/3.[129,130] This promotes the recruitment of other repressive factors, including methyltransferases and histone deacetylase, which function together to promote silencing.[131,132] The high density of H3K9me3 within heterochromatin suggests that Tip60 may be preferentially activated by DSBs generated within heterochromatin, which would be consistent with reports that DSB repair in heterochromatin requires ATM.[8] However, as discussed in the previous section, there is no evidence to support the idea that ATM is only activated within specific chromatin domains.

The majority of the H3K9me3 in heterochromatin is associated with HP1. In this situation, HP1 and Tip60 may compete for binding to H3K9me3, with the potential that occupation of H3K9me3 sites by HP1 may preclude Tip60 binding and therefore negatively impact the DDR within heterochromatin. Recent work has pointed to a potential mechanism for removing HP1 from the chromatin at DSBs. This work demonstrated that DNA damage leads to the phosphorylation of the chromodomain of HP1β by the CKII kinase.[133] Phosphorylation of HP1's chromodomain blocks the binding of HP1 to H3K9me3, leading to the rapid release of HP1β from the chromatin.[51,133] This suggested a simple mechanism in which DNA damage leads to the rapid phosphorylation and release of HP1β from H3K9me3 at DSBs.[51,133] This mechanism would therefore generate domains of H3K9me3 which would then be available to interact with Tip60's chromodomain. However, this simple idea is more complex than originally thought. A recent study reported that HP1 proteins are actively recruited to sites of DNA damage, and that this recruitment was independent of their ability to associate with H3K9me3.[134,135] Instead, HP1 proteins were localized to the damaged chromatin via their chromoshadow domain. Indeed, the original work detailing release of HP1 from the chromatin after damage noted that the ejected HP1 rebound to H3K9me3 at later times.[133] The dynamics of HP1 binding to H3K9me3 are therefore complex and subject to regulation at multiple levels.

Recent work indicates that, in addition to HP1, ATM can phosphorylate kap1, a previously identified heterochromatin-binding protein.[136] Phosphorylation of kap1 by ATM is required for global relaxation of chromatin structure[136] and for DSB repair within heterochromatin.[8,137] DSB repair was demonstrably slower in heterochromatic regions, and unrepaired DSBs persisted at heterochromatin boundaries in the absence of either ATM or kap-1 phosphorylation[8,137] (reviewed in Ref. 7). The key conclusion from these studies is that phosphorylation of kap-1 by ATM opens up the heterochromatin structure and that this process promotes both access to and repair of DSBs. Interestingly, it is yet not clear how phosphorylation of kap1 impacts heterochromatin structure as phospho-kap1 remains associated with the chromatin.[8] Phosphorylation may alter kap1 association with other repressive proteins, such as HDACs and methyltransferases,[131,132,138] shifting the balance toward a less repressed chromatin structure. However, given that two sets of heterochromatin-binding proteins, the HP1 family and the kap1 repressor, are phosphorylated in response to DNA damage and are either released or relocated on the chromatin,[8,51,133,134,136,137] it is clear that reorganization of heterochromatin and changes in accessibility to H3K9me3 are fundamental aspects of the cells' immediate response to DSBs. Reorganization of the chromatin and, in particular, altering the interactions between HP1, H3K9me3, and kap1, may introduce flexibility into the chromatin and increase access of Tip60 to preexisting H3K9me3, as well as promoting *de novo* methylation of H3K9me3. Probing ATM activation and function within clearly defined heterochromatin domains will continue to provide key insights into this process.

D. Chromatin Dynamics and ATM Activation

Figure 1 outlines a general scheme for activation of ATM by Tip60, taking into account contributions from HP1, the MRN complex, H3K9me3, and the Tip60 acetyltransferase. Generation of DSBs leads to the rapid recruitment of the MRN complex to the break. In addition, HP1 associated with adjacent

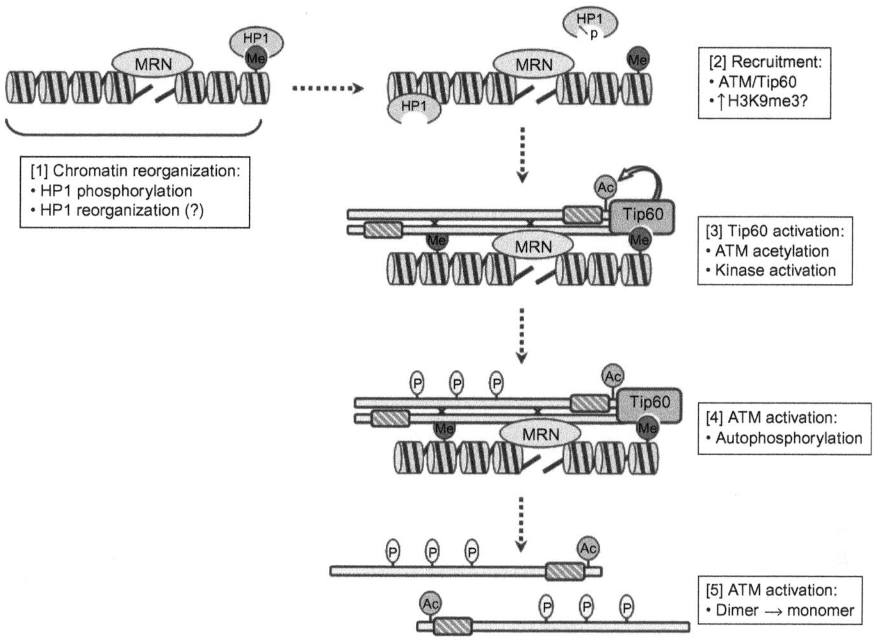

FIG. 1. *A mechanism for ATM activation.* [1] Following DSB production, MRN is recruited to DSB. [2] HP1 proteins are released from H3K9me3, and either retained on the chromatin or released to the nucleoplasm by a process involving phosphorylation of HP1 by the CK2 kinase.[133] It is not known whether MRN participates in this process. Inducible methylation of H3K9 by unknown methyltransferases may also occur at this point. [3] The inactive ATM–Tip60 complex is recruited to the DSB by MRN, facilitating interactions between Tip60's chromodomain and H3K9me3. This activates Tip60 acetyltransferase activity, leading to acetylation of the PRD of ATM. [4] Interaction between MRN and ATM, in combination with acetylation of ATM by Tip60, activates ATM's kinase activity, promoting autophosphorylation of ATM. [5] Separation of inactive ATM dimers into the fully active ATM monomers. Me, methylation; Ac, acetylation; P, phosphorylation; MRN, Mre11–Rad50–nbs1; HP1, heterochromatin-binding protein. (See Color Insert.)

H3K9me3 is phosphorylated and either ejected from the chromatin or relocated to nearby chromatin locations. This primary reorganization of the local chromatin architecture will create domains of H3K9me3 free of associated HP1. In addition, there is the potential that DSBs also lead to the inducible methylation of H3K9 through recruitment of specific methyltransferases, which may be particularly important for chromatin regions which normally lack this modification, such as euchromatin. However, these early steps in the DSB repair response are poorly characterized and further work will be required to unravel the complexity of these issues. In particular, it is important to resolve the issue of how HP1 binding is altered during repair, and to determine if this process is mediated by the MRN complex. In addition, determining if H3K9me3 levels are increased at DSB and identifying the methyltransferase involved are of primary importance. Finally, comparing ATM activation by DSBs in heterochromatin and euchromatin will provide key information concerning the role of reorganization of heterochromatin in the DDR response.

Following the reorganization of the chromatin, the ATM–Tip60 complex is targeted to the chromatin through interaction with the MRN (Fig. 1). This promotes interaction between Tip60's chromodomain and H3K9me3 on adjacent nucleosomes, activating Tip60's acetyltransferase activity. Tip60 then acetylates the PRD domain of ATM, promoting activation of ATM's kinase, a process which also requires input from the MRN complex. Subsequently, ATM undergoes autophosphorylation on multiple sites, promoting conversion from the inactive dimer to the fully active monomeric form. As a significant fraction of active ATM is mobile within the nucleoplasm, ATM acetylation may also function to maintain ATM in the fully active, monomeric form, preventing inactivation of ATM by phosphatases and HDACs. The continued exploration of the complexities of ATM regulation will provide new insights into the molecular events that occur on the chromatin in response to DSBs.

V. Conclusions and Implications for Cancer Therapy

Tip60 functions in many chromatin-dependent transactions, including DSB repair, transcription, and regulation of the androgen receptor. The binding of Tip60's chromodomain to H3K9me3 may therefore be a common mechanism to regulate Tip60 activity, indicating that the chromodomain is the key transducer of epigenetic information on gene promoters regulated by Tip60. Further, Tip60 is known to be inactivated in breast[82] and other cancers.[78–81] Loss of Tip60 would be predicted to impact ATM activation, leading to impaired DSB repair and increasing the risk of a mutagenic event. Further, as Tip60 activation relies on H3K9me3, alterations in global histone methylation patterns in tumor cells may impact Tip60 activity and its ability to regulate DNA repair. In fact, studies have demonstrated altered or decreased levels of H3K9me2/3 in prostate, pancreatic, lung, breast, and other cancers.[139–144] Further, genetic inactivation of several histone methyltransferases, including Dot1, Suv4-20h, and the H3K9 methyltransferases Suv39h1 and Suv39h2, leads to aberrant histone methylation, genomic instability, checkpoint activation, and aberrant DNA repair.[113,145–148] Taken together, these studies indicate that alterations in the global levels of histone methylation, including changes in H3K9me3 levels, can lead to increased genomic instability and may be a contributing factor to the etiology and progression of cancer. In particular, reduced levels of H3K9me3 in tumor cells may provide a mechanism for limiting ATM activity and allowing the tumor to maintain or tolerate increased genomic instability. Further, it implies that DNA damage occurring in regions of low/absent H3K9me3 may be repaired with lower efficiency than in regions of high H3K9me3 density. Overall, this indicates that histone methylation patterns may play a key role in controlling both genomic stability and the sensitivity of tumor cells to DNA damage.

Finally, histone methylation provides the potential for novel therapeutic strategies to be employed against tumors. For example, tumors with reduced levels of H3K9me3 and impaired ATM activation may be more sensitive to specific types of chemotherapy or to radiotherapy. Epigenetic therapies aimed at inhibiting H3K9 methyltransferases (to decrease H3K9me3) or H3K9me3 demethylases (to increase H3K9 methylation) may be effective at sensitizing tumor cells to radiation therapy and/or protecting normal tissue from radiation damage. Understanding the dynamics of histone methylation, and the crucial role that epigenetic markers play in DSB repair and ATM activation, will provide new therapeutic approaches for the prevention and treatment of cancer.

Acknowledgments

We thank A. D'Andrea and members of the Price laboratory for the critical discussions and reading of the manuscript. This work was supported by grants from the NCI (CA64585 and CA93602) and the DOD Breast Cancer Program to B. D. P.

References

1. Jackson SP, Bartek J. The DNA-damage response in human biology and disease. *Nature* 2009;**461**:1071–8.
2. Lieber MR. The mechanism of double-strand DNA break repair by the nonhomologous DNA end-joining pathway. *Annu Rev Biochem* 2010;**79**:181–211.
3. Kass EM, Jasin M. Collaboration and competition between DNA double-strand break repair pathways. *FEBS Lett* 2010;**584**:3703–8.
4. Heyer WD, Ehmsen KT, Liu J. Regulation of homologous recombination in eukaryotes. *Annu Rev Genet* 2010;**44**:113–39.
5. Xu Y, Price BD. Chromatin dynamics and the repair of DNA double strand breaks. *Cell Cycle* 2011;**10**:261–7.
6. Sun Y, Jiang X, Price BD. Tip60: connecting chromatin to DNA damage signaling. *Cell Cycle* 2010;**9**:930–6.
7. Goodarzi AA, Noon AT, Jeggo PA. The impact of heterochromatin on DSB repair. *Biochem Soc Trans* 2009;**37**:569–76.
8. Goodarzi AA, Noon AT, Deckbar D, Ziv Y, Shiloh Y, Lobrich M, et al. ATM signaling facilitates repair of DNA double-strand breaks associated with heterochromatin. *Mol Cell* 2008;**31**:167–77.
9. Lavin MF. Ataxia-telangiectasia: from a rare disorder to a paradigm for cell signalling and cancer. *Nat Rev Mol Cell Biol* 2008;**9**:759–69.

10. Taylor AM, Byrd PJ. Molecular pathology of ataxia telangiectasia. *J Clin Pathol* 2005;**58**:1009–15.
11. Taylor AM, Harnden DG, Arlett CF, Harcourt SA, Lehmann AR, Stevens S, et al. Ataxia telangiectasia: a human mutation with abnormal radiation sensitivity. *Nature* 1975;**258**:427–9.
12. Lavin MF, Birrell G, Chen P, Kozlov S, Scott S, Gueven N. ATM signaling and genomic stability in response to DNA damage. *Mutat Res* 2005;**569**:123–32.
13. Meyn MS. Ataxia-telangiectasia, cancer and the pathobiology of the ATM gene. *Clin Genet* 1999;**55**:289–304.
14. Savitsky K, Sfez S, Tagle DA, Ziv Y, Sartiel A, Collins FS, et al. The complete sequence of the coding region of the ATM gene reveals similarity to cell cycle regulators in different species. *Hum Mol Genet* 1995;**4**:2025–32.
15. Savitsky K, Bar-Shira A, Gilad S, Rotman G, Ziv Y, Vanagaite L, et al. A single ataxia telangiectasia gene with a product similar to PI-3 kinase. *Science* 1995;**268**:1749–53.
16. Lempiainen H, Halazonetis TD. Emerging common themes in regulation of PIKKs and PI3Ks. *EMBO J* 2009;**28**:3067–73.
17. Gilad S, Khosravi R, Shkedy D, Uziel T, Ziv Y, Savitsky K, et al. Predominance of null mutations in ataxia-telangiectasia. *Hum Mol Genet* 1996;**5**:433–9.
18. Byrd PJ, McConville CM, Cooper P, Parkhill J, Stankovic T, McGuire GM, et al. Mutations revealed by sequencing the 5' half of the gene for ataxia telangiectasia. *Hum Mol Genet* 1996;**5**:145–9.
19. Canman CE, Lim DS, Cimprich KA, Taya Y, Tamai K, Sakaguchi K, et al. Activation of the ATM kinase by ionizing radiation and phosphorylation of p53. *Science* 1998;**281**:1677–9.
20. Matsuoka S, Ballif BA, Smogorzewska A, McDonald 3rd ER, Hurov KE, Luo J, et al. ATM and ATR substrate analysis reveals extensive protein networks responsive to DNA damage. *Science* 2007;**316**:1160–6.
21. Bonner WM, Redon CE, Dickey JS, Nakamura AJ, Sedelnikova OA, Solier S, et al. Gamma-H2AX and cancer. *Nat Rev Cancer* 2008;**8**:957–67.
22. Rogakou EP, Boon C, Redon C, Bonner WM. Megabase chromatin domains involved in DNA double-strand breaks in vivo. *J Cell Biol* 1999;**146**:905–16.
23. Stucki M, Clapperton JA, Mohammad D, Yaffe MB, Smerdon SJ, Jackson SP. MDC1 directly binds phosphorylated histone H2AX to regulate cellular responses to DNA double-strand breaks. *Cell* 2005;**123**:1213–26.
24. Stucki M, Jackson SP. gammaH2AX and MDC1: anchoring the DNA-damage-response machinery to broken chromosomes. *DNA Repair (Amst)* 2006;**5**:534–43.
25. Huen MS, Grant R, Manke I, Minn K, Yu X, Yaffe MB, et al. RNF8 transduces the DNA-damage signal via histone ubiquitylation and checkpoint protein assembly. *Cell* 2007;**131**: 901–14.
26. Kolas NK, Chapman JR, Nakada S, Ylanko J, Chahwan R, Sweeney FD, et al. Orchestration of the DNA-damage response by the RNF8 ubiquitin ligase. *Science* 2007;**318**:1637–40.
27. Sobhian B, Shao G, Lilli DR, Culhane AC, Moreau LA, Xia B, et al. RAP80 targets BRCA1 to specific ubiquitin structures at DNA damage sites. *Science* 2007;**316**:1198–202.
28. Cortez D, Wang Y, Qin J, Elledge SJ. Requirement of ATM-dependent phosphorylation of brca1 in the DNA damage response to double-strand breaks. *Science* 1999;**286**:1162–6.
29. Bakkenist CJ, Kastan MB. DNA damage activates ATM through intermolecular autophosphorylation and dimer dissociation. *Nature* 2003;**421**:499–506.
30. Kozlov SV, Graham ME, Peng C, Chen P, Robinson PJ, Lavin MF. Involvement of novel autophosphorylation sites in ATM activation. *EMBO J* 2006;**25**:3504–14.
31. Kozlov SV, Graham ME, Jakob B, Tobias F, Kijas AW, Tanuji M, et al. Autophosphorylation and ATM activation: additional sites add to the complexity. *J Biol Chem* 2011;**286**:9107–19.

32. Sun Y, Xu Y, Roy K, Price BD. DNA damage-induced acetylation of lysine 3016 of ATM activates ATM kinase activity. *Mol Cell Biol* 2007;**27**:8502–9.
33. Daniel JA, Pellegrini M, Lee JH, Paull TT, Feigenbaum L, Nussenzweig A. Multiple autophosphorylation sites are dispensable for murine ATM activation in vivo. *J Cell Biol* 2008;**183**: 777–83.
34. Pellegrini M, Celeste A, Difilippantonio S, Guo R, Wang W, Feigenbaum L, et al. Autophosphorylation at serine 1987 is dispensable for murine Atm activation in vivo. *Nature* 2006;**443**: 222–5.
35. Eilam R, Peter Y, Elson A, Rotman G, Shiloh Y, Groner Y, et al. Selective loss of dopaminergic nigro-striatal neurons in brains of Atm-deficient mice. *Proc Natl Acad Sci USA* 1998;**95**:12653–6.
36. Xu Y, Ashley T, Brainerd EE, Bronson RT, Meyn MS, Baltimore D. Targeted disruption of ATM leads to growth retardation, chromosomal fragmentation during meiosis, immune defects, and thymic lymphoma. *Genes Dev* 1996;**10**:2411–22.
37. Dupre A, Boyer-Chatenet L, Gautier J. Two-step activation of ATM by DNA and the Mre11-Rad50-Nbs1 complex. *Nat Struct Mol Biol* 2006;**13**:451–7.
38. Lee JH, Paull TT. ATM activation by DNA double-strand breaks through the Mre11-Rad50-Nbs1 complex. *Science* 2005;**308**:551–4.
39. Lee JH, Paull TT. Activation and regulation of ATM kinase activity in response to DNA double-strand breaks. *Oncogene* 2007;**26**:7741–8.
40. Uziel T, Lerenthal Y, Moyal L, Andegeko Y, Mittelman L, Shiloh Y. Requirement of the MRN complex for ATM activation by DNA damage. *EMBO J* 2003;**22**:5612–21.
41. Difilippantonio S, Celeste A, Fernandez-Capetillo O, Chen HT, Reina San Martin B, Van Laethem F, et al. Role of Nbs1 in the activation of the Atm kinase revealed in humanized mouse models. *Nat Cell Biol* 2005;**7**:675–85.
42. Cerosaletti K, Wright J, Concannon P. Active role for nibrin in the kinetics of atm activation. *Mol Cell Biol* 2006;**26**:1691–9.
43. Falck J, Coates J, Jackson SP. Conserved modes of recruitment of ATM, ATR and DNA-PKcs to sites of DNA damage. *Nature* 2005;**434**:605–11.
44. Stracker TH, Petrini JHJ. The MRE11 complex: starting from the ends. *Nat Rev Mol Cell Biol* 2011;**12**:90–103.
45. Xie A, Kwok A, Scully R. Role of mammalian Mre11 in classical and alternative nonhomologous end joining. *Nat Struct Mol Biol* 2009;**16**:814–8.
46. Taylor AM, Groom A, Byrd PJ. Ataxia-telangiectasia-like disorder (ATLD)—its clinical presentation and molecular basis. *DNA Repair (Amst)* 2004;**3**:1219–25.
47. McKinnon PJ, Caldecott KW. DNA strand break repair and human genetic disease. *Annu Rev Genomics Hum Genet* 2007;**8**:37–55.
48. You Z, Chahwan C, Bailis J, Hunter T, Russell P. ATM activation and its recruitment to damaged DNA require binding to the C terminus of Nbs1. *Mol Cell Biol* 2005;**25**:5363–79.
49. Lee JH, Paull TT. Direct activation of the ATM protein kinase by the Mre11/Rad50/Nbs1 complex. *Science* 2004;**304**:93–6.
50. Lim DS, Kim ST, Xu B, Maser RS, Lin J, Petrini JH, et al. ATM phosphorylates p95/nbs1 in an S-phase checkpoint pathway. *Nature* 2000;**404**:613–7.
51. Sun Y, Jiang X, Xu Y, Ayrapetov MK, Moreau LA, Whetstine JR, et al. Histone H3 methylation links DNA damage detection to activation of the tumour suppressor Tip60. *Nat Cell Biol* 2009;**11**:1376–82.
52. Kitagawa R, Bakkenist CJ, McKinnon PJ, Kastan MB. Phosphorylation of SMC1 is a critical downstream event in the ATM-NBS1-BRCA1 pathway. *Genes Dev* 2004;**18**:1423–38.
53. Horejsi Z, Falck J, Bakkenist CJ, Kastan MB, Lukas J, Bartek J. Distinct functional domains of Nbs1 modulate the timing and magnitude of ATM activation after low doses of ionizing radiation. *Oncogene* 2004;**23**:3122–7.

54. Kim ST, Xu B, Kastan MB. Involvement of the cohesin protein, Smc1, in Atm-dependent and independent responses to DNA damage. *Genes Dev* 2002;**16**:560–70.
55. Kanu N, Behrens A. ATMINistrating ATM signalling: regulation of ATM by ATMIN. *Cell Cycle* 2008;**7**:3483–6.
56. Kanu N, Behrens A. ATMIN defines an NBS1-independent pathway of ATM signalling. *EMBO J* 2007;**26**:2933–41.
57. Soutoglou E, Misteli T. Activation of the cellular DNA damage response in the absence of DNA lesions. *Science* 2008;**320**:1507–10.
58. Murga M, Jaco I, Fan Y, Soria R, Martinez-Pastor B, Cuadrado M, et al. Global chromatin compaction limits the strength of the DNA damage response. *J Cell Biol* 2007;**178**:1101–8.
59. Kim YC, Gerlitz G, Furusawa T, Catez F, Nussenzweig A, Oh KS, et al. Activation of ATM depends on chromatin interactions occurring before induction of DNA damage. *Nat Cell Biol* 2009;**11**:92–6.
60. Sun Y, Jiang X, Chen S, Fernandes N, Price BD. A role for the Tip60 histone acetyltransferase in the acetylation and activation of ATM. *Proc Natl Acad Sci USA* 2005;**102**:13182–7.
61. Squatrito M, Gorrini C, Amati B. Tip60 in DNA damage response and growth control: many tricks in one HAT. *Trends Cell Biol* 2006;**16**:433–42.
62. Fazzio TG, Huff JT, Panning B. An RNAi screen of chromatin proteins identifies Tip60-p400 as a regulator of embryonic stem cell identity. *Cell* 2008;**134**:162–74.
63. Carrozza MJ, Utley RT, Workman JL, Cote J. The diverse functions of histone acetyltransferase complexes. *Trends Genet* 2003;**19**:321–9.
64. Marmorstein R. Structure of histone acetyltransferases. *J Mol Biol* 2001;**311**:433–44.
65. Kimura A, Horikoshi M. Tip60 acetylates six lysines of a specific class in core histones in vitro. *Genes Cells* 1998;**3**:789–800.
66. Kusch T, Florens L, Macdonald WH, Swanson SK, Glaser RL, Yates Iii JR, et al. Acetylation by Tip60 is required for selective histone variant exchange at DNA lesions. *Science* 2004;**306**:2084–7.
67. Bird AW, Yu DY, Pray-Grant MG, Qiu Q, Harmon KE, Megee PC, et al. Acetylation of histone H4 by Esa1 is required for DNA double-strand break repair. *Nature* 2002;**419**:411–5.
68. Doyon Y, Selleck W, Lane WS, Tan S, Cote J. Structural and functional conservation of the NuA4 histone acetyltransferase complex from yeast to humans. *Mol Cell Biol* 2004;**24**: 1884–96.
69. Gaughan L, Logan IR, Cook S, Neal DE, Robson CN. Tip60 and histone deacetylase 1 regulate androgen receptor activity through changes to the acetylation status of the receptor. *J Biol Chem* 2002;**277**:25904–13.
70. Tang Y, Luo J, Zhang W, Gu W. Tip60-dependent acetylation of p53 modulates the decision between cell-cycle arrest and apoptosis. *Mol Cell* 2006;**24**:827–39.
71. Sykes SM, Mellert HS, Holbert MA, Li K, Marmorstein R, Lane WS, et al. Acetylation of the p53 DNA-binding domain regulates apoptosis induction. *Mol Cell* 2006;**24**:841–51.
72. Lin YY, Lu JY, Zhang J, Walter W, Dang W, Wan J, et al. Protein acetylation microarray reveals that NuA4 controls key metabolic target regulating gluconeogenesis. *Cell* 2009;**136**:1073–84.
73. Sun Y, Jiang X, Chen S, Price BD. Inhibition of histone acetyltransferase activity by anacardic acid sensitizes tumor cells to ionizing radiation. *FEBS Lett* 2006;**580**:4353–6.
74. Lehner B, Crombie C, Tischler J, Fortunato A, Fraser AG. Systematic mapping of genetic interactions in Caenorhabditis elegans identifies common modifiers of diverse signaling pathways. *Nat Genet* 2006;**38**:896–903.
75. Hu Y, Fisher JB, Koprowski S, McAllister D, Kim MS, Lough J. Homozygous disruption of the Tip60 gene causes early embryonic lethality. *Dev Dyn* 2009;**238**:2912–21.
76. Ikura T, Ogryzko VV, Grigoriev M, Groisman R, Wang J, Horikoshi M, et al. Involvement of the TIP60 histone acetylase complex in DNA repair and apoptosis. *Cell* 2000;**102**:463–73.

77. Murr R, Loizou JI, Yang YG, Cuenin C, Li H, Wang ZQ, et al. Histone acetylation by Trrap-Tip60 modulates loading of repair proteins and repair of DNA double-strand breaks. *Nat Cell Biol* 2006;**8**:91–9.
78. Me LL, Vidal F, Gallardo D, Diaz-Fuertes M, Rojo F, Cuatrecasas M, et al. New p53 related genes in human tumors: significant downregulation in colon and lung carcinomas. *Oncol Rep* 2006;**16**:603–8.
79. Halkidou K, Gnanapragasam VJ, Mehta PB, Logan IR, Brady ME, Cook S, et al. Expression of Tip60, an androgen receptor coactivator, and its role in prostate cancer development. *Oncogene* 2003;**22**:2466–77.
80. Mattera L, Escaffit F, Pillaire MJ, Selves J, Tyteca S, Hoffmann JS, et al. The p400/Tip60 ratio is critical for colorectal cancer cell proliferation through DNA damage response pathways. *Oncogene* 2009;**28**:1506–17.
81. Sakuraba K, Yasuda T, Sakata M, Kitamura YH, Shirahata A, Goto T, et al. Down-regulation of Tip60 gene as a potential marker for the malignancy of colorectal cancer. *Anticancer Res* 2009;**29**:3953–5.
82. Gorrini C, Squatrito M, Luise C, Syed N, Perna D, Wark L, et al. Tip60 is a haplo-insufficient tumour suppressor required for an oncogene-induced DNA damage response. *Nature* 2007;**448**:1063–7.
83. Doyon Y, Cote J. The highly conserved and multifunctional NuA4 HAT complex. *Curr Opin Genet Dev* 2004;**14**:147–54.
84. Downs JA, Allard S, Jobin-Robitaille O, Javaheri A, Auger A, Bouchard N, et al. Binding of chromatin-modifying activities to phosphorylated histone H2A at DNA damage sites. *Mol Cell* 2004;**16**:979–90.
85. Xu Y, Sun Y, Jiang X, Ayrapetov MK, Moskwa P, Yang S, et al. The p400 ATPase regulates nucleosome stability and chromatin ubiquitination during DNA repair. *J Cell Biol* 2010;**191**:31–43.
86. Ikura T, Tashiro S, Kakino A, Shima H, Jacob N, Amunugama R, et al. DNA damage-dependent acetylation and ubiquitination of H2AX enhances chromatin dynamics. *Mol Cell Biol* 2007;**27**:7028–40.
87. Jha S, Shibata E, Dutta A. Human Rvb1/Tip49 is required for the histone acetyltransferase activity of Tip60/NuA4 and for the downregulation of phosphorylation on H2AX after DNA damage. *Mol Cell Biol* 2008;**28**:2690–700.
88. Li H, Balajee AS, Su T, Cen B, Hei TK, Weinstein IB. The HINT1 tumor suppressor regulates both gamma-H2AX and ATM in response to DNA damage. *J Cell Biol* 2008;**183**:253–65.
89. Eymin B, Claverie P, Salon C, Leduc C, Col E, Brambilla E, et al. p14ARF activates a Tip60-dependent and p53-independent ATM/ATR/CHK pathway in response to genotoxic stress. *Mol Cell Biol* 2006;**26**:4339–50.
90. Jiang X, Sun Y, Chen S, Roy K, Price BD. The FATC domains of PIKK proteins are functionally equivalent and participate in the Tip60-dependent activation of DNA-PKcs and ATM. *J Biol Chem* 2006;**281**:15741–6.
91. Abraham RT. PI 3-kinase related kinases: 'big' players in stress-induced signaling pathways. *DNA Repair (Amst)* 2004;**3**:883–7.
92. Bosotti R, Isacchi A, Sonnhammer EL. FAT: a novel domain in PIK-related kinases. *Trends Biochem Sci* 2000;**25**:225–7.
93. Mordes DA, Glick GG, Zhao R, Cortez D. TopBP1 activates ATR through ATRIP and a PIKK regulatory domain. *Genes Dev* 2008;**22**:1478–89.
94. Mordes DA, Cortez D. Activation of ATR and related PIKKs. *Cell Cycle* 2008;**7**:2809–12.
95. Beamish HJ, Jessberger R, Riballo E, Priestley A, Blunt T, Kysela B, et al. The C-terminal conserved domain of DNA-PKcs, missing in the SCID mouse, is required for kinase activity. *Nucleic Acids Res* 2000;**28**:1506–13.

96. Priestley A, Beamish HJ, Gell D, Amatucci AG, Muhlmann-Diaz MC, Singleton BK, et al. Molecular and biochemical characterisation of DNA-dependent protein kinase-defective rodent mutant irs-20. *Nucleic Acids Res* 1998;**26**:1965–73.
97. Takahashi T, Hara K, Inoue H, Kawa Y, Tokunaga C, Hidayat S, et al. Carboxyl-terminal region conserved among phosphoinositide-kinase-related kinases is indispensable for mTOR function in vivo and in vitro. *Genes Cells* 2000;**5**:765–75.
98. Sekulic A, Hudson CC, Homme JL, Yin P, Otterness DM, Karnitz LM, et al. A direct linkage between the phosphoinositide 3-kinase-AKT signaling pathway and the mammalian target of rapamycin in mitogen-stimulated and transformed cells. *Cancer Res* 2000;**60**:3504–13.
99. Lemercier C, Legube G, Caron C, Louwagie M, Garin J, Trouche D, et al. Tip60 acetyltransferase activity is controlled by phosphorylation. *J Biol Chem* 2003;**278**:4713–8.
100. Wang J, Chen J. SIRT1 regulates autoacetylation and histone acetyltransferase activity of TIP60. *J Biol Chem* 2010;**285**:11458–64.
101. Taverna SD, Li H, Ruthenburg AJ, Allis CD, Patel DJ. How chromatin-binding modules interpret histone modifications: lessons from professional pocket pickers. *Nat Struct Mol Biol* 2007;**14**:1025–40.
102. Bernstein BE, Meissner A, Lander ES. The mammalian epigenome. *Cell* 2007;**128**:669–81.
103. Kouzarides T. Chromatin modifications and their function. *Cell* 2007;**128**:693–705.
104. Regha K, Sloane MA, Huang R, Pauler FM, Warczok KE, Melikant B, et al. Active and repressive chromatin are interspersed without spreading in an imprinted gene cluster in the mammalian genome. *Mol Cell* 2007;**27**:353–66.
105. Barski A, Cuddapah S, Cui K, Roh TY, Schones DE, Wang Z, et al. High-resolution profiling of histone methylations in the human genome. *Cell* 2007;**129**:823–37.
106. Heintzman ND, Stuart RK, Hon G, Fu Y, Ching CW, Hawkins RD, et al. Distinct and predictive chromatin signatures of transcriptional promoters and enhancers in the human genome. *Nat Genet* 2007;**39**:311–8.
107. Guenther MG, Levine SS, Boyer LA, Jaenisch R, Young RA. A chromatin landmark and transcription initiation at most promoters in human cells. *Cell* 2007;**130**:77–88.
108. Mikkelsen TS, Ku M, Jaffe DB, Issac B, Lieberman E, Giannoukos G, et al. Genome-wide maps of chromatin state in pluripotent and lineage-committed cells. *Nature* 2007;**448**:553–60.
109. Bhaumik SR, Smith E, Shilatifard A. Covalent modifications of histones during development and disease pathogenesis. *Nat Struct Mol Biol* 2007;**14**:1008–16.
110. Trievel RC. Structure and function of histone methyltransferases. *Crit Rev Eukaryot Gene Expr* 2004;**14**:147–69.
111. Shi Y, Whetstine JR. Dynamic regulation of histone lysine methylation by demethylases. *Mol Cell* 2007;**25**:1–14.
112. Whetstine JR, Nottke A, Lan F, Huarte M, Smolikov S, Chen Z, et al. Reversal of histone lysine trimethylation by the JMJD2 family of histone demethylases. *Cell* 2006;**125**:467–81.
113. Peters AH, O'Carroll D, Scherthan H, Mechtler K, Sauer S, Schofer C, et al. Loss of the Suv39h histone methyltransferases impairs mammalian heterochromatin and genome stability. *Cell* 2001;**107**:323–37.
114. Robert F, Hardy S, Nagy Z, Baldeyron C, Murr R, Dery U, et al. The transcriptional histone acetyltransferase cofactor TRRAP associates with the MRN repair complex and plays a role in DNA double-strand break repair. *Mol Cell Biol* 2006;**26**:402–12.
115. Chailleux C, Tyteca S, Papin C, Boudsocq F, Puget N, Courilleau C, et al. Physical interaction between the histone acetyl transferase Tip60 and the DNA double-strand breaks sensor MRN complex. *Biochem J* 2010;**426**:365–71.
116. Goodarzi AA, Jonnalagadda JC, Douglas P, Young D, Ye R, Moorhead GB, et al. Autophosphorylation of ataxia-telangiectasia mutated is regulated by protein phosphatase 2A. *EMBO J* 2004;**23**:4451–61.

117. Guo Z, Kozlov S, Lavin MF, Person MD, Paull TT. ATM activation by oxidative stress. *Science* 2010;**330**:517–21.
118. Fernandes ND, Sun Y, Price BD. Activation of the kinase activity of ATM by retinoic acid is required for CREB-dependent differentiation of neuroblastoma cells. *J Biol Chem* 2007;**282**:16577–84.
119. Yang DQ, Halaby MJ, Li Y, Hibma JC, Burn P. Cytoplasmic ATM protein kinase: an emerging therapeutic target for diabetes, cancer and neuronal degeneration. *Drug Discov Today* 2011;**16**:332–8.
120. Halaby MJ, Hibma JC, He J, Yang DQ. ATM protein kinase mediates full activation of Akt and regulates glucose transporter 4 translocation by insulin in muscle cells. *Cell Signal* 2008;**20**:1555–63.
121. Peng JC, Karpen GH. Epigenetic regulation of heterochromatic DNA stability. *Curr Opin Genet Dev* 2008;**18**:204–11.
122. Cheutin T, McNairn AJ, Jenuwein T, Gilbert DM, Singh PB, Misteli T. Maintenance of stable heterochromatin domains by dynamic HP1 binding. *Science* 2003;**299**:721–5.
123. Vakoc CR, Sachdeva MM, Wang H, Blobel GA. Profile of histone lysine methylation across transcribed mammalian chromatin. *Mol Cell Biol* 2006;**26**:9185–95.
124. Tjeertes JV, Miller KM, Jackson SP. Screen for DNA-damage-responsive histone modifications identifies H3K9Ac and H3K56Ac in human cells. *EMBO J* 2009;**28**:1878–89.
125. Fnu S, Williamson EA, De Haro LP, Brenneman M, Wray J, Shaheen M, et al. Methylation of histone H3 lysine 36 enhances DNA repair by nonhomologous end-joining. *Proc Natl Acad Sci USA* 2011;**108**:540–5.
126. Pei H, Zhang L, Luo K, Qin Y, Chesi M, Fei F, et al. MMSET regulates histone H4K20 methylation and 53BP1 accumulation at DNA damage sites. *Nature* 2011;**470**:124–8.
127. de Wit E, van Steensel B. Chromatin domains in higher eukaryotes: insights from genome-wide mapping studies. *Chromosoma* 2009;**118**:25–36.
128. Pokholok DK, Harbison CT, Levine S, Cole M, Hannett NM, Lee TI, et al. Genome-wide map of nucleosome acetylation and methylation in yeast. *Cell* 2005;**122**:517–27.
129. Jacobs SA, Khorasanizadeh S. Structure of HP1 chromodomain bound to a lysine 9-methylated histone H3 tail. *Science* 2002;**295**:2080–3.
130. Nielsen PR, Nietlispach D, Mott HR, Callaghan J, Bannister A, Kouzarides T, et al. Structure of the HP1 chromodomain bound to histone H3 methylated at lysine 9. *Nature* 2002;**416**: 103–7.
131. Li X, Lin HH, Chen H, Xu X, Shih HM, Ann DK. SUMOylation of the transcriptional co-repressor KAP1 is regulated by the serine and threonine phosphatase PP1. *Sci Signal* 2010;**3**:ra32.
132. Zeng L, Yap KL, Ivanov AV, Wang X, Mujtaba S, Plotnikova O, et al. Structural insights into human KAP1 PHD finger-bromodomain and its role in gene silencing. *Nat Struct Mol Biol* 2008;**15**:626–33.
133. Ayoub N, Jeyasekharan AD, Bernal JA, Venkitaraman AR. HP1-beta mobilization promotes chromatin changes that initiate the DNA damage response. *Nature* 2008;**453**:682–6.
134. Luijsterburg MS, Dinant C, Lans H, Stap J, Wiernasz E, Lagerwerf S, et al. Heterochromatin protein 1 is recruited to various types of DNA damage. *J Cell Biol* 2009;**185**:577–86.
135. Ball Jr. AR, Yokomori K. Revisiting the role of heterochromatin protein 1 in DNA repair. *J Cell Biol* 2009;**185**:573–5.
136. Ziv Y, Bielopolski D, Galanty Y, Lukas C, Taya Y, Schultz DC, et al. Chromatin relaxation in response to DNA double-strand breaks is modulated by a novel ATM- and KAP-1 dependent pathway. *Nat Cell Biol* 2006;**8**:870–6.
137. Noon AT, Shibata A, Rief N, Lobrich M, Stewart GS, Jeggo PA, et al. 53BP1-dependent robust localized KAP-1 phosphorylation is essential for heterochromatic DNA double-strand break repair. *Nat Cell Biol* 2010;**12**:177–84.

138. Li X, Lee YK, Jeng JC, Yen Y, Schultz DC, Shih HM, et al. Role for KAP1 serine 824 phosphorylation and sumoylation/desumoylation switch in regulating KAP1-mediated transcriptional repression. *J Biol Chem* 2007;**282**:36177–89.
139. Park YS, Jin MY, Kim YJ, Yook JH, Kim BS, Jang SJ. The global histone modification pattern correlates with cancer recurrence and overall survival in gastric adenocarcinoma. *Ann Surg Oncol* 2008;**15**:1968–76.
140. Northcott PA, Nakahara Y, Wu X, Feuk L, Ellison DW, Croul S, et al. Multiple recurrent genetic events converge on control of histone lysine methylation in medulloblastoma. *Nat Genet* 2009;**41**:465–72.
141. Fraga MF, Ballestar E, Villar-Garea A, Boix-Chornet M, Espada J, Schotta G, et al. Loss of acetylation at Lys16 and trimethylation at Lys20 of histone H4 is a common hallmark of human cancer. *Nat Genet* 2005;**37**:391–400.
142. Manuyakorn A, Paulus R, Farrell J, Dawson NA, Tze S, Cheung-Lau G, et al. Cellular histone modification patterns predict prognosis and treatment response in resectable pancreatic adenocarcinoma: results from RTOG 9704. *J Clin Oncol* 2010;**28**:1358–65.
143. Ellinger J, Kahl P, von der Gathen J, Rogenhofer S, Heukamp LC, Gutgemann I, et al. Global levels of histone modifications predict prostate cancer recurrence. *Prostate* 2010;**70**:61–9.
144. Seligson DB, Horvath S, McBrian MA, Mah V, Yu H, Tze S, et al. Global levels of histone modifications predict prognosis in different cancers. *Am J Pathol* 2009;**174**:1619–28.
145. Giannattasio M, Lazzaro F, Plevani P, Muzi-Falconi M. The DNA damage checkpoint response requires histone H2B ubiquitination by Rad6-Bre1 and H3 methylation by Dot1. *J Biol Chem* 2005;**280**:9879–86.
146. Wysocki R, Javaheri A, Allard S, Sha F, Cote J, Kron SJ. Role of Dot1-dependent histone H3 methylation in G1 and S phase DNA damage checkpoint functions of Rad9. *Mol Cell Biol* 2005;**25**:8430–43.
147. Tardat M, Murr R, Herceg Z, Sardet C, Julien E. PR-Set7-dependent lysine methylation ensures genome replication and stability through S phase. *J Cell Biol* 2007;**179**:1413–26.
148. Sakaguchi A, Steward R. Aberrant monomethylation of histone H4 lysine 20 activates the DNA damage checkpoint in Drosophila melanogaster. *J Cell Biol* 2007;**176**:155–62.

Exploiting Synthetic Lethal Interactions Between DNA Damage Signaling, Checkpoint Control, and p53 for Targeted Cancer Therapy

SANDRA MORANDELL AND
MICHAEL B. YAFFE

Departments of Biology and Biological Engineering, David H. Koch Institute for Integrative Cancer Research, Massachusetts Institute of Technology, Cambridge, Massachusetts, USA

DNA damage signaling and checkpoint control pathways are among the most commonly mutated networks in human tumors. Emerging data suggest that synthetic lethal interactions between mutated oncogenes or tumor suppressor genes with molecules involved in the DNA damage response and DNA repair pathways can be therapeutically exploited to preferentially kill cancer cells. In this review, we discuss the concept of synthetic lethality with a focus on *p53*, a commonly lost tumor suppressor gene, in the context of DNA damage signaling. We describe several recent examples in which this concept was successfully applied to target tumor cells in culture or in mouse models, as well as in human cancer patients.

Progress in Molecular Biology
and Translational Science, Vol. 110
DOI: 10.1016/B978-0-12-387665-2.00011-0

I. Introduction

A. Targeting Cancer Genes in the Context of DNA Damage

Genomic instability is a critical component of tumor development and progression from in situ lesions to invasive cancers. As a consequence, cancer cells acquire the ability to tolerate an increased amount of DNA damage compared to noncancer cells. This is often achieved by dampening one or more DNA damage repair pathways combined with suppression of DNA damage signaling networks that control cell cycle arrest in the presence of genotoxic stress.[1] These mutated or partially compromised DNA damage pathways, while directly contributing to cancer development, also furnish an "Achilles' heel" for anticancer therapy.

Many cytotoxic chemotherapies as well as ionizing radiation therapy used in anticancer treatment target rapidly growing cells by introducing lesions into their chromosomal DNA, or by inhibiting DNA replication. Even though cancer cells show an enhanced sensitivity to these agents, the targets of these conventional cancer drugs are present in both normal and cancer cells. It is well accepted that the off-target effects of DNA damage in nontumor cells (the so-called bystander effect) is responsible for most dose-limiting toxicities.[2] As therapeutic success in killing cancer cells depends on separating tumor toxicity from normal tissue toxicity, the use of additional agents that specifically affect other signaling or repair pathways that may be compromised in tumors can be exploited to achieve synthetic lethal effects. These synthetic lethal effects can be due to inherent underlying mutations within the cancer cells, or they can be induced by specifically targeting pathways which tumor cells have become dependent on (i.e., poly(ADP-ribose) polymerase (PARP) in BRCA1/2-defective tumors). Thus, novel targeted anticancer therapies can be developed by focusing on the oncogenic context of cancer cells, and making use of unique tumor properties that are not shared by normal tissue.[3] Current clinical problems such as the limited success of specific chemotherapeutic agents in only a small subset of cancer types and the evolution of drug resistance highlight the need for treatments with wider therapeutic windows.

During tumor development, cancer cells accumulate a variety of genetic lesions. As a consequence of these dynamic changes in the genome and the consequent rewiring of their signaling networks, cancer cells acquire properties such as unlimited replicative potential, self-sufficiency in growth signals, insensitivity to antiproliferative and apoptotic signals, and the potential to sustain angiogenesis as well as to invade tissue and metastasize.[4] These typical cancer phenotypes are often caused, at least in part, by either the gain of function of tumor oncogenes or by the loss of tumor suppressor genes.[5]

The acquired mutations often affect a subset of conserved pathways, including DNA damage repair and cell cycle control, in addition to apoptosis, integrin signaling, regulation of cell adhesion and invasion, small GTPase-dependent signaling, TGFβ-signaling, Wnt/Notch signaling, and Hedgehog signaling.[6] As these pathways become rewired and subject to new mechanisms of regulation, they become promising targets for anticancer therapy.

Some of the critical "driving" genetic lesions within these oncogenic signaling pathways can be targeted with small molecules. Inhibition of mutant or amplified ErbB receptors (Erlotinib, Gefitinib, Trastuzumab)[7,8] or BCR-Abl (Imatinib),[9] for example, is among the most successful strategies for targeting critical signaling components in cancer cells. Other classes of oncogenes, especially nonkinases such as Myc or Ras, have proven to be less "druggable."[10,11] In addition, therapeutic restoration of commonly lost tumor suppressor genes, including *p53*, *Rb*, *BRCA1/2*, *p16*^{INK4A}, and *ATM*, has not yet become a feasible option for anticancer therapy. Importantly, many of these genes play a role in the DNA damage response and DNA repair pathways, which are among the most frequently compromised networks in human cancers.[12–20] Therefore, targeting oncogenic signaling pathways in combination with DNA damage treatments may offer improved efficacy compared to treatment with pathway inhibitors alone.

B. The DNA Damage Response

In response to DNA damage, cells activate complex signaling networks that mediate DNA repair and cell cycle arrest, or if the damage is extensive, they trigger apoptosis[2,21] (Fig. 1A). The DNA damage response is initiated by the activation of the PI(3)K (phosphatidylinositol-3-OH-kinase)-like kinases ATM (ataxia telangiectasia mutated), ATR (ATM and Rad related), and DNA-PKcs (DNA-dependent protein kinase catalytic subunit). These kinases recruit repair machinery directly to sites of DNA damage while also halting progression through the cell cycle by activating the effector kinases Chk2 and Chk1.[22–24] While the ATM/Chk2 pathway responds primarily to DNA double-strand breaks (DSBs), the ATR/Chk1 module is activated by exposure of DNA single strands at breaks or sites of bulky DNA base adducts. Recently, we have identified the stress-activated p38MAPK/MK2 (MAPKAPK2) pathway as a third checkpoint regulator that is activated downstream of ATM and ATR upon DNA damage.[25,26] The networks regulated by these checkpoint effector kinases affect a variety of cellular outcomes including cell cycle arrest, DNA repair, chromatin assembly, transcriptional and posttranscriptional regulation of gene expression, and cell death[2,27] (Fig. 1A).

The activation of ATM/ATR-Chk1/Chk2/MK2-controlled checkpoints leads to a delay in cell cycle progression through the G1, S, or G2 phase (Fig. 1B). The transcription factor p53 is a major effector of DNA kinase pathways[28] and mediates arrest in G1 mainly through the upregulation of

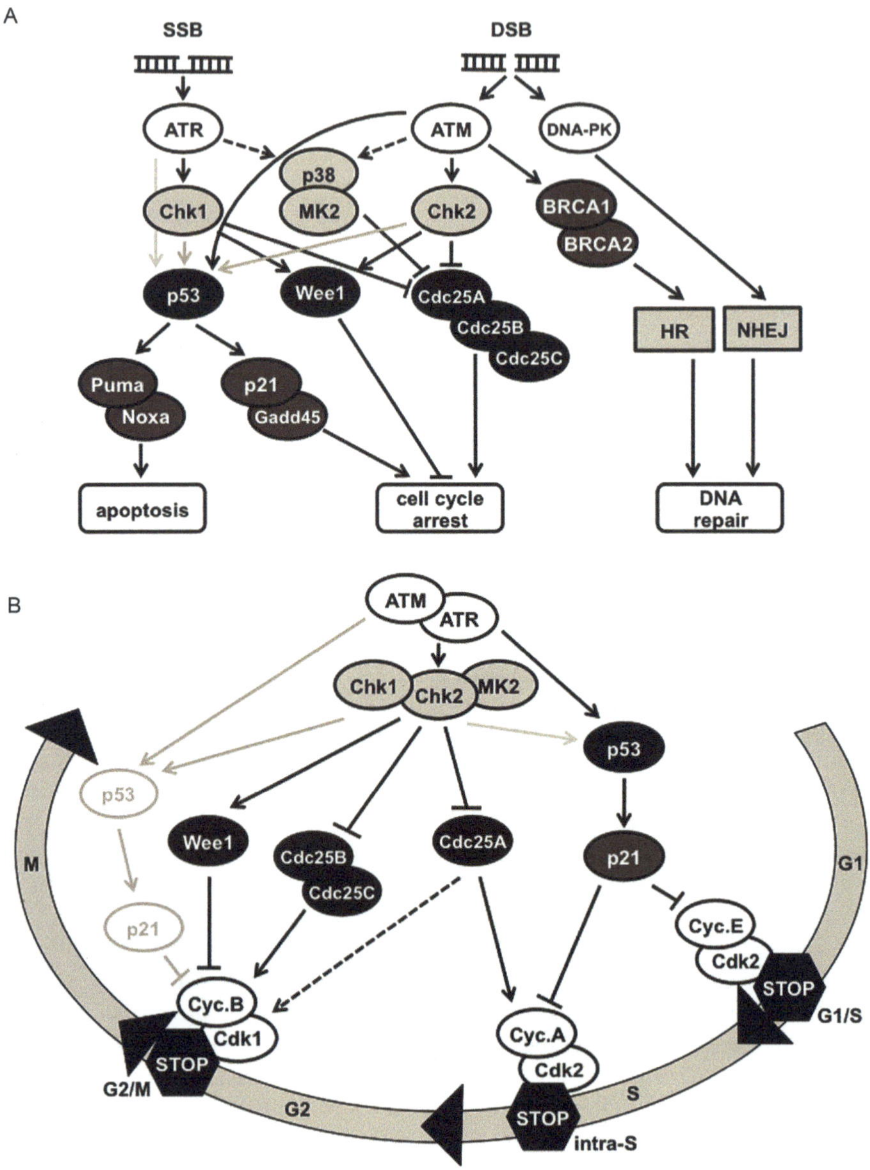

FIG. 1. DNA damage signaling, checkpoint control, and the role of p53. (A) DNA damage signaling pathways. In response to genotoxic stress by chemotherapeutic agents, cells activate complex signaling networks that control DNA repair and cell cycle arrest or initiate apoptosis if the damage is extensive. DNA double-strand breaks (DSBs) primarily activate ATM, which in turn phosphorylates its downstream effectors Chk2 and p53. Induction of cell cycle-regulatory p53 target genes like *p21* or *Gadd45α* results in cell cycle arrest, while the activation of proapoptotic p53

p21CIP1/WAF1.[29] p21 inhibits the G1/S-promoting cyclin E/Cdk2 kinase and, as a consequence, the ability for DNA synthesis, that is, the duplication of damaged DNA. Alternatively, if the DNA damage is extensive, p53 triggers apoptotic cell death through intrinsic and extrinsic pathways, by activating target genes such as *Puma* and *Noxa*[30,31] (Fig. 1A). In addition to p53, checkpoints are controlled by negatively regulating the Cdc25 family of phosphatases, direct substrates of the effector kinases Chk1 and Chk2, as well as MK2.[24] Cdc25 proteins regulate G1/S and G2/M transitions through the removal of inhibitory phosphorylations on Cyclin/Cdk complexes (Fig. 1B). During the intra-S checkpoint, phosphorylation of Cdc25A by Chk1 leads to rapid degradation of the phosphatase by the 26S proteasome.[21,32–34] Cdc25B and Cdc25C are inhibited to enforce the G2/M checkpoint, where phosphorylation by DDR effector kinases leads to 14-3-3 binding in the cytosol, resulting in reduced catalytic activity and sequestration away from their substrates, the Cyclin/Cdk complexes.[32,35,36] The increased selectivity of cancer cells to DNA-damaging agents indicates that they have rewired their signaling networks and established new dependencies between cell cycle checkpoint and survival pathways that do not exist, or are less pronounced, in normal cells. The G1/S boundary, for example, is the predominant checkpoint upon DNA damage in normal mammalian cells. Most tumor cells, however, show a loss of function in the tumor suppressor gene *p53* or its regulatory pathways,[29] leading to a selective loss of this G1 checkpoint. Therefore, cancer cells are often completely dependent on the S and G2/M checkpoints to arrest cell cycle after genotoxic

---

target genes *Puma* and *Noxa* following DNA damage is responsible for cell death. The ATR/Chk1 module is activated by DNA single-strand breaks (SSBs) and bulky lesions. The p38MAPK/MK2 pathway is an additional checkpoint regulator that is activated downstream of both ATM and ATR upon DNA damage and is especially important in the context of p53 deficiency. Targets of the effector kinases Chk1, Chk2, and MK2 include Cdc25 phosphatases and Wee1 kinase, which are regulators of cell cycle arrest. BRCA1/2 are components of the homologous recombination (HR) repair pathway. DNA-PKcs is a third PI(3)K-like kinase mainly involved in the activation of DNA damage repair pathways including nonhomologous end-joining (NHEJ). (B) Cell cycle checkpoint mechanisms. Genotoxic and replicative stress activate checkpoints to delay progression from one cell cycle phase to the next. Activation of p53 by DNA damage response kinases mediates arrest at the G1 (Gap1)/S (synthesis) and intra-S transitions, mainly through the upregulation of p21. p21 in turn inhibits the cell cycle promoting cyclin/Cdk kinase complexes. In addition, checkpoints are regulated by Cdc25 phosphatases, which are inhibited by the effector kinases Chk1, Chk2, and MK2. Cdc25 proteins drive S-phase and G2 (Gap2)/M (Mitosis) entry through removal of inhibitory phosphorylations on Cdk/Cyclin complexes. The Chk1 target Wee1 is responsible for generating these inhibitory phosphorylations on Cdk kinases. Many tumor cells are defective in p53 function, leading to a selective loss of the G1/S checkpoint. Therefore, cancer cells are often dependent on the intra-S and G2/M checkpoints to arrest the cell cycle after genotoxic stress. For this reason, interference with these two checkpoints has become an interesting and promising strategy to sensitize G1-deficient cancer cells to DNA-damaging therapy.

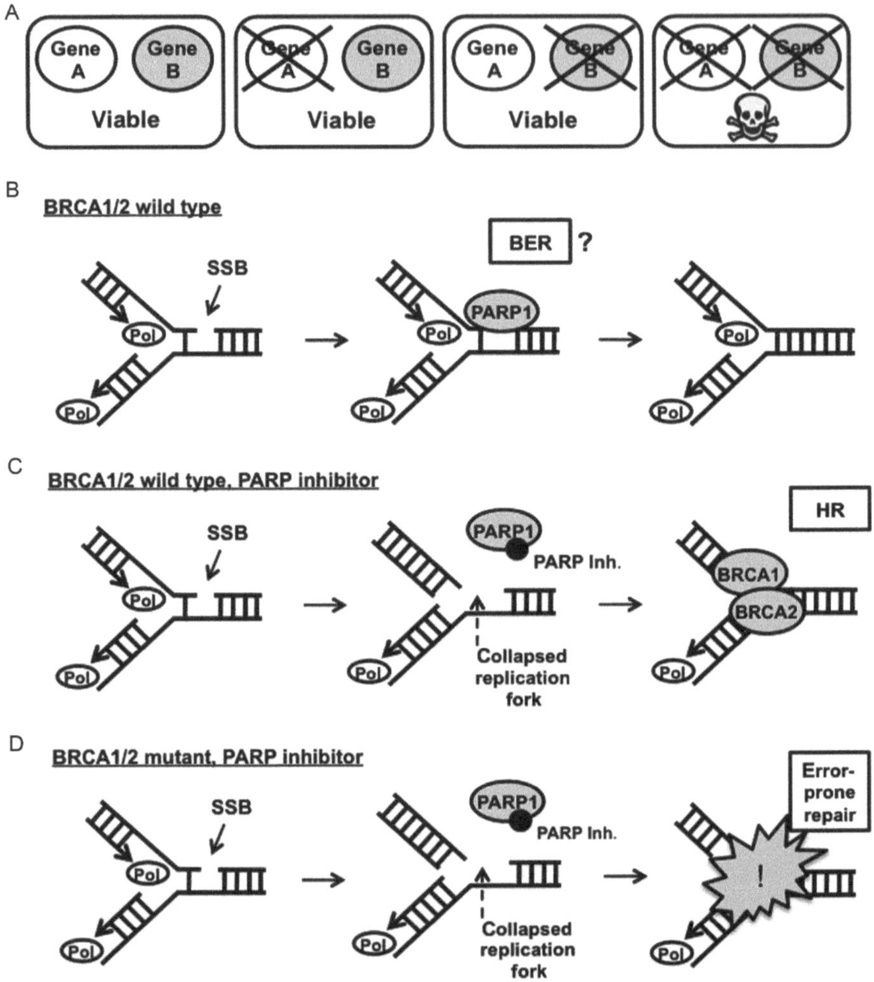

FIG. 2. Synthetic lethality in cancer treatment. (A) The concept of synthetic lethality. Two genes are in a synthetic lethal relationship when combined inactivation results in greater cell death than that predicted from the additive effects of each gene inactivation in isolation. (B)–(D) Exploiting the synthetic lethal relationship between PARP1 and BRCA1/2 for the targeted treatment of HR-deficient human tumors. Homologous recombination (HR) is used as a DNA repair pathway to resolve double-strand breaks (DSBs), for example, those resulting from replication fork collapse. (B) In normal cells, PARP1 is involved in the repair of base modifications using base excision repair (BER) prior to S-phase entry, preventing replication forks from colliding with single-strand breaks. (C) Pharmacological inhibition of the BER component PARP1 results in unrepaired single-strand breaks (SSBs), which lead to DSBs when the replication forks progress during the S-phase. In BRCA1/2-proficient cells, the newly synthesized sister chromatid can serve as a

stress. For this reason, interference with these latter two checkpoints has become an interesting and promising strategy to sensitize G1-deficient cancer cells to DNA damage therapy.

II. The Concept of Synthetic Lethality in Cancer Therapy

Advances in the understanding of the pathways affected by these cancer-specific mutations are beginning to lead to novel therapeutic approaches aimed at killing cancer cells based on the concept of synthetic lethality. Synthetic lethal interactions are defined as two genetic alterations that, when occurring together, cause cell death, while neither mutation alone is lethal (Fig. 2A). This phenomenon was originally revealed through experiments in yeast and fruit flies,[38,39] but variations on this basic concept have clear application to cancer therapy.[40,41] Some of the changes in the molecular nature of cancer cell signaling may serve as sensitizing mutations, leading to a dependency for survival on certain molecules or pathway combinations that are not necessary in normal cells, providing interesting therapeutic opportunities if these pathway combinations can be selectively targeted and/or engaged.

Although synthetic lethality has been classically described for genetic alterations, it can also be applied to other cellular perturbations such as the action of small molecules or environmental factors such as DNA-damaging ionizing radiation. Therefore, pharmacological inhibition of one gene product that is synthetic lethal in combination with a preexisting cancer mutation will selectively target and kill the tumor cell, while healthy cells not carrying the cancer-specific mutation tolerate the treatment remaining largely unharmed. In addition, synthetic lethality-based therapies provide the possibility to exploit treatments that endogenously manipulate molecular targets that are not easily accessible for inhibition by small molecules, in addition to exploiting loss-of-function mutations in tumor suppressor genes to enhance the effects of DNA-damaging agents.[42]

III. Synthetic Lethality Between PARP1 and BRCA1/2 as a Model for Enhancing DNA Damage-Induced Cell Death

The potential for using synthetic lethality in the context of DNA damage signaling as a principle for anticancer therapy has been demonstrated by recent findings on the interaction between *PARP1* and *BRCA1/2*.[43,44] These genes are

template for HR-mediated repair. (D) HR-mediated repair of DSBs induced by PARP inhibition is not available in BRCA1/2-deficient cancer cells. Instead, breaks are resolved by error-prone repair pathways, such as NHEJ and SSA, leading to progressive genomic instability and ultimately cell death as a consequence of mitotic catastrophe (modified from Ref. 37).

critical components of two different DNA damage repair pathways. PARP1 is best appreciated for its role in facilitating base excision repair (BER), a mechanism used to fix chemically induced base mutations. This involves the temporary creation of single-strand DNA breaks and the possible exposure of modest lengths of single-stranded DNA[45] (Fig. 2B). Exactly how PARP1 participates in BER at the molecular level remains unclear, but known substrates of PARP1 include histones and PARP1 itself. BRCA1 and 2 are major components of the error-free homologous recombination (HR) repair for DNA DSBs. Deficiency in BRCA genes leads to high susceptibility for breast and ovarian cancer, as cells become dependent on the repair of DSBs by alternative mechanisms, including nonhomologous end-joining (NHEJ) or single-strand annealing (SSA). These repair mechanisms are error-prone, which ultimately results in increased genomic instability.[46] Studies by Bryant et al. and Farmer et al. demonstrated that the inhibition of PARP1-dependent repair mechanisms ultimately leads to the formation of DNA lesions that are dependent on HR for resolution.[43,44] In normal cells with intact HR mechanisms, this results in only minor survival differences after PARP inhibition (Fig. 2C). In BRCA1- or 2-deficient cells, however, there is a dramatic increase of cell death in response to PARP1 inhibitors caused by the absence of HR activity, most likely resulting in genomic instability that cannot be rescued by alternative repair mechanisms (Fig. 2D). PARP1 function is observed in response to a wide variety of DNA lesions, and the extent to which the synthetic lethal interaction between PARP1 and BRCA1/2 is dependent on BER is not completely clear.

Nonetheless, these findings indicate that the inhibition of PARP1 can and should be used as a therapeutic strategy to target *BRCA1*- or *2*-deficient cancers. For this reason, PARP inhibitors are currently being evaluated in clinical trials; an initial phase I clinical trial using the PARP1 inhibitor Olaparib showed antitumor activity in cancer associated with the *BRCA1* or *2* mutation and reported a few adverse effects in patients.[47] Additional studies with Olaparib confirmed efficacy and tolerability in patients[48] as well as antitumor activity in *BRCA1/2* mutation ovarian cancer, which was associated with increased sensitivity to platinum-based chemotherapy.[49] PARP1 inhibition in a phase II trial has now been shown to result in a favorable therapeutic index for BRCA-deficient breast cancer patients.[50] In addition to Olaparib, a variety of other PARP inhibitors such as Inaparib (BSI-201), AG014699, ABT-888, CEP-9722, and INO-1001 have been developed and are currently being tested in different studies.[51–53] The best clinical results seem to be emerging from the inhibitors that have the lowest IC_{50} values, indicative of strong PARP binding (A. Ashworth, personal communication). At present, there are more than 50 clinical trials ongoing worldwide for different tumor types, many in combination with DNA-damaging chemotherapy.[54]

The results from these trials have to be interpreted carefully, particularly with respect to long-term follow-up. PARP1-deficient mice show a predisposition to epidermal hyperplasia[55] and mammary carcinomas,[56] indicating that long-term treatment with PARP inhibitors might cause severe adverse effects. In addition, studies on BRCA2-deficient cancer cells treated in culture indicate that secondary mutations in the BRCA2 gene restore HR function and, as a consequence, resistance to PARP inhibitors.[57,58] A deeper understanding of the specific role of PARP in cancer cells and the consideration of factors such as the levels of PARP-1 expression in tumors, and the optimal dosing schedule for PARP inhibitors/chemotherapy combinations may improve the selection of patients for PARP inhibitory therapy. Nevertheless, the success of the BRCA/PARP studies has established a paradigm for synthetic lethality as a functionally new concept in anticancer therapy.

IV. Synthetic Lethality in the Context of p53 Mutations

Synthetic lethal interactions not only occur between two genetic mutations or by single-gene mutations in combination with inhibition of other gene products by small molecules, but also become apparent when cells are exposed to environmental factors such as extrinsic cellular stress by DNA-damaging ionizing radiation or cytotoxic chemotherapy. The loss of one tumor suppressor gene alone may not change the survival of cancer cells in response to genotoxic stress as compared to normal cells. Inhibition of a second pathway that engages a synthetic lethal interaction with the lost tumor suppressor, however, might result in a dramatic difference in the chemosensitivity and enhanced chemotherapy/radiation-induced cell death in tumors following DNA damage, but not necessarily under unstressed conditions prior to the damage.

As mentioned earlier, the DNA damage response and DNA repair pathways are among the most commonly perturbed networks in human cancers.[12–20] Several recent studies have therefore identified specific components of the DNA damage response pathway as promising candidates for targeting synthetic lethal interactions with commonly lost tumor suppressor genes such as *p53* in the context of genotoxic therapy. The focus of many of these studies has been on the kinases regulating the intra-S and G2/M checkpoints, as kinases are well established as druggable targets.

A. Synthetic Lethality Between p53 and the ATR/Chk1 Signaling Pathway

The ATR/Chk1 signaling pathway is activated mainly by DNA single-strand breaks (DSSs) and stalled replication forks. These kinases function in the regulation of cell cycle arrest following genotoxic stress, but are also critical components of essentially every S-phase as a consequence of inherent errors that frequently arise at sites of replication, particularly when cells experience oncogene-driven replication stress. ATR and Chk1 prevent new replication origins from firing during the S-phase[59,60] and are involved in the intra-S and G2/M checkpoints where Chk1 phosphorylation of Cdc25A facilitates its proteosomal degradation and Chk1 phosphorylation of Cdc25B/C induces 14-3-3 binding.[32,33,61] In p53-deficient cells, knockdown of Chk1 has been shown to induce loss of both these cell cycle checkpoints.[62]

1. Inhibition of Chk1

Because Chk1 and its upstream activator ATR are essential in mammals,[63–65] functional studies of this pathway have been limited mainly to RNAi-mediated knockdown approaches or the use of inhibitors. These inhibitors are often not specific for the kinase of interest, although more selective inhibitors are currently in development. The largest body of clinical data to date comes from studies with UCN01 (7-hydroxystaurosporine), a molecule originally isolated as an inhibitor of protein kinase C (PKC) and also known to inhibit a number of other kinases including Chk1, MK2, phosphoinositide-dependent kinase-1 (Pdk1), and several cyclin-dependent kinases (Cdks).[26,66–69] In combination with DNA-damaging agents or during replicative stress, UCN-01 causes a bypass of intra-S and G2 checkpoints in cells with defective p53 function,[70–72] which phenocopies the loss of Chk1 using RNAi.[62] As UCN-01 also potently inhibits MK2, some of its efficacy may result from the simultaneous disruption of two critical checkpoint pathways in p53-defective cells.[26]

In phase I clinical trials, UCN-01 used either as a single agent or in combination with several DNA-damaging agents including Irinotecan (CPT-11), an analog of the topoisomerase I inhibitor camptothecin,[73] showed initial promise. Two patients with p53-deficient triple-negative breast cancers responded dramatically to therapy, confirming preclinical data that show a synergistic effect between UCN-01 and DNA-damaging agents in tumor cells with defective p53 function.[70–72] However, due to adverse toxicities and poor pharmacokinetics, UCN-01 has not been further clinically developed.

In recent years, novel inhibitors with a higher specificity for Chk1 have been tested successfully in preclinical models and are currently being evaluated in clinical phase I studies in combination with DNA-damaging agents or antimetabolites, which interfere with DNA replication (reviewed in Ref. 74).[54] These newer inhibitors, including AZD7762, PF477736, and SCH900776, are more selective for Chk1 than UCN-01, although none of them is completely exempt from inhibition of other kinases.

2. A Focus on ATR

Recently, a synthetic lethal interaction between *p53* and *ATR*, the upstream activator of Chk1, has been described using a genetic model system in which a human cancer cell line was modified to carry one of the *ATR* mutations known to cause Seckel syndrome.[75] This mutation results in severely decreased levels of ATR expression due to aberrant splicing.[15,76,77] Cells with defective levels of ATR signaling showed a highly increased sensitivity to many DNA-damaging agents and antimetabolites.[78] The authors of this study demonstrated that loss of ATR expression selectively enhanced the response of *p53*-deficient cells to treatment with the DNA-cross-linking agent cisplatin. In contrast, restoration of p53 function in these ATR-deficient cells suppressed treatment-dependent apoptosis, similar to the chemoprotective effect that was independently observed in cells with functional p53 signaling in an ATM-deficient background.[79]

Using a cell-based high-throughput screen in which inhibition of replication stress was used as a readout for ATR activity, Toledo and colleagues recently identified two potent compounds that abrogate ATR signaling in the nanomolar range.[80] One of these compounds, NVP-BEZ235, had previously been recognized as a dual inhibitor of phosphatidylinositol-3-OH-kinase (PI3K) and mTOR, and is currently being tested in the context of anticancer therapy in clinical trials.[81] Another screen led to the identification of VE-821, a novel compound with strong selectivity for ATR and little cross-reactivity toward other PI(3)K-like kinases including mTOR.[82,83] Tests for the synergy of ATR inhibition in combination with DNA-damaging therapy in a panel of normal and cancer cell lines revealed that the combinatorial treatment was most effective in inducing apoptosis in the context of ATM or p53 deficiency.

To date, no data related to the inhibition of ATR in combination with DNA-damaging agents in mouse models of cancer have been published. However, a recent study focusing on the suppression of ATR in the context of oncogenic Ras transformation[84] highlights the importance of this signaling pathway in the maintenance of genomic stability and raises important concerns about the possible dangers of the use of ATR/Chk1-inhibiting compounds in the clinic. The phenotypic outcome of ATR suppression in the used model systems was highly dose-dependent: haploinsufficiency of *ATR* promoted oncogenic Ras-induced tumorigenesis in *p53*-heterozygous knockout mice due to increased genomic instability. However, abrogation of ATR signaling in cell lines expressing oncogenic Ras increased genomic instability to intolerable levels, resulting in suppressed proliferation and mitotic catastrophe. Haploinsufficiency-induced tumorigenesis has also been described for *Chk1*-heterozygous mice in a breast cancer model,[85] further elevating concerns about the therapeutic value of ATR/Chk1 inhibitors in cancer patients. Consistent with this, human mutations or suppression of the ATR/Chk1 pathway has been linked to different tumor types.[86–88] The fact that disruption of *ATR* or *Chk1* in mice is embryonically lethal[63–65] adds to the concerns about the safety of ATR or Chk1 inhibitors as drugs. Both ATR and Chk1 play a major role during a normal cell cycle by monitoring the integrity of replication forks.[59] Even in the absence of exogenously caused DNA damage, their function is critical to maintain genomic stability, which might cause dangerous side effects in healthy tissues exposed to Chk1 or ATR inhibitors, especially during long-term or repeated treatment. In addition, efficacy of these therapies might be highly dependent on the genetic context of the tumors including their p53 status.

B. Synthetic Lethality Between p53 and ATM/Chk2

Our lab and others have described a similar synthetic lethality between p53 and the ATM/Chk2 pathway.[79] Previous studies in *p53* knockout mouse embryonic fibroblasts (MEFs) suggested that a combined loss of ATM and p53 increased sensitivity to topoisomerase inhibitors when compared to ATM-expressing cells.[89] We observed that loss of either ATM or Chk2 in both human and murine cell lines strongly sensitized p53-deficient cells to DNA-damaging treatment with doxorubicin, resulting in increased cell death.[79] The interaction between ATM/Chk2 and p53 caused the loss of a functional G2/M cell cycle checkpoint and drove cells into mitotic catastrophe.

Surprisingly, however, loss of ATM or Chk2 in p53-proficient cells caused exactly the opposite effect, making the cancer cells resistant to genotoxic stress. Both sets of results—synthetic lethality in p53-defective tumor cells and synthetic viability in p53-proficient tumor cells—were also confirmed *in vivo*. Knockdown of ATM in H-Ras^{V12}-driven, p53-deficient allograft tumors in nude mice increased their sensitivity to chemotherapy, as it did in syngeneic transplants of Eμ-myc-driven p53-defective lymphomas. In contrast, p53-proficient tumors in which ATM was downregulated were more resistant to treatment than control tumors with normal ATM levels in both the xenograft and lymphoma models.[79] This effect could be explained when the expression of p53 target genes upon genotoxic stress was analyzed. While cell cycle checkpoint molecules such as p21 or Gadd45α were normally induced in the absence or presence of ATM, we found that loss of ATM in p53-proficient cells selectively suppressed the upregulation of proapoptotic target genes such as *Noxa* and *Puma*. Notably, the functional cell cycle arrest without the ability to induce cell death in p53 wild-type animals seemed to specifically protect the ATM-deficient p53-expressing cells from doxorubicin treatment (Fig. 3).

The results could be directly compared with genetic data on the ATM/Chk2 and p53 status in human cancer patients. Analysis of the 10-year survival data from a large cohort of breast cancer patients after DNA-damaging chemotherapy supports the observations about the lethal or protective interactions between ATM/Chk2 and p53, showing a critical dependence on the p53 status

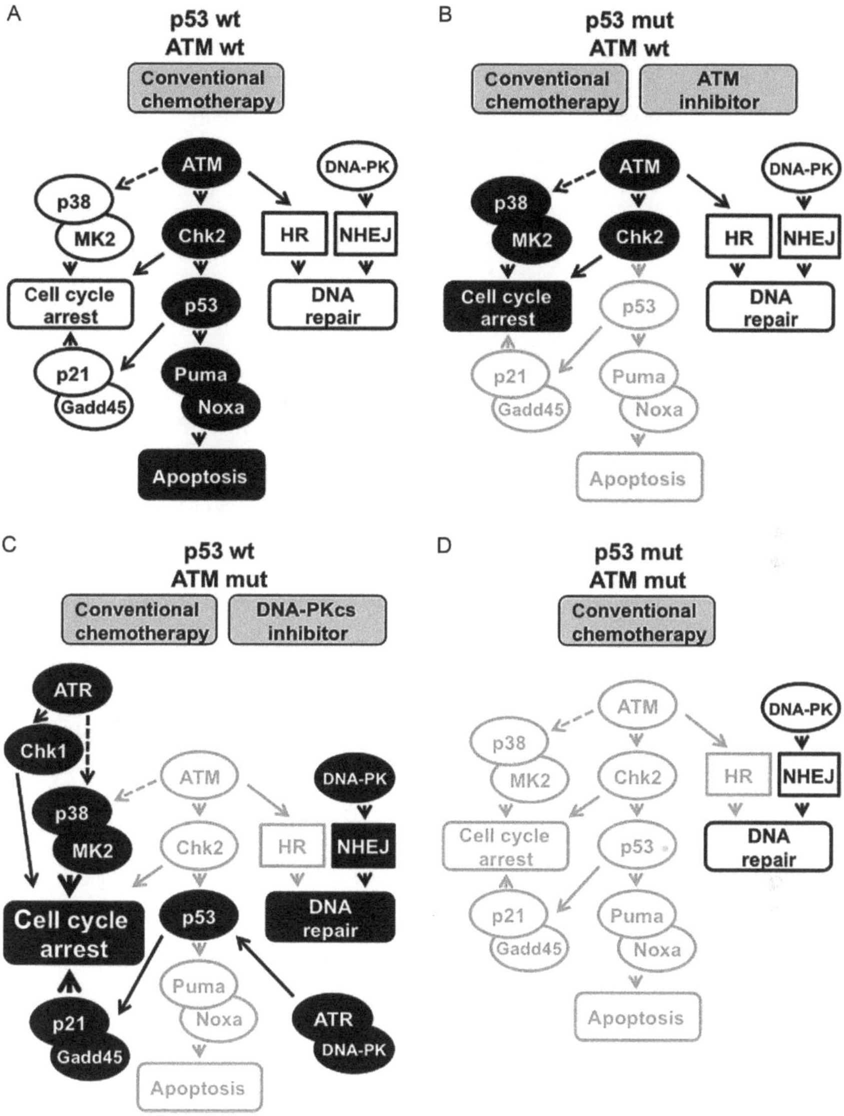

FIG. 3. Exploiting synthetic lethal interactions for the treatment of human cancer. DNA damage signaling networks commonly show extensive rewiring in tumor cells. (A) In p53- and ATM-proficient (wt) cancer cells, DNA-damaging agents induce apoptosis. Conventional DNA-damaging chemotherapy should be effective due to a functional proapoptotic ATM/Chk2-p53-Puma/Noxa signaling pathway. (B) Loss-of-function mutations (mut) of *p53* in cancer cells largely abrogate apoptosis induction in response to genotoxic stress. In these cells, ATM signaling induces a robust cell cycle arrest, and ATM-mediated homologous recombination (HR) repair remains intact

of cells. The combined status of ATM and p53 was found to be a key determinant for survival of patients after treatment: patients with tumors that were deficient for either ATM or p53 alone had a significantly reduced survival compared to patients with p53- and ATM-proficient tumors. Moreover, when tumors had lost ATM in the presence of functional p53, those patients had the poorest prognosis despite equivalent therapy. In contrast, patients with tumors that had lost both ATM and p53 had a strikingly increased survival, although the combination of these two mutations was highly underrepresented.[79] These results again highlight that loss of ATM or p53 in isolation promotes resistance against chemotherapy, while their combined loss or inactivation leads to increased chemosensitivity (Fig. 3).

C. Sensitizing Chemo-Resistant Tumors that Have p53 Function: A Role for DNA-PKcs

The data discussed suggest that inhibition of ATM or Chk2 in a p53-deficient background is a powerful method for sensitizing tumors to DNA-damaging therapy by synthetic lethality. However, targeting of ATM in a p53 wild-type setting causes exactly the opposite phenotype, conferring therapeutic resistance, a phenomenon that was reflected in the decreased patient survival. These data raised the possibility that loss of ATM in a p53-deficient background might lead to new potentially synthetic lethal interaction in the resistant tumors that could be used to kill these cells more efficiently by DNA-damaging therapy. Indeed, the blunted response to chemotherapy

in p53-deficient cancer cells. This redirecting of DNA damage-induced ATM signaling promotes cellular survival in response to DNA damage. Treatment of p53-deficient tumors should include a combination of conventional DNA-damaging chemotherapeutics and ATM inhibitors. (C) Loss of ATM reduces the induction of the proapoptotic p53 target genes *Puma* and *Noxa* following DNA damage. However, induction of cell cycle-regulatory p53 target genes such as *p21* or *Gadd45α* remains intact, allowing ATM-deficient cancer cells to have enhanced cell cycle arrest after genotoxic stress. As these cells are not dead, they can continue to proliferate once the DNA damage is resolved. These tumors, however, depend on the DNA-PKcs-mediated nonhomologous end-joining (NHEJ) pathway to repair chemotherapy-induced DSBs and maintain genomic stability. Abolishing DNA-PKcs signaling in ATM-deficient cells leads to a strong increase in sensitivity to DNA-damaging chemotherapy. Treatment of ATM-depleted tumors with retained p53 function should include a combination of conventional DNA-damaging chemotherapy and DNA-PKcs inhibitors. (D) Cancer cells with a combined loss of p53 and ATM lack functional cell cycle checkpoints in response to genotoxic stress. This inability to stop progression through the cell cycle despite the presence of DNA damage ultimately results in mitotic catastrophe. p53- and ATM-deficient cancer cells should be exquisitely sensitive to treatment with conventional DNA-damaging chemotherapy. However, these cells are likely to be deselected during tumor evolution as a consequence of their presumed sensitivity to oncogene-induced replication stress (modified from Ref. 37).

observed in the ATM-deficient, p53-proficient tumors could be reversed by inhibition of another PI(3)K-like kinase involved in the DNA damage response, DNA-PKcs. This finding could be explained by the distinct roles of ATM and DNA-PKcs in the repair of DNA DSBs. ATM is involved in HR-mediated error-free repair[90–95] while DNA-PKcs is an important component of the more error-prone NHEJ-mediated repair.[96–99] The most likely explanation for the observed synthetic lethality between ATM and DNA-PK in p53-proficient cells was that loss of both kinases eliminated the two major pathways required for DSB repair. In the absence of ATM, the p53-deficient cells were defective in performing HR, and instead exhibited an increased reliance on alternative DSB repair mechanisms, including NHEJ (Fig. 3).

This synthetic lethal interaction between ATM and DNA-PKcs in the presence of chemotherapy-induced DSBs was also confirmed *in vivo*, where suppression of DNA-PKcs was found to resensitize ATM-depleted, p53-proficient Eμ-myc-driven lymphomas to doxorubicin. Interestingly, the inhibition of DNA-PKcs-dependent signaling had no sensitizing effect on ATM wild-type cells or p53-deficient cells, regardless of their ATM status. These observations are consistent with reports that the combined germline deficiency of ATM and DNA-PKcs in mice results in early embryonic lethality.[100] These data strongly support a role for NHEJ as a backup mechanism for DSB repair when HR is lost in p53-proficient cells, analogous to the role of HR in the context of PARP inhibition, as discussed earlier. Whether these observations could ever be translated into clinical practice remains unclear, as small-molecule inhibition of both ATM and DNA-PK would be expected to have very high toxicity. DNA-PK inhibitors, however, might prove very useful in the subset of cancer patients whose tumors have functional p53 but defective ATM signaling as a consequence of genetic mutations or epigenetic silencing.

D. Synthetic Lethality Between p53 and MK2

In addition to the canonical DNA damage signaling modules ATM/Chk2 and ATR/Chk1, genotoxic stress also leads to the activation of the p38/MK2 pathway.[24–26,101–105] Activation of the p38/MK2 complex upon genotoxic stress is dependent on both ATM and ATR[103,106] and functions in parallel and independently of Chk1 or Chk2 activation. MK2 is a direct substrate of p38, a MAPK that plays a central role in cellular stress responses. The p38 signaling pathway is activated by a variety of stimuli besides DNA damage, including cytokines, reactive oxygen species, osmotic stress, heat shock, and various toxins.[107]

Our lab found that the p38/MK2 pathway is critical for arresting the cell cycle after ultraviolet (UV)-induced damage.[25] MK2-depleted cells were defective in both the G1/S and G2/M checkpoints, and they became more sensitive to DNA damage-induced cell death as a consequence of mitotic catastrophe, indicating that MK2 is the crucial downstream effector of p38MAPK in the checkpoint response. To further investigate the role of MK2 in cell survival after genotoxic stress in a more cancer-specific context, studies were performed in cells following cisplatin-induced intrastrand cross-links as well as after DNA breaks induced by topoisomerase inhibitors such as doxorubicin,[26] two agents commonly used in anticancer therapy. Depletion of MK2 in p53-proficient and p53-deficient MEFs revealed that the function of this pathway is especially important for cell cycle arrest and prevention of apoptosis in the context of defective p53 function. MK2 activity was dispensable for survival in p53 wild-type cells after both cisplatin and doxorubicin treatment. In p53-deficient cells, however, loss of MK2 resulted in a markedly increased sensitivity to DNA-damaging chemotherapy, particularly at low doses of these genotoxic agents.

Studies on allograft tumors in mice derived from transformed H-Ras^{V12} and p53-deficient MEFs confirmed these results. When the tumors were examined for chemosensitivity, knockdown of MK2 was found to result in a dramatically increased tumor regression following systemic treatment with cisplatin or doxorubicin *in vivo*. This synthetic lethality between p53 and MK2 could be explained by the fact that p53-deficient cells lacking MK2 activity could initiate, but were unable to maintain, functional G1/S and G2/M cell cycle checkpoints at late times after DNA damage as a consequence of the reduced levels of Cdc25A, B, and C phosphorylation.[26] Furthermore, studies on the subcellular localization of MK2, compared with Chk1, explained the requirement for both effector checkpoint kinases.[27] Chk1 activity arises early after damage and is largely localized to the nuclear compartment. In contrast, MK2 activity is prolonged at times long after Chk1 activity has ceased, and the active form of MK2 is primarily found in the cytoplasm. These results suggested that in p53-proficient cells, damage-induced activation of p53 maintains the cell cycle checkpoint when Chk1 activity fades, and that in the absence of p53, MK2 subsumes this function. These results suggest that MK2 might be an ideal therapeutic target for use in chemosensitizing p53-defective tumors to specific types of chemotherapy. A variety of small-molecule inhibitors of MK2 are in various stages of development, although their use in cancer treatment trials is still unclear.

1. Inhibition of the Chk1 Target Wee1

In addition to the major DNA damage response kinases ATM/Chk2, MK2, and ATR/Chk1, other molecules involved in maintaining the G2/M cell cycle checkpoint arrest have recently been identified as potential druggable targets for a synthetic lethal approach to treat p53-deficient tumors with higher efficiency. One of them is Wee1, a protein kinase activated by Chk1 in DNA damage response and replication stress signaling. Wee1 is directly responsible for the inhibitory phosphorylation of Cdk1 and Cdk2, and, therefore, it is the complementary counterpart to the Cdk/cyclin complex-activating protein phosphatases Cdc25A, B, and C, which remove these phosphorylations.[108–111] Like its upstream kinases ATR and Chk1, Wee1 is essential for mammalian development, and its conditional deletion in mice leads to growth defects and apoptosis due to aneuploidy and DNA damage.[112] Several studies using knock-down of Wee1[113,114] or Wee1-specific inhibitors indicate that loss of Wee1 function is a promising strategy to abrogate the G2/M arrest in combination with DNA-damaging therapy.

Various compounds that inhibit Wee1 have recently been described in the literature, including PD0166285[115–117] and MK-1775, each of which shows a different cross-reactivity to other kinases (reviewed in Ref. 118). The more selective compound MK-1775 has successfully been used in combination with DNA-damaging agents in mouse xenograf models.[119–122] Importantly, the efficiency of checkpoint abrogation appears to correlate with *p53* deficiency in many of the tumor types and human cell lines tested. Its clinical application is currently being tested in phase I and phase II studies. Preliminary results show good tolerability and promising anticancer activity.[123]

V. Potential for Future Therapies

A. Novel Players in the Genotoxic Stress Response: Posttranscriptional Regulation of Selective mRNAs Upon DNA Damage

Many of the changes in cellular mechanisms that occur upon DNA damage, such as delay in cell cycle progression, activation and recruitment of the DNA damage repair machinery, and others, can be explained by posttranslational modifications of proteins within the DNA damage signaling network through phosphorylation, ubiquitinylation, or sumoylation.[23] Recently, however, the posttranslational control of mRNA stability, splicing, and translation in response to genotoxic stress has emerged as an additional aspect of regulation in the cellular DNA damage response.

DNA damage triggers a transient repression of transcriptional activity in eukaryotic cells.[124,125] Despite this transcriptional block, cells up- and down-regulate a large portion of their mRNAs following exposure to a variety of DNA-damaging agents.[126,127] A major contributor to these changes in gene expression is the posttranscriptional regulation of mRNA levels. Global analysis of changes in mRNA levels upon UV irradiation of lung carcinoma cells, for example, showed that mRNA turnover is responsible for more than half of the observed regulatory events.[128] Stability, splicing, and translation of mRNA are all regulated by RNA-binding proteins (RBPs). Through direct interaction with their target mRNAs, these RBPs control nuclear export, half-life, accessibility by the translation machinery, and spliceosome targeting.[101,129–132] Recently, several independent screens for DNA damage modulators identified RBPs as important components of the DNA damage response. In a phospho-proteomic mass spectrometry screen searching for direct targets of the DDR kinases ATM, ATR, and DNA-PKcs in response to ionizing radiation, molecules involved in mRNA regulation constituted a large subset of the identified kinase substrates.[133] Moreover, when looking for spontaneous DNA damage in a genome-wide RNAi screen, mRNA-binding proteins and molecules involved in RNA processing were the most prominent group of identified factors.[134] Another genome-wide siRNA screen that was performed to search for novel players in DNA damage signaling in response to ionizing radiation also identified proteins regulating RNA posttranslational modifications as one of the most significant "hits."[135]

Our understanding of how these mRNA regulatory factors are connected to DNA damage signaling pathways is in its infancy. It is likely that the activity of RBPs is directly regulated by kinases such as those implicated by the screen by Matsuoka et al. for ATM and ATR.[133] Our lab recently identified RBPs as a major target for the p38MAPK/MK2 pathway upon genotoxic stress involved in the regulation of Gadd45α,[27] a cell cycle checkpoint protein that is induced after genotoxic stress.[136,137] In addition to limited transcriptional control, Gadd45α levels are regulated posttranscriptionally.[138] The RBPs AUF1 and TIAR, which negatively regulate Gadd45α mRNA and protein levels, rapidly dissociate from its mRNA upon DNA damage, leading to increased mRNA stability and protein accumulation.[139] The inhibitory function of TIAR is released through its direct phosphorylation by p38 MAPK.[27] In addition, MK2 directly phosphorylates two other regulators, hnRNPA0[140] and PARN, that were identified as novel Gadd45α mRNA-binding proteins.[27] The resulting accumulation of Gadd45α mRNA and protein functions in a positive feed-forward loop that prolongs p38/MK2 activity in later times after DNA damage, a function that appears to be especially important in p53-defective cells[27] (reviewed in Ref. 141).

A future challenge will be to identify additional critical mRNA targets in the DNA damage response, the RBPs that control them, and the regulation of these interactions between mRNA and mRNA regulatory proteins by DNA damage response signaling pathways. Posttranscriptional regulatory mechanisms are just emerging as a critical component in a cell's response to genotoxic stress. Understanding this response in more detail, and with a special focus on its oncogenic context, might open a new window of opportunity for the identification of synthetic lethal interactions with cancer-specific mutations, as well as identifying a novel class of molecules for the design and development of small-molecule inhibitors.

B. Future Directions

A limited number of well-studied synthetic lethal interactions have been identified in cancer cells. These findings give hope for the translation of successful basic research into improvements in the clinical treatment of cancers. Unbiased screening approaches will be necessary to discover new synthetic lethal combinations. High-throughput technologies such as genome-wide RNAi approaches or screening of chemical compound libraries should enable us to explore the promising concept of synthetic lethality in greater detail. New sequencing techniques allow gathering of more information about mutations in cancer cell lines and primary tumor samples.

Many ongoing studies discussed earlier focus on the loss of function in tumor suppressor genes such as BRCA or p53 deficiency. However, there is great potential for the discovery of lethal interactions with other genetic lesions, for example, gain-of-function mutations as described for p53.[142] Mutations in *p53* can lead to elevated activation of the Rho-pathway and thereby increase proliferation and evasion potential in tumor cells.[143,144] Just as for loss-of-function mutations, this gain of novel properties specific to cancer cells could offer alternative possibilities for p53-targeted approaches. In addition, the principles of synthetic lethality can be applied for common "non-druggable" oncogenes, such as *Myc* amplifications or *K-ras* mutations, some of the most commonly mutated oncogenes. Recent RNAi screens for synthetic lethal interactors with oncogenic Ras found that *K-ras* mutated tumor cells are hypersensitive to the loss of polo-like kinase 1 (Plk1) and STK33,[145,146] which could be potentially druggable targets.

Synthetic lethality has emerged as an important concept to understand the function of genes, and the interdependence between pathways following rewiring of signaling networks in tumor cells. It forms the basis for new strategies in targeted anticancer therapy that focus on the oncogenic context of cancer cells and make use of unique tumor properties that are not shared by normal tissue. Currently, many of these novel approaches exploiting synthetic lethal genes are highly effective in preclinical studies, some holding great promise for a successful validation in the clinic.

References

1. Kennedy RD, D'Andrea AD. DNA repair pathways in clinical practice: lessons from pediatric cancer susceptibility syndromes. *J Clin Oncol* 2006;**24**:3799–808.
2. Kastan MB, Bartek J. Cell-cycle checkpoints and cancer. *Nature* 2004;**432**:316–23.
3. Luo J, Solimini NL, Elledge SJ. Principles of cancer therapy: oncogene and non-oncogene addiction. *Cell* 2009;**136**:823–37.
4. Hanahan D, Weinberg RA. The hallmarks of cancer. *Cell* 2000;**100**:57–70.

5. Hahn WC, Weinberg RA. Modelling the molecular circuitry of cancer. *Nat Rev Cancer* 2002;**2**:331–41.
6. Jones S, Zhang X, Parsons DW, Lin JC, Leary RJ, Angenendt P, et al. Core signaling pathways in human pancreatic cancers revealed by global genomic analyses. *Science* 2008;**321**:1801–6.
7. Roberts PJ, Der CJ. Targeting the Raf-MEK-ERK mitogen-activated protein kinase cascade for the treatment of cancer. *Oncogene* 2007;**26**:3291–310.
8. Sharma SV, Bell DW, Settleman J, Haber DA. Epidermal growth factor receptor mutations in lung cancer. *Nat Rev Cancer* 2007;**7**:169–81.
9. Druker BJ, Tamura S, Buchdunger E, Ohno S, Segal GM, Fanning S, et al. Effects of a selective inhibitor of the Abl tyrosine kinase on the growth of Bcr-Abl positive cells. *Nat Med* 1996;**2**:561–6.
10. Cox AD, Der CJ. Ras family signaling: therapeutic targeting. *Cancer Biol Ther* 2002;**1**:599–606.
11. Fritz G. Targeting the mevalonate pathway for improved anticancer therapy. *Curr Cancer Drug Targets* 2009;**9**:626–38.
12. Frebourg T, Friend SH. Cancer risks from germline p53 mutations. *J Clin Invest* 1992;**90**:1637–41.
13. Lavin MF, Shiloh Y. The genetic defect in ataxia-telangiectasia. *Annu Rev Immunol* 1997;**15**:177–202.
14. Lehmann AR. DNA repair-deficient diseases, xeroderma pigmentosum, Cockayne syndrome and trichothiodystrophy. *Biochimie* 2003;**85**:1101–11.
15. O'Driscoll M, Ruiz-Perez VL, Woods CG, Jeggo PA, Goodship JA. A splicing mutation affecting expression of ataxia-telangiectasia and Rad3-related protein (ATR) results in Seckel syndrome. *Nat Genet* 2003;**33**:497–501.
16. Shiloh Y. ATM and related protein kinases: safeguarding genome integrity. *Nat Rev Cancer* 2003;**3**:155–68.
17. Taylor AM, Groom A, Byrd PJ. Ataxia-telangiectasia-like disorder (ATLD)-its clinical presentation and molecular basis. *DNA Repair (Amst)* 2004;**3**:1219–25.
18. Bartkova J, Horejsi Z, Koed K, Kramer A, Tort F, Zieger K, et al. DNA damage response as a candidate anti-cancer barrier in early human tumorigenesis. *Nature* 2005;**434**:864–70.
19. Nevanlinna H, Bartek J. The CHEK2 gene and inherited breast cancer susceptibility. *Oncogene* 2006;**25**:5912–9.
20. Fackenthal JD, Olopade OI. Breast cancer risk associated with BRCA1 and BRCA2 in diverse populations. *Nat Rev Cancer* 2007;**7**:937–48.
21. Zhou BB, Elledge SJ. The DNA damage response: putting checkpoints in perspective. *Nature* 2000;**408**:433–9.
22. Bartek J, Lukas J. Chk1 and Chk2 kinases in checkpoint control and cancer. *Cancer Cell* 2003;**3**:421–9.
23. Harper JW, Elledge SJ. The DNA damage response: ten years after. *Mol Cell* 2007;**28**:739–45.
24. Reinhardt HC, Yaffe MB. Kinases that control the cell cycle in response to DNA damage: Chk1, Chk2, and MK2. *Curr Opin Cell Biol* 2009;**21**:245–55.
25. Manke IA, Nguyen A, Lim D, Stewart MQ, Elia AE, Yaffe MB. MAPKAP kinase-2 is a cell cycle checkpoint kinase that regulates the G2/M transition and S phase progression in response to UV irradiation. *Mol Cell* 2005;**17**:37–48.
26. Reinhardt HC, Aslanian AS, Lees JA, Yaffe MB. p53-deficient cells rely on ATM- and ATR-mediated checkpoint signaling through the p38MAPK/MK2 pathway for survival after DNA damage. *Cancer Cell* 2007;**11**:175–89.
27. Reinhardt HC, Hasskamp P, Schmedding I, Morandell S, van Vugt MA, Wang X, et al. DNA damage activates a spatially distinct late cytoplasmic cell-cycle checkpoint network controlled by MK2-mediated RNA stabilization. *Mol Cell* 2010;**40**:34–49.

28. Harris SL, Levine AJ. The p53 pathway: positive and negative feedback loops. *Oncogene* 2005;**24**:2899–908.
29. Vogelstein B, Lane D, Levine AJ. Surfing the p53 network. *Nature* 2000;**408**:307–10.
30. Fridman JS, Lowe SW. Control of apoptosis by p53. *Oncogene* 2003;**22**:9030–40.
31. Vousden KH, Lu X. Live or let die: the cell's response to p53. *Nat Rev Cancer* 2002;**2**:594–604.
32. Donzelli M, Draetta GF. Regulating mammalian checkpoints through Cdc25 inactivation. *EMBO Rep* 2003;**4**:671–7.
33. Jin J, Shirogane T, Xu L, Nalepa G, Qin J, Elledge SJ, et al. SCFbeta-TRCP links Chk1 signaling to degradation of the Cdc25A protein phosphatase. *Genes Dev* 2003;**17**:3062–74.
34. Sorensen CS, Syljuasen RG, Falck J, Schroeder T, Ronnstrand L, Khanna KK, et al. Chk1 regulates the S phase checkpoint by coupling the physiological turnover and ionizing radiation-induced accelerated proteolysis of Cdc25A. *Cancer Cell* 2003;**3**:247–58.
35. Zeng Y, Piwnica-Worms H. DNA damage and replication checkpoints in fission yeast require nuclear exclusion of the Cdc25 phosphatase via 14-3-3 binding. *Mol Cell Biol* 1999;**19**:7410–9.
36. Gardino AK, Yaffe MB. 14-3-3 proteins as signaling integration points for cell cycle control and apoptosis. *Semin Cell Dev Biol* 2011;**22**:688–95.
37. Reinhardt HC, Jiang H, Hemann MT, Yaffe MB. Exploiting synthetic lethal interactions for targeted cancer therapy. *Cell Cycle* 2009;**8**:3112–9.
38. Dobzhansky T. Genetics of Natural Populations. Xiii. Recombination and Variability in Populations of Drosophila Pseudoobscura. *Genetics* 1946;**31**:269–90.
39. Hartman JL, Garvik B, Hartwell L. Principles for the buffering of genetic variation. *Science* 2001;**291**:1001–4.
40. Kaelin Jr. WG. The concept of synthetic lethality in the context of anticancer therapy. *Nat Rev Cancer* 2005;**5**:689–98.
41. Hartwell LH, Szankasi P, Roberts CJ, Murray AW, Friend SH. Integrating genetic approaches into the discovery of anticancer drugs. *Science* 1997;**278**:1064–8.
42. Bommi-Reddy A, Almeciga I, Sawyer J, Geisen C, Li W, Harlow E, et al. Kinase requirements in human cells: III. Altered kinase requirements in VHL-/- cancer cells detected in a pilot synthetic lethal screen. *Proc Natl Acad Sci U S A* 2008;**105**:16484–9.
43. Bryant HE, Schultz N, Thomas HD, Parker KM, Flower D, Lopez E, et al. Specific killing of BRCA2-deficient tumours with inhibitors of poly(ADP-ribose) polymerase. *Nature* 2005;**434**:913–7.
44. Farmer H, McCabe N, Lord CJ, Tutt AN, Johnson DA, Richardson TB, et al. Targeting the DNA repair defect in BRCA mutant cells as a therapeutic strategy. *Nature* 2005;**434**:917–21.
45. Hoeijmakers JH. Genome maintenance mechanisms for preventing cancer. *Nature* 2001;**411**:366–74.
46. Martin SA, Lord CJ, Ashworth A. DNA repair deficiency as a therapeutic target in cancer. *Curr Opin Genet Dev* 2008;**18**:80–6.
47. Fong PC, Boss DS, Yap TA, Tutt A, Wu P, Mergui-Roelvink M, et al. Inhibition of poly(ADP-ribose) polymerase in tumors from BRCA mutation carriers. *N Engl J Med* 2009;**361**:123–34.
48. Audeh MW, Carmichael J, Penson RT, Friedlander M, Powell B, Bell-McGuinn KM, et al. Oral poly(ADP-ribose) polymerase inhibitor olaparib in patients with BRCA1 or BRCA2 mutations and recurrent ovarian cancer: a proof-of-concept trial. *Lancet* 2010;**376**:245–51.
49. Fong PC, Yap TA, Boss DS, Carden CP, Mergui-Roelvink M, Gourley C, et al. Poly(ADP)-ribose polymerase inhibition: frequent durable responses in BRCA carrier ovarian cancer correlating with platinum-free interval. *J Clin Oncol* 2010;**28**:2512–9.
50. Tutt A, Robson M, Garber JE, Domchek SM, Audeh MW, Weitzel JN, et al. Oral poly(ADP-ribose) polymerase inhibitor olaparib in patients with BRCA1 or BRCA2 mutations and advanced breast cancer: a proof-of-concept trial. *Lancet* 2010;**376**:235–44.

51. Sandhu SK, Yap TA, de Bono JS. Poly(ADP-ribose) polymerase inhibitors in cancer treatment: a clinical perspective. *Eur J Cancer* 2010;**46**:9–20.
52. Iglehart JD, Silver DP. Synthetic lethality—a new direction in cancer-drug development. *N Engl J Med* 2009;**361**:189–91.
53. Ashworth A. A synthetic lethal therapeutic approach: poly(ADP) ribose polymerase inhibitors for the treatment of cancers deficient in DNA double-strand break repair. *J Clin Oncol* 2008;**26**:3785–90.
54. ClinicalTrials.gov, A service of the U.S. National Institute of Health; (n.d.); retrieved from www.clinicaltrials.gov.
55. Wang ZQ, Auer B, Stingl L, Berghammer H, Haidacher D, Schweiger M, et al. Mice lacking ADPRT and poly(ADP-ribosyl)ation develop normally but are susceptible to skin disease. *Genes Dev* 1995;**9**:509–20.
56. Tong WM, Yang YG, Cao WH, Galendo D, Frappart L, Shen Y, et al. Poly(ADP-ribose) polymerase-1 plays a role in suppressing mammary tumourigenesis in mice. *Oncogene* 2007;**26**:3857–67.
57. Edwards SL, Brough R, Lord CJ, Natrajan R, Vatcheva R, Levine DA, et al. Resistance to therapy caused by intragenic deletion in BRCA2. *Nature* 2008;**451**:1111–5.
58. Sakai W, Swisher EM, Karlan BY, Agarwal MK, Higgins J, Friedman C, et al. Secondary mutations as a mechanism of cisplatin resistance in BRCA2-mutated cancers. *Nature* 2008; **451**:1116–20.
59. Maya-Mendoza A, Petermann E, Gillespie DA, Caldecott KW, Jackson DA. Chk1 regulates the density of active replication origins during the vertebrate S phase. *EMBO J* 2007;**26**:2719–31.
60. Cimprich KA, Cortez D. ATR: an essential regulator of genome integrity. *Nat Rev Mol Cell Biol* 2008;**9**:616–27.
61. Boutros R, Lobjois V, Ducommun B. CDC25 phosphatases in cancer cells: key players? Good targets? *Nat Rev Cancer* 2007;**7**:495–507.
62. Zhao H, Watkins JL, Piwnica-Worms H. Disruption of the checkpoint kinase 1/cell division cycle 25A pathway abrogates ionizing radiation-induced S and G2 checkpoints. *Proc Natl Acad Sci U S A* 2002;**99**:14795–800.
63. Brown EJ, Baltimore D. ATR disruption leads to chromosomal fragmentation and early embryonic lethality. *Genes Dev* 2000;**14**:397–402.
64. Liu Q, Guntuku S, Cui XS, Matsuoka S, Cortez D, Tamai K, et al. Chk1 is an essential kinase that is regulated by Atr and required for the G(2)/M DNA damage checkpoint. *Genes Dev* 2000;**14**:1448–59.
65. Takai H, Tominaga K, Motoyama N, Minamishima YA, Nagahama H, Tsukiyama T, et al. Aberrant cell cycle checkpoint function and early embryonic death in Chk1(-/-) mice. *Genes Dev* 2000;**14**:1439–47.
66. Takahashi I, Kobayashi E, Nakano H, Murakata C, Saitoh H, Suzuki K, et al. Potent selective inhibition of 7-O-methyl UCN-01 against protein kinase C. *J Pharmacol Exp Ther* 1990;**255**:1218–21.
67. Graves PR, Yu L, Schwarz JK, Gales J, Sausville EA, O'Connor PM, et al. The Chk1 protein kinase and the Cdc25C regulatory pathways are targets of the anticancer agent UCN-01. *J Biol Chem* 2000;**275**:5600–5.
68. Busby EC, Leistritz DF, Abraham RT, Karnitz LM, Sarkaria JN. The radiosensitizing agent 7-hydroxystaurosporine (UCN-01) inhibits the DNA damage checkpoint kinase hChk1. *Cancer Res* 2000;**60**:2108–12.
69. Sato S, Fujita N, Tsuruo T. Interference with PDK1-Akt survival signaling pathway by UCN-01 (7-hydroxystaurosporine). *Oncogene* 2002;**21**:1727–38.
70. Bunch RT, Eastman A. Enhancement of cisplatin-induced cytotoxicity by 7-hydroxystaurosporine (UCN-01), a new G2-checkpoint inhibitor. *Clin Cancer Res* 1996;**2**:791–7.

71. Levesque AA, Fanous AA, Poh A, Eastman A. Defective p53 signaling in p53 wild-type tumors attenuates p21waf1 induction and cyclin B repression rendering them sensitive to Chk1 inhibitors that abrogate DNA damage-induced S and G2 arrest. *Mol Cancer Ther* 2008;**7**:252–62.
72. Wang Q, Fan S, Eastman A, Worland PJ, Sausville EA, O'Connor PM. UCN-01: a potent abrogator of G2 checkpoint function in cancer cells with disrupted p53. *J Natl Cancer Inst* 1996;**88**:956–65.
73. Fracasso PM, Williams KJ, Chen RC, Picus J, Ma CX, Ellis MJ, et al. A Phase 1 study of UCN-01 in combination with irinotecan in patients with resistant solid tumor malignancies. *Cancer Chemother Pharmacol* 2011;**67**:1225–37.
74. Ma CX, Janetka JW, Piwnica-Worms H. Death by releasing the breaks: CHK1 inhibitors as cancer therapeutics. *Trends Mol Med* 2011;**17**:88–96.
75. Sangster-Guity N, Conrad BH, Papadopoulos N, Bunz F. ATR mediates cisplatin resistance in a p53 genotype-specific manner. *Oncogene* 2011;**30**:2526–33.
76. Hurley PJ, Wilsker D, Bunz F. Human cancer cells require ATR for cell cycle progression following exposure to ionizing radiation. *Oncogene* 2007;**26**:2535–42.
77. Alderton GK, Joenje H, Varon R, Borglum AD, Jeggo PA, O'Driscoll M. Seckel syndrome exhibits cellular features demonstrating defects in the ATR-signalling pathway. *Hum Mol Genet* 2004;**13**:3127–38.
78. Wilsker D, Bunz F. Loss of ataxia telangiectasia mutated- and Rad3-related function potentiates the effects of chemotherapeutic drugs on cancer cell survival. *Mol Cancer Ther* 2007;**6**:1406–13.
79. Jiang H, Reinhardt HC, Bartkova J, Tommiska J, Blomqvist C, Nevanlinna H, et al. The combined status of ATM and p53 link tumor development with therapeutic response. *Genes Dev* 2009;**23**:1895–909.
80. Toledo LI, Murga M, Zur R, Soria R, Rodriguez A, Martinez S, et al. A cell-based screen identifies ATR inhibitors with synthetic lethal properties for cancer-associated mutations. *Nat Struct Mol Biol* 2011;**18**:721–7.
81. Maira SM, Stauffer F, Brueggen J, Furet P, Schnell C, Fritsch C, et al. Identification and characterization of NVP-BEZ235, a new orally available dual phosphatidylinositol 3-kinase/mammalian target of rapamycin inhibitor with potent in vivo antitumor activity. *Mol Cancer Ther* 2008;**7**:1851–63.
82. Charrier JD, Durrant SJ, Golec JM, Kay DP, Knegtel RM, MacCormick S, et al. Discovery of potent and selective inhibitors of ataxia telangiectasia mutated and Rad3 related (ATR) protein kinase as potential anticancer agents. *J Med Chem* 2011;**54**:2320–30.
83. Reaper PM, Griffiths MR, Long JM, Charrier JD, Maccormick S, Charlton PA, et al. Selective killing of ATM- or p53-deficient cancer cells through inhibition of ATR. *Nat Chem Biol* 2011;**7**:428–30.
84. Gilad O, Nabet BY, Ragland RL, Schoppy DW, Smith KD, Durham AC, et al. Combining ATR suppression with oncogenic Ras synergistically increases genomic instability, causing synthetic lethality or tumorigenesis in a dosage-dependent manner. *Cancer Res* 2010;**70**:9693–702.
85. Lam MH, Liu Q, Elledge SJ, Rosen JM. Chk1 is haploinsufficient for multiple functions critical to tumor suppression. *Cancer Cell* 2004;**6**:45–59.
86. Menoyo A, Alazzouzi H, Espin E, Armengol M, Yamamoto H, Schwartz Jr. S. Somatic mutations in the DNA damage-response genes ATR and CHK1 in sporadic stomach tumors with microsatellite instability. *Cancer Res* 2001;**61**:7727–30.
87. Vassileva V, Millar A, Briollais L, Chapman W, Bapat B. Genes involved in DNA repair are mutational targets in endometrial cancers with microsatellite instability. *Cancer Res* 2002;**62**:4095–9.

88. Lewis KA, Bakkum-Gamez J, Loewen R, French AJ, Thibodeau SN, Cliby WA. Mutations in the ataxia telangiectasia and rad3-related-checkpoint kinase 1 DNA damage response axis in colon cancers. *Genes Chromosomes Cancer* 2007;**46**:1061–8.
89. Fedier A, Schlamminger M, Schwarz VA, Haller U, Howell SB, Fink D. Loss of atm sensitises p53-deficient cells to topoisomerase poisons and antimetabolites. *Ann Oncol* 2003;**14**:938–45.
90. Morrison C, Sonoda E, Takao N, Shinohara A, Yamamoto K, Takeda S. The controlling role of ATM in homologous recombinational repair of DNA damage. *EMBO J* 2000;**19**:463–71.
91. Kuhne M, Riballo E, Rief N, Rothkamm K, Jeggo PA, Lobrich M. A double-strand break repair defect in ATM-deficient cells contributes to radiosensitivity. *Cancer Res* 2004;**64**:500–8.
92. Dar ME, Winters TA, Jorgensen TJ. Identification of defective illegitimate recombinational repair of oxidatively-induced DNA double-strand breaks in ataxia-telangiectasia cells. *Mutat Res* 1997;**384**:169–79.
93. Bredemeyer AL, Sharma GG, Huang CY, Helmink BA, Walker LM, Khor KC, et al. ATM stabilizes DNA double-strand-break complexes during V(D)J recombination. *Nature* 2006;**442**:466–70.
94. Riballo E, Kuhne M, Rief N, Doherty A, Smith GC, Recio MJ, et al. A pathway of double-strand break rejoining dependent upon ATM, Artemis, and proteins locating to gamma-H2AX foci. *Mol Cell* 2004;**16**:715–24.
95. Yuan SS, Chang HL, Lee EY. Ionizing radiation-induced Rad51 nuclear focus formation is cell cycle-regulated and defective in both ATM(-/-) and c-Abl(-/-) cells. *Mutat Res* 2003;**525**:85–92.
96. DeFazio LG, Stansel RM, Griffith JD, Chu G. Synapsis of DNA ends by DNA-dependent protein kinase. *EMBO J* 2002;**21**:3192–200.
97. Lees-Miller SP, Meek K. Repair of DNA double strand breaks by non-homologous end joining. *Biochimie* 2003;**85**:1161–73.
98. Spagnolo L, Rivera-Calzada A, Pearl LH, Llorca O. Three-dimensional structure of the human DNA-PKcs/Ku70/Ku80 complex assembled on DNA and its implications for DNA DSB repair. *Mol Cell* 2006;**22**:511–9.
99. Smith GC, Jackson SP. The DNA-dependent protein kinase. *Genes Dev* 1999;**13**:916–34.
100. Gurley KE, Kemp CJ. Synthetic lethality between mutation in Atm and DNA-PK(cs) during murine embryogenesis. *Curr Biol* 2001;**11**:191–4.
101. Lafarga V, Cuadrado A, Lopez de Silanes I, Bengoechea R, Fernandez-Capetillo O, Nebreda AR. p38 Mitogen-activated protein kinase- and HuR-dependent stabilization of p21(Cip1) mRNA mediates the G(1)/S checkpoint. *Mol Cell Biol* 2009;**29**:4341–51.
102. Bulavin DV, Higashimoto Y, Popoff IJ, Gaarde WA, Basrur V, Potapova O, et al. Initiation of a G2/M checkpoint after ultraviolet radiation requires p38 kinase. *Nature* 2001;**411**:102–7.
103. Raman M, Earnest S, Zhang K, Zhao Y, Cobb MH. TAO kinases mediate activation of p38 in response to DNA damage. *EMBO J* 2007;**26**:2005–14.
104. Hirose Y, Katayama M, Stokoe D, Haas-Kogan DA, Berger MS, Pieper RO. The p38 mitogen-activated protein kinase pathway links the DNA mismatch repair system to the G2 checkpoint and to resistance to chemotherapeutic DNA-methylating agents. *Mol Cell Biol* 2003;**23**:8306–15.
105. Hirose Y, Katayama M, Berger MS, Pieper RO. Cooperative function of Chk1 and p38 pathways in activating G2 arrest following exposure to temozolomide. *J Neurosurg* 2004;**100**:1060–5.
106. Perfettini JL, Nardacci R, Bourouba M, Subra F, Gros L, Seror C, et al. Critical involvement of the ATM-dependent DNA damage response in the apoptotic demise of HIV-1-elicited syncytia. *PLoS One* 2008;**3**:e2458.
107. Kyriakis JM, Avruch J. Mammalian mitogen-activated protein kinase signal transduction pathways activated by stress and inflammation. *Physiol Rev* 2001;**81**:807–69.
108. Parker LL, Piwnica-Worms H. Inactivation of the p34cdc2-cyclin B complex by the human WEE1 tyrosine kinase. *Science* 1992;**257**:1955–7.

109. Rowley R, Hudson J, Young PG. The wee1 protein kinase is required for radiation-induced mitotic delay. *Nature* 1992;**356**:353–5.
110. Heald R, McLoughlin M, McKeon F. Human wee1 maintains mitotic timing by protecting the nucleus from cytoplasmically activated Cdc2 kinase. *Cell* 1993;**74**:463–74.
111. Lee J, Kumagai A, Dunphy WG. Positive regulation of Wee1 by Chk1 and 14-3-3 proteins. *Mol Biol Cell* 2001;**12**:551–63.
112. Tominaga Y, Li C, Wang RH, Deng CX. Murine Wee1 plays a critical role in cell cycle regulation and pre-implantation stages of embryonic development. *Int J Biol Sci* 2006;**2**:161–70.
113. Wang Y, Decker SJ, Sebolt-Leopold J. Knockdown of Chk1, Wee1 and Myt1 by RNA interference abrogates G2 checkpoint and induces apoptosis. *Cancer Biol Ther* 2004;**3**:305–13.
114. Mir SE, De Witt Hamer PC, Krawczyk PM, Balaj L, Claes A, Niers JM, et al. In silico analysis of kinase expression identifies WEE1 as a gatekeeper against mitotic catastrophe in glioblastoma. *Cancer Cell* 2010;**18**:244–57.
115. Panek RL, Lu GH, Klutchko SR, Batley BL, Dahring TK, Hamby JM, et al. In vitro pharmacological characterization of PD 166285, a new nanomolar potent and broadly active protein tyrosine kinase inhibitor. *J Pharmacol Exp Ther* 1997;**283**:1433–44.
116. Wang Y, Li J, Booher RN, Kraker A, Lawrence T, Leopold WR, et al. Radiosensitization of p53 mutant cells by PD0166285, a novel G(2) checkpoint abrogator. *Cancer Res* 2001; **61**:8211–7.
117. PosthumaDeBoer J, Wurdinger T, Graat HC, van Beusechem VW, Helder MN, van Royen BJ, et al. WEE1 inhibition sensitizes osteosarcoma to radiotherapy. *BMC Cancer* 2011;**11**:156.
118. De Witt Hamer PC, Mir SE, Noske D, Van Noorden CJ, Wurdinger T. WEE1 kinase targeting combined with DNA-damaging cancer therapy catalyzes mitotic catastrophe. *Clin Cancer Res* 2011;**17**:4200–7.
119. Mizuarai S, Yamanaka K, Itadani H, Arai T, Nishibata T, Hirai H, et al. Discovery of gene expression-based pharmacodynamic biomarker for a p53 context-specific anti-tumor drug Wee1 inhibitor. *Mol Cancer* 2009;**8**:34.
120. Hirai H, Iwasawa Y, Okada M, Arai T, Nishibata T, Kobayashi M, et al. Small-molecule inhibition of Wee1 kinase by MK-1775 selectively sensitizes p53-deficient tumor cells to DNA-damaging agents. *Mol Cancer Ther* 2009;**8**:2992–3000.
121. Hirai H, Arai T, Okada M, Nishibata T, Kobayashi M, Sakai N, et al. MK-1775, a small molecule Wee1 inhibitor, enhances anti-tumor efficacy of various DNA-damaging agents, including 5-fluorouracil. *Cancer Biol Ther* 2010;**9**:514–22.
122. Rajeshkumar NV, De Oliveira E, Ottenhof N, Watters J, Brooks D, Demuth T, et al. MK-1775, a potent Wee1 inhibitor, synergizes with gemcitabine to achieve tumor regressions, selectively in p53-deficient pancreatic cancer xenografts. *Clin Cancer Res* 2011;**17**:2799–806.
123. Leijen S, Beijnen JH, Schellens JH. Abrogation of the G2 checkpoint by inhibition of Wee-1 kinase results in sensitization of p53-deficient tumor cells to DNA-damaging agents. *Curr Clin Pharmacol* 2010;**5**:186–91.
124. Vichi P, Coin F, Renaud JP, Vermeulen W, Hoeijmakers JH, Moras D, et al. Cisplatin- and UV-damaged DNA lure the basal transcription factor TFIID/TBP. *EMBO J* 1997;**16**:7444–56.
125. Rockx DA, Mason R, van Hoffen A, Barton MC, Citterio E, Bregman DB, et al. UV-induced inhibition of transcription involves repression of transcription initiation and phosphorylation of RNA polymerase II. *Proc Natl Acad Sci U S A* 2000;**97**:10503–8.
126. Rieger KE, Chu G. Portrait of transcriptional responses to ultraviolet and ionizing radiation in human cells. *Nucleic Acids Res* 2004;**32**:4786–803.
127. Gasch AP, Huang M, Metzner S, Botstein D, Elledge SJ, Brown PO. Genomic expression responses to DNA-damaging agents and the regulatory role of the yeast ATR homolog Mec1p. *Mol Biol Cell* 2001;**12**:2987–3003.

128. Fan J, Yang X, Wang W, Wood 3rd WH, Becker KG, Gorospe M. Global analysis of stress-regulated mRNA turnover by using cDNA arrays. *Proc Natl Acad Sci U S A* 2002;**99**:10611–6.
129. Abdelmohsen K, Kuwano Y, Kim HH, Gorospe M. Posttranscriptional gene regulation by RNA-binding proteins during oxidative stress: implications for cellular senescence. *Biol Chem* 2008;**389**:243–55.
130. Lunde BM, Moore C, Varani G. RNA-binding proteins: modular design for efficient function. *Nat Rev Mol Cell Biol* 2007;**8**:479–90.
131. Mazan-Mamczarz K, Galban S, Lopez de Silanes I, Martindale JL, Atasoy U, Keene JD, et al. RNA-binding protein HuR enhances p53 translation in response to ultraviolet light irradiation. *Proc Natl Acad Sci U S A* 2003;**100**:8354–9.
132. Wang W, Furneaux H, Cheng H, Caldwell MC, Hutter D, Liu Y, et al. HuR regulates p21 mRNA stabilization by UV light. *Mol Cell Biol* 2000;**20**:760–9.
133. Matsuoka S, Ballif BA, Smogorzewska A, McDonald 3rd ER, Hurov KE, Luo J, et al. ATM and ATR substrate analysis reveals extensive protein networks responsive to DNA damage. *Science* 2007;**316**:1160–6.
134. Paulsen RD, Soni DV, Wollman R, Hahn AT, Yee MC, Guan A, et al. A genome-wide siRNA screen reveals diverse cellular processes and pathways that mediate genome stability. *Mol Cell* 2009;**35**:228–39.
135. Hurov KE, Cotta-Ramusino C, Elledge SJ. A genetic screen identifies the Triple T complex required for DNA damage signaling and ATM and ATR stability. *Genes Dev* 2010;**24**:1939–50.
136. Hildesheim J, Bulavin DV, Anver MR, Alvord WG, Hollander MC, Vardanian L, et al. Gadd45a protects against UV irradiation-induced skin tumors, and promotes apoptosis and stress signaling via MAPK and p53. *Cancer Res* 2002;**62**:7305–15.
137. Wang XW, Zhan Q, Coursen JD, Khan MA, Kontny HU, Yu L, et al. GADD45 induction of a G2/M cell cycle checkpoint. *Proc Natl Acad Sci U S A* 1999;**96**:3706–11.
138. Jackman J, Alamo Jr. I, Fornace Jr. AJ. Genotoxic stress confers preferential and coordinate messenger RNA stability on the five gadd genes. *Cancer Res* 1994;**54**:5656–62.
139. Lal A, Abdelmohsen K, Pullmann R, Kawai T, Galban S, Yang X, et al. Posttranscriptional derepression of GADD45alpha by genotoxic stress. *Mol Cell* 2006;**22**:117–28.
140. Rousseau S, Morrice N, Peggie M, Campbell DG, Gaestel M, Cohen P. Inhibition of SAPK2a/p38 prevents hnRNP A0 phosphorylation by MAPKAP-K2 and its interaction with cytokine mRNAs. *EMBO J* 2002;**21**:6505–14.
141. Reinhardt HC, Cannell IG, Morandell S, Yaffe MB. Is post-transcriptional stabilization, splicing and translation of selective mRNAs a key to the DNA damage response? *Cell Cycle* 2011;**10**:23–7.
142. Brosh R, Rotter V. When mutants gain new powers: news from the mutant p53 field. *Nat Rev Cancer* 2009;**9**:701–13.
143. Guo F, Zheng Y. Rho family GTPases cooperate with p53 deletion to promote primary mouse embryonic fibroblast cell invasion. *Oncogene* 2004;**23**:5577–85.
144. Mizuarai S, Yamanaka K, Kotani H. Mutant p53 induces the GEF-H1 oncogene, a guanine nucleotide exchange factor-H1 for RhoA, resulting in accelerated cell proliferation in tumor cells. *Cancer Res* 2006;**66**:6319–26.
145. Luo J, Emanuele MJ, Li D, Creighton CJ, Schlabach MR, Westbrook TF, et al. A genome-wide RNAi screen identifies multiple synthetic lethal interactions with the Ras oncogene. *Cell* 2009;**137**:835–48.
146. Scholl C, Frohling S, Dunn IF, Schinzel AC, Barbie DA, Kim SY, et al. Synthetic lethal interaction between oncogenic KRAS dependency and STK33 suppression in human cancer cells. *Cell* 2009;**137**:821–34.

Index

Note: Page numbers followed by "*f*" indicate figures, and "*t*" indicate tables.

A

B

C

D

U

X

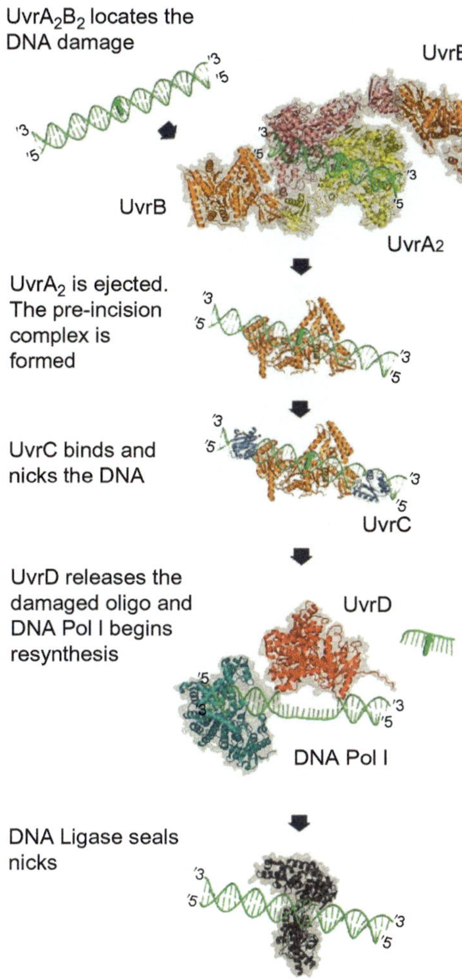

KAD AND VAN HOUTEN, FIG. 1.

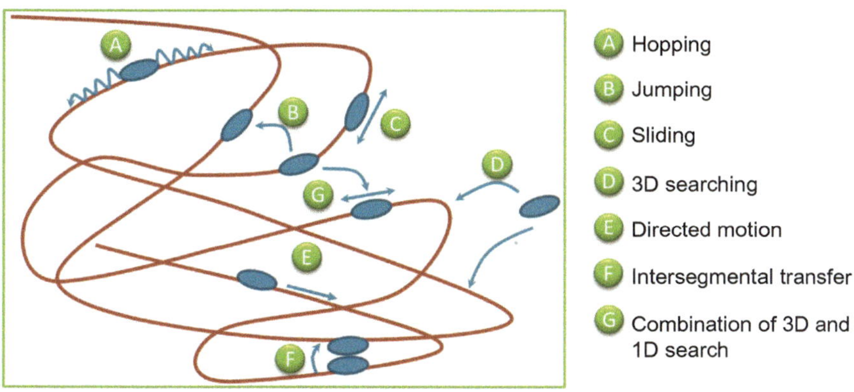

KAD AND VAN HOUTEN, FIG. 2.

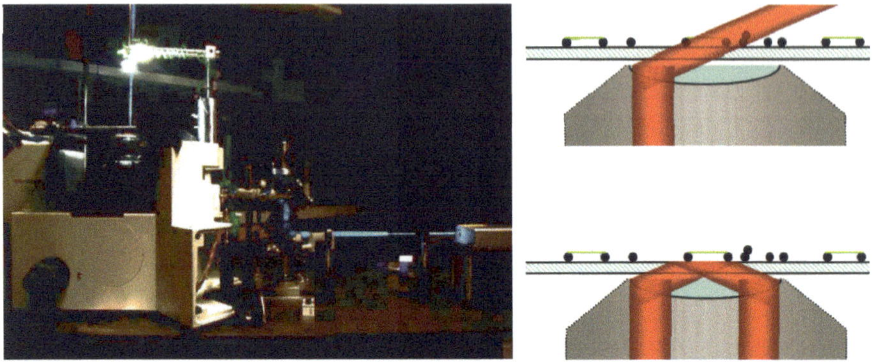

KAD AND VAN HOUTEN, FIG. 3.

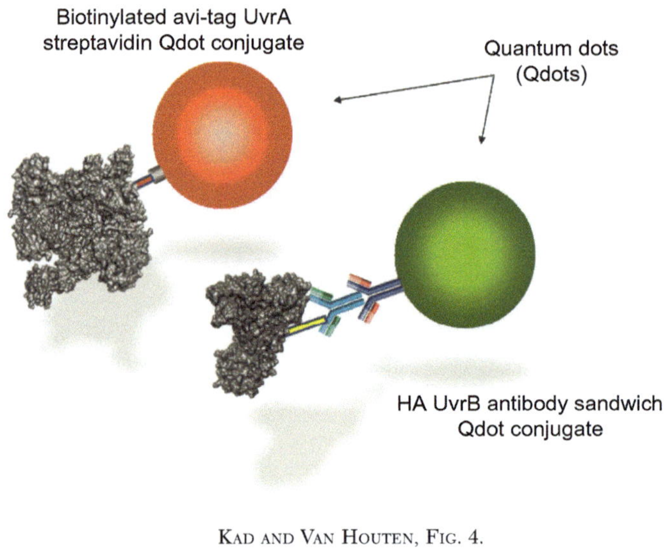

KAD AND VAN HOUTEN, FIG. 4.

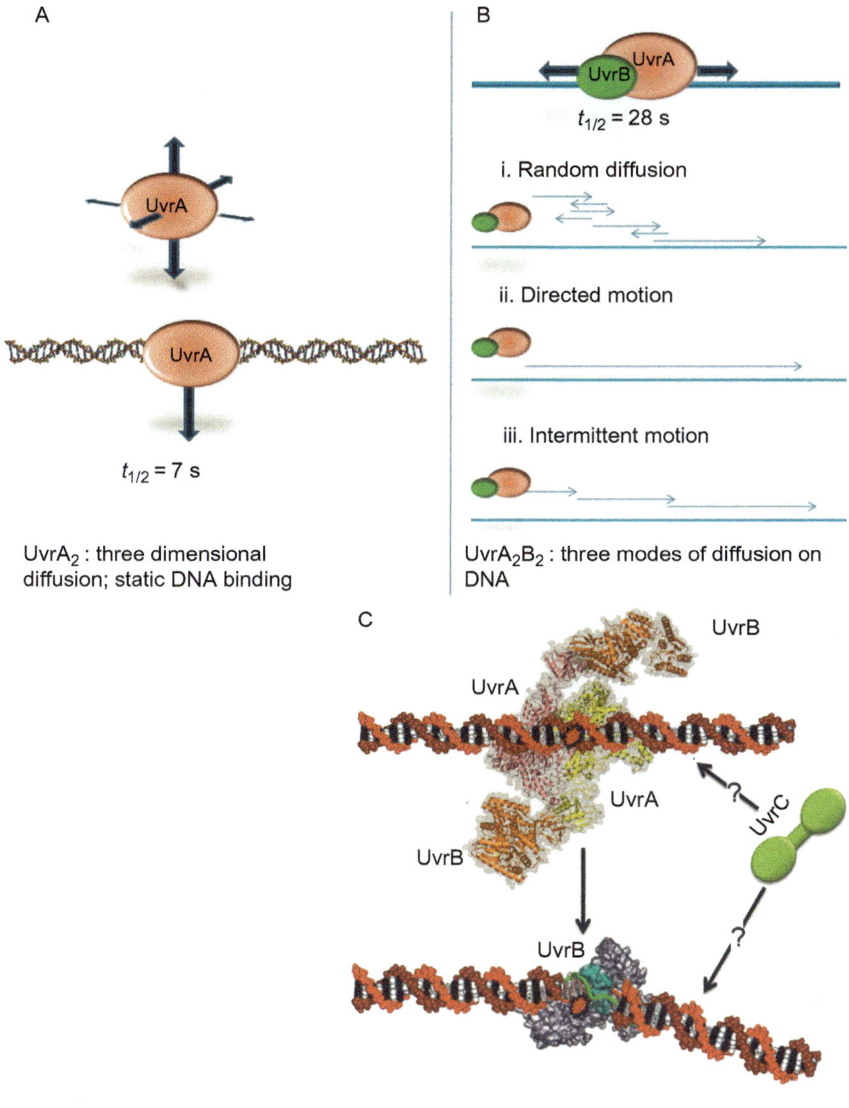

KAD AND VAN HOUTEN, FIG. 5.

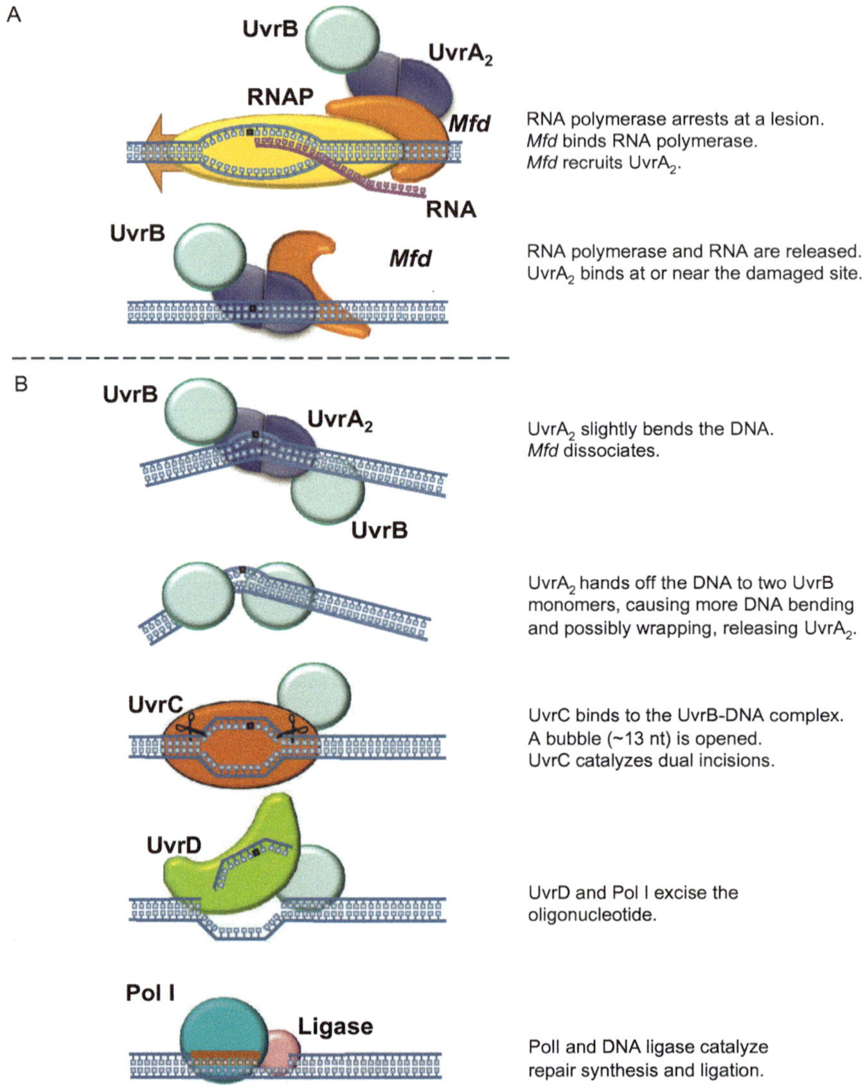

GANESAN *ET AL.*, FIG. 1.

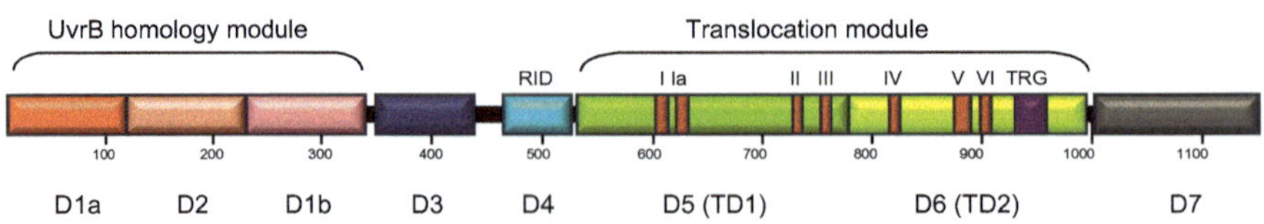

GANESAN *ET AL.*, FIG. 2.

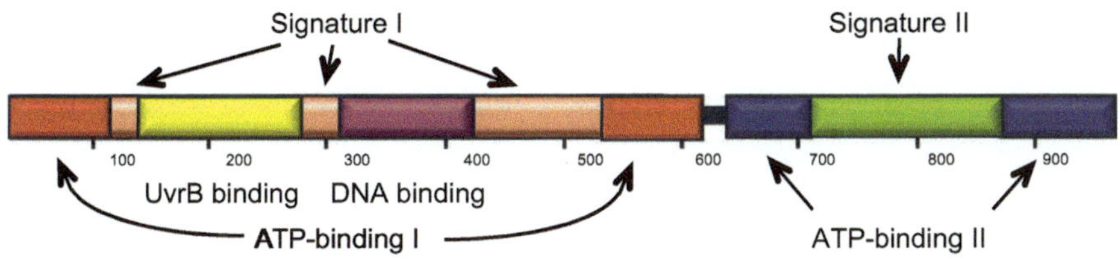

Ganesan *et al.*, Fig. 3.

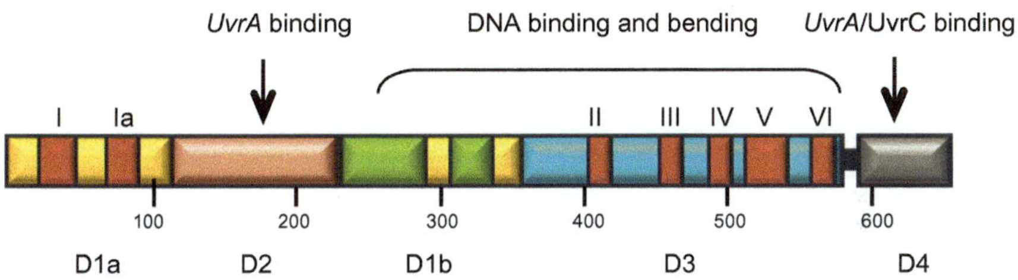

Ganesan *et al.*, Fig. 4.

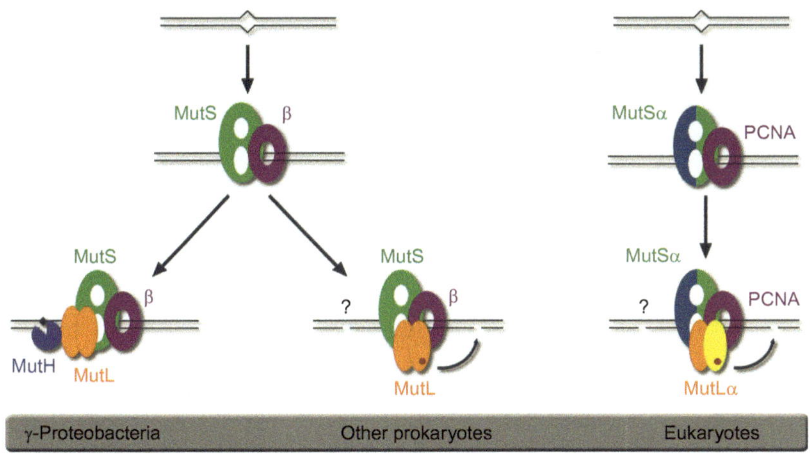

Alba Guarné, Fig. 1.

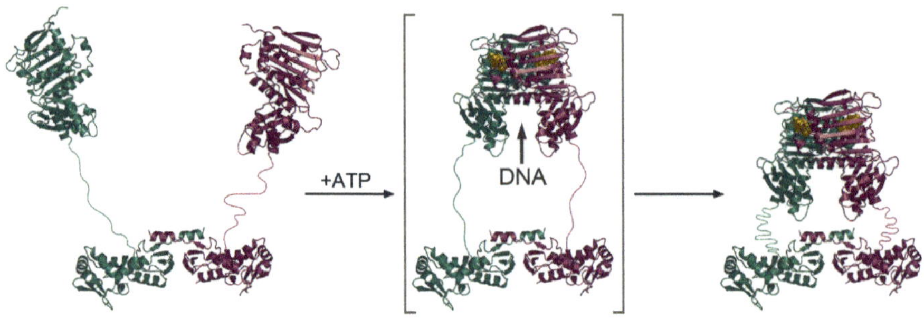

Alba Guarné, Fig. 2.

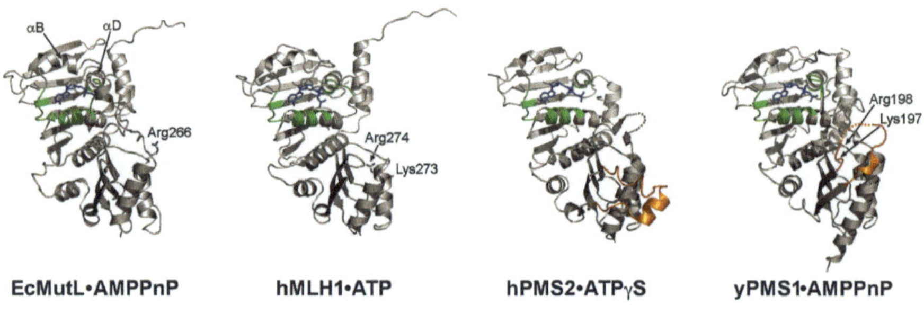

Alba Guarné, Fig. 3.

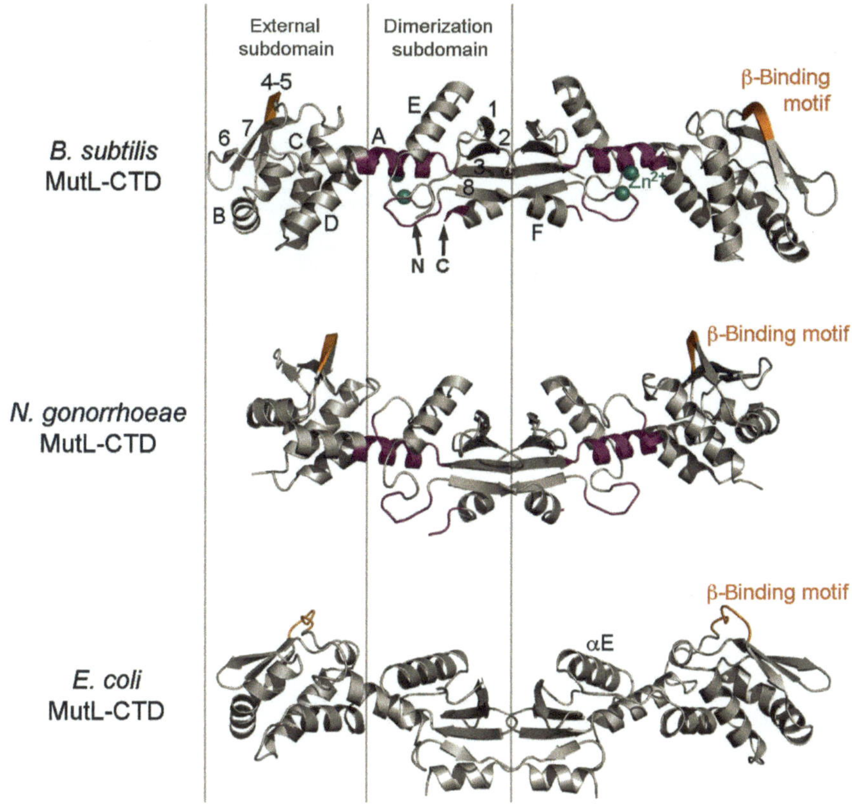

ALBA GUARNÉ, FIG. 4.

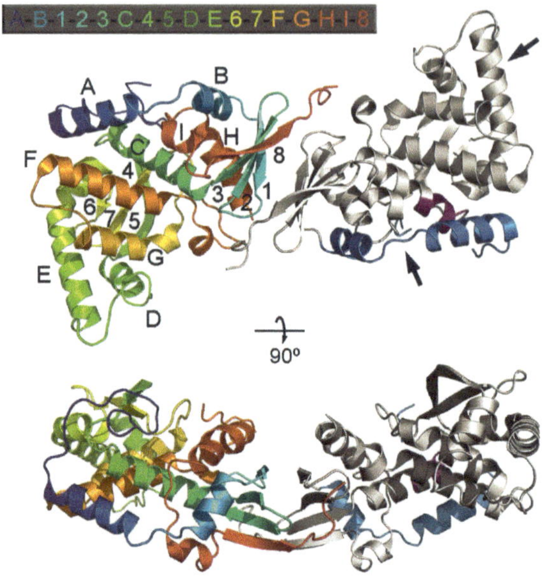

ALBA GUARNÉ, FIG. 5.

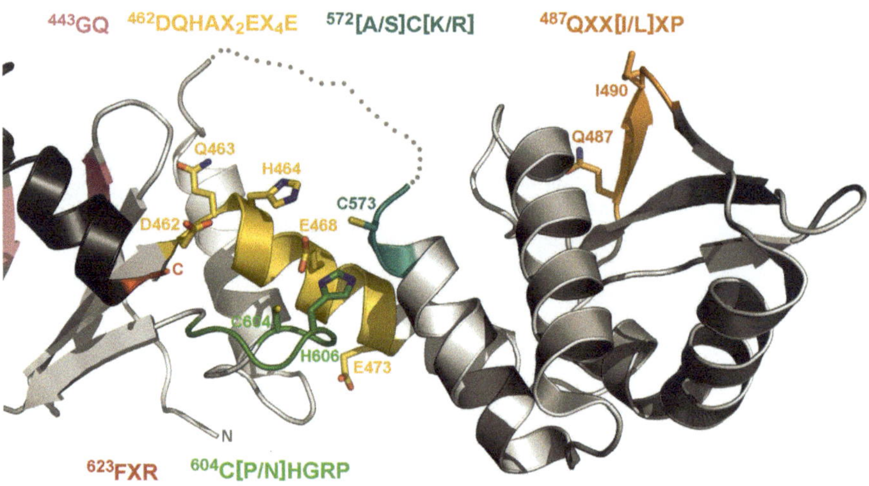

ALBA GUARNÉ, FIG. 6.

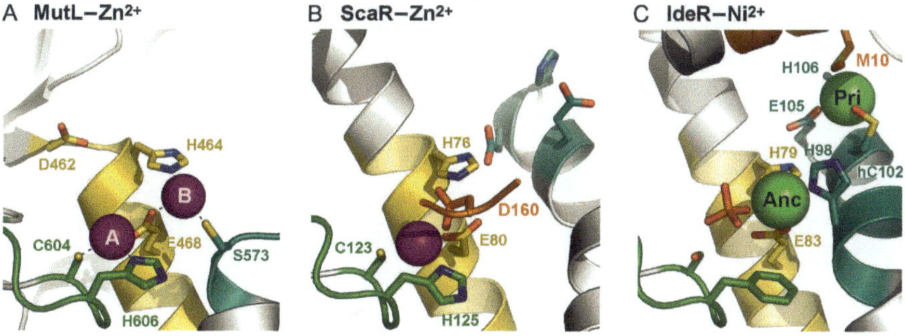

ALBA GUARNÉ, FIG. 7.

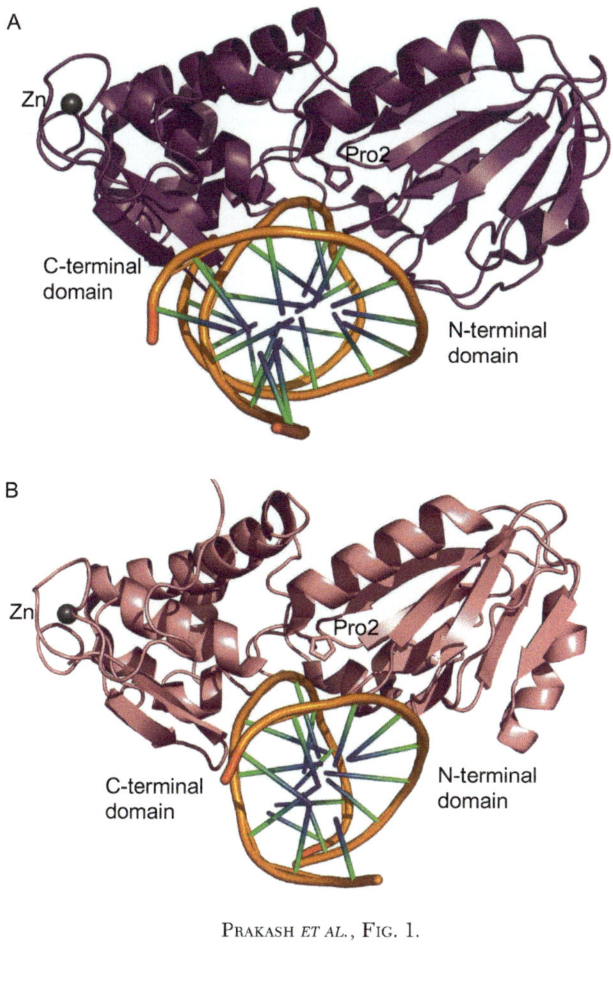

PRAKASH *ET AL.*, FIG. 1.

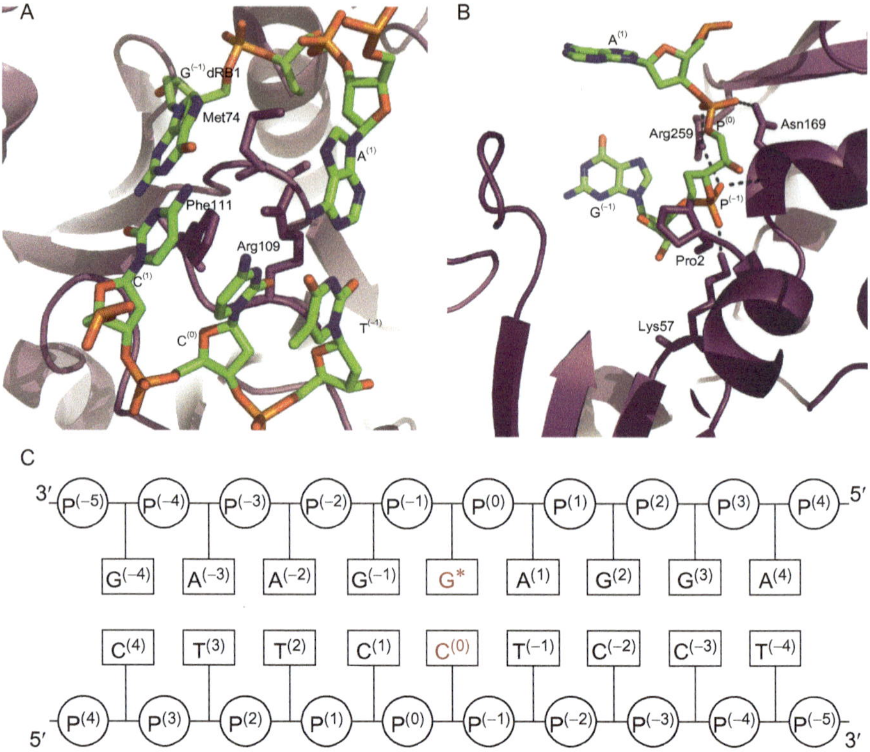

PRAKASH *ET AL.*, FIG. 2.

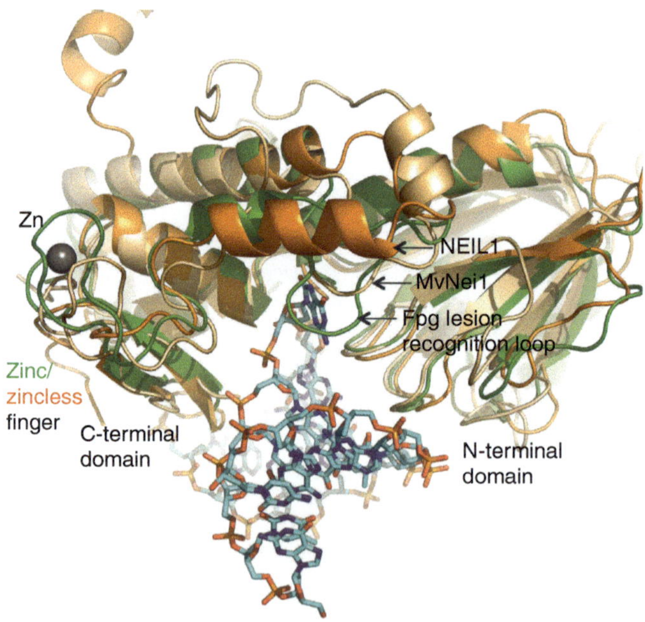

PRAKASH *ET AL.*, FIG. 3.

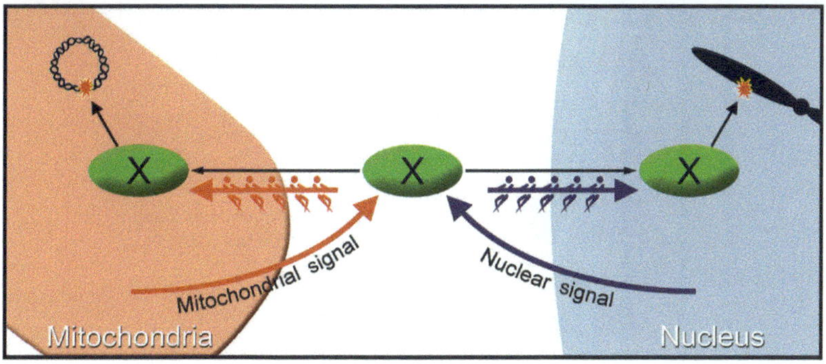

SWARTZLANDER *ET AL.*, FIG. 1.

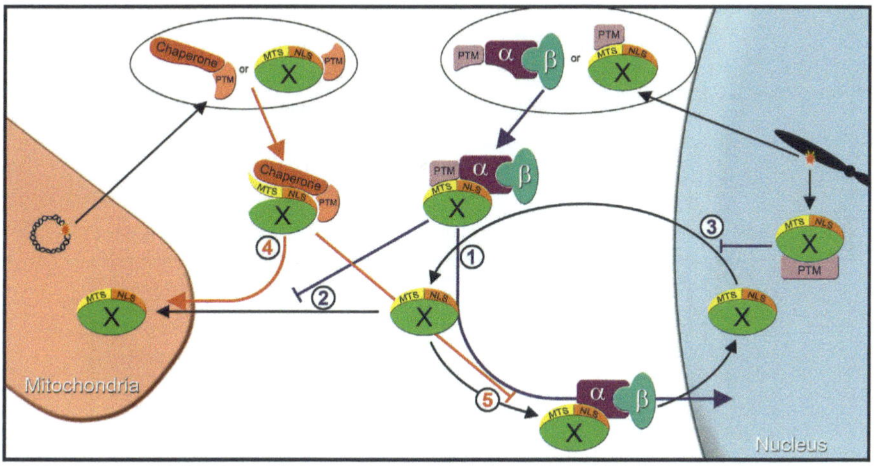

SWARTZLANDER *ET AL.*, FIG. 2.

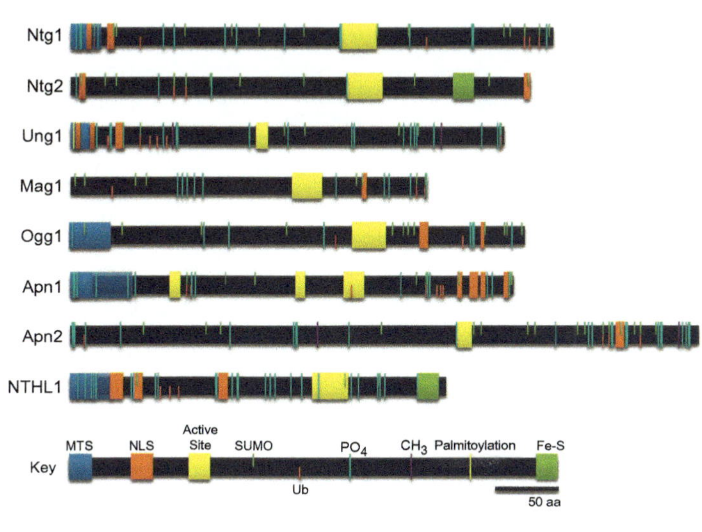

SWARTZLANDER *ET AL.*, FIG. 3.

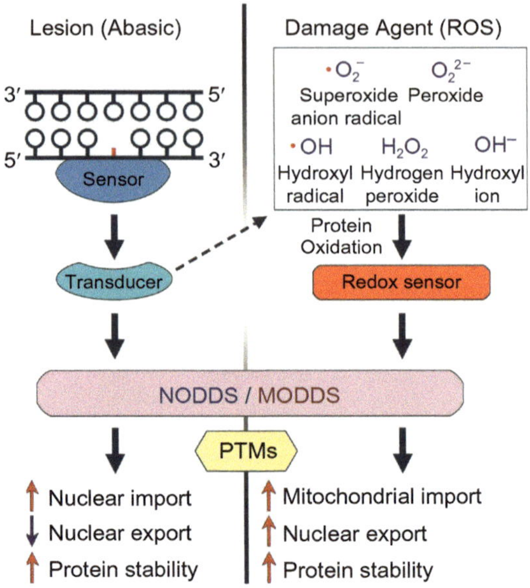

SWARTZLANDER *ET AL.*, FIG. 4.

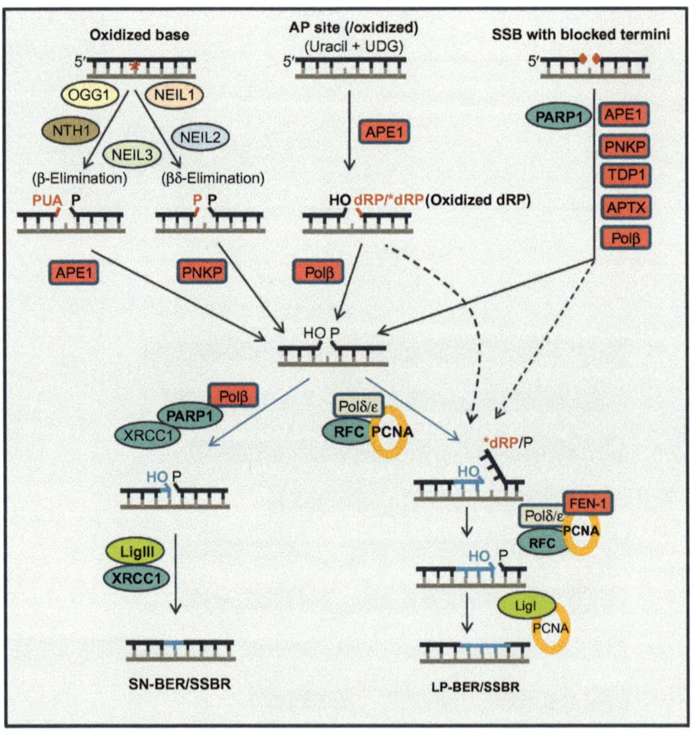

HEGDE *ET AL.*, FIG. 1.

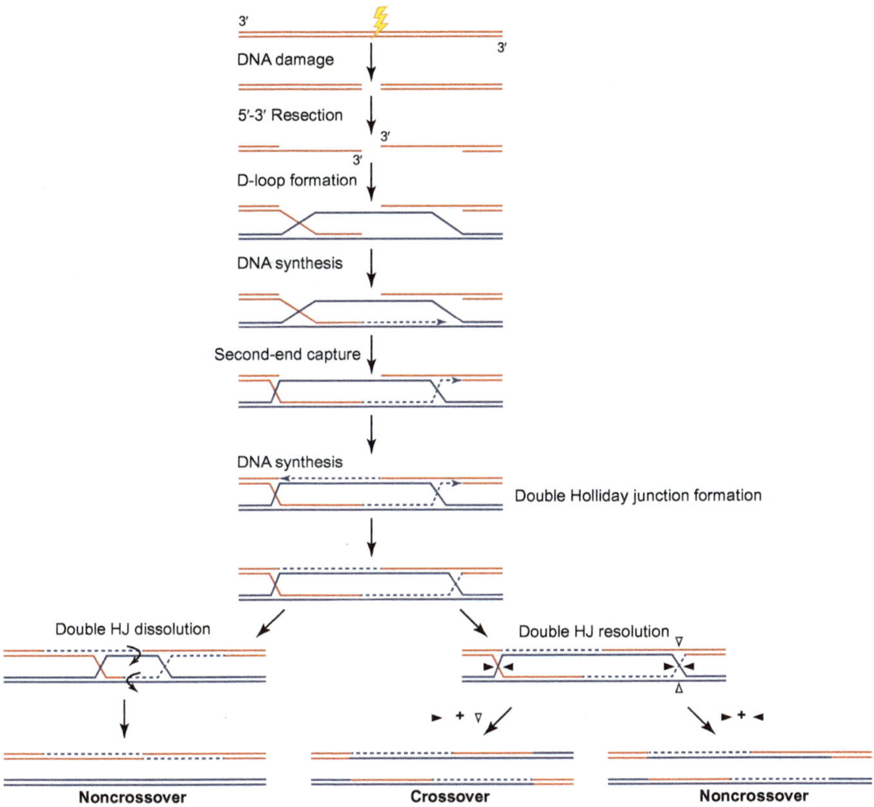

AMUNUGAMA AND FISHEL, FIG. 1.

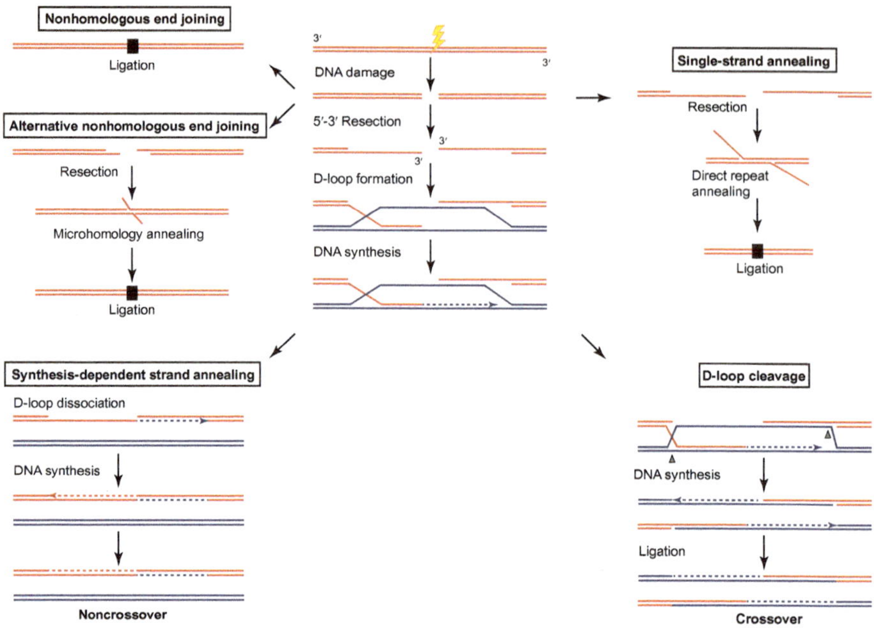

Amunugama and Fishel, Fig. 2.

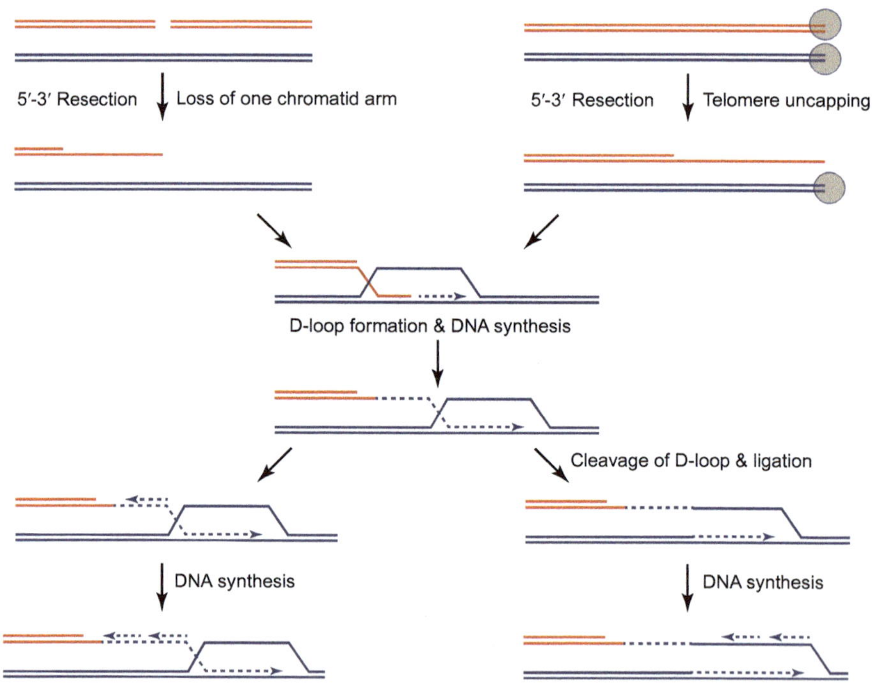

Amunugama and Fishel, Fig. 3.

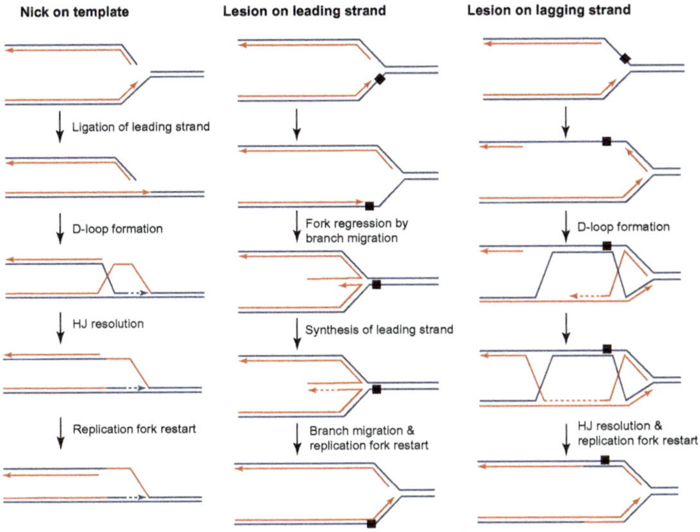

AMUNUGAMA AND FISHEL, FIG. 4.

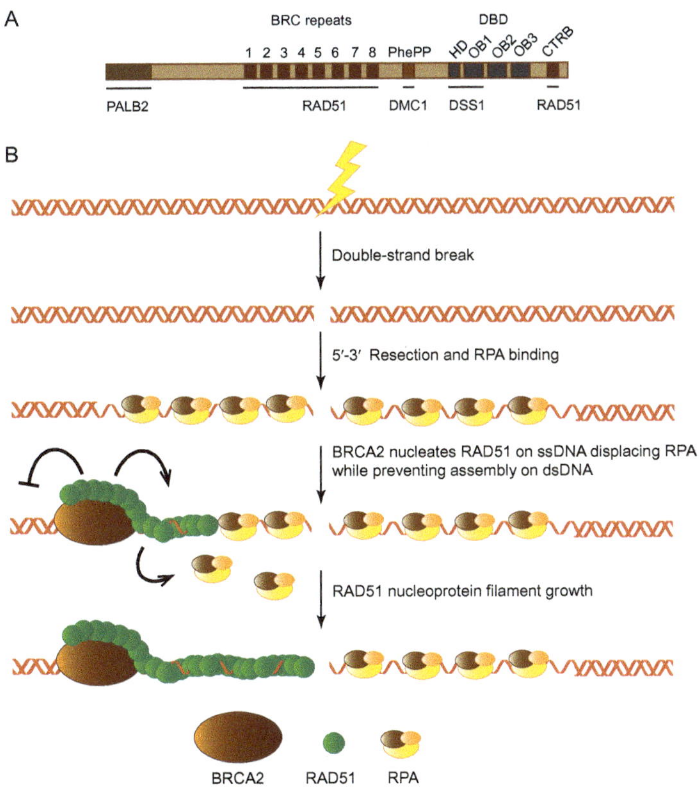

AMUNUGAMA AND FISHEL, FIG. 5.

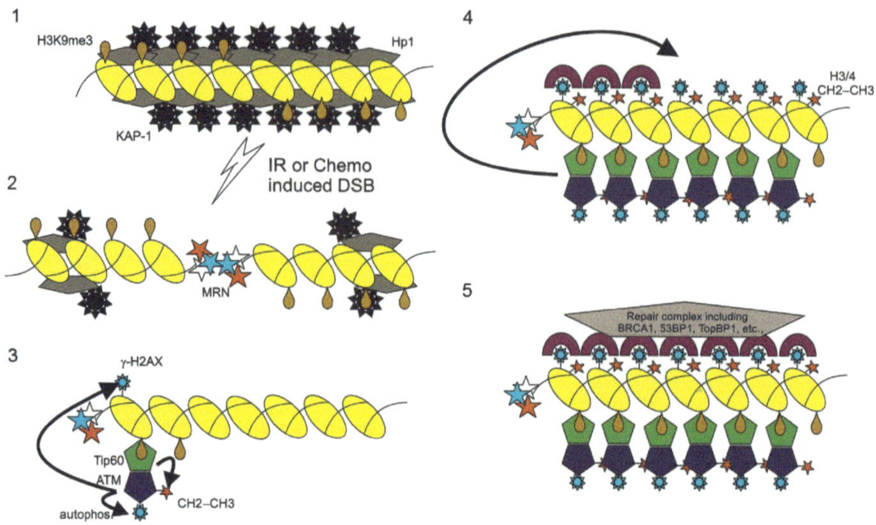

WILLIAMSON *ET AL.*, FIG. 1.

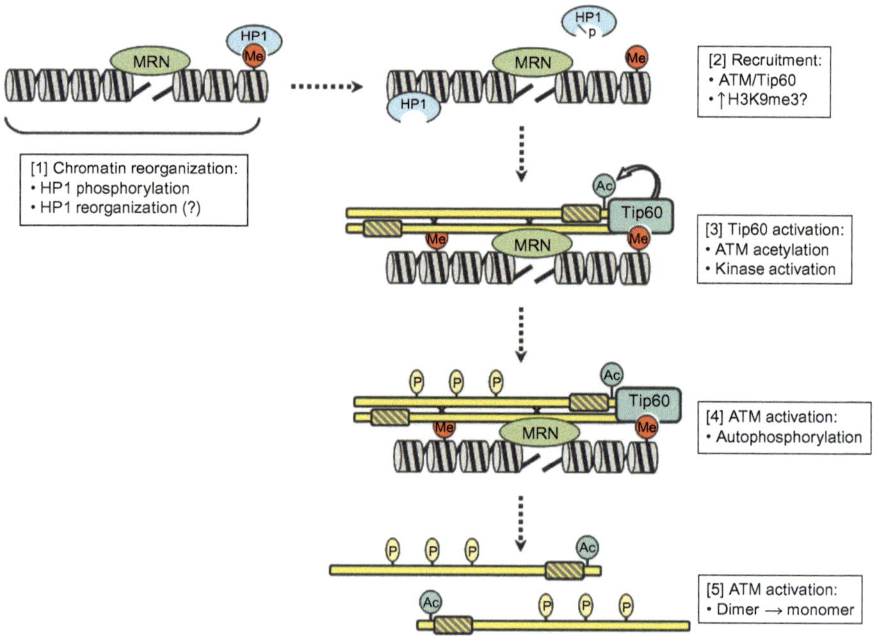

XU *ET AL.*, FIG. 1.

FSC
www.fsc.org
MIX
Papier | Fördert gute Waldnutzung
FSC® C083411

Kolibri 360 GmbH
Ferdinand-Jühlke-Straße 7
99095 Erfurt, Deutschland
produktsicherheit@kolibri360.de